THE SOCIAL BIOLOGY OF
ROPALIDIA MARGINATA

The Social Biology of
Ropalidia marginata

TOWARD UNDERSTANDING
THE EVOLUTION OF
EUSOCIALITY

Raghavendra Gadagkar

HARVARD UNIVERSITY PRESS

Cambridge, Massachusetts ❋ London, England ❋ 2001

Learning Resources
Centre

Library of Congress Cataloging-in-Publication Data

Gadagkar, Raghavendra.
  The social biology of Ropalidia marginata :
  toward understanding the evolution of eusociality / Raghavendra Gadagkar.
    p. cm.
  Includes bibliographical references (p.).
  ISBN 0-674-00611-9 (hardcover : alk. paper)
  1. Ropalidia marginata—Behavior.
  2. Insect societies.   I. Title
  QL568.V5 G34 2001
  595.79'8—dc21
  2001024420

To my students and collaborators,
past, present, and future

CONTENTS

PREFACE

My most pleasant childhood memories are of collecting insects and tadpoles and watching them metamorphose and staring at drops of pond water under my uncle's microscope in his veterinary hospital. My interest in insects persisted through my teenage years, and eventually I decided to study zoology. While in college, I read with fascination *King Solomon's Ring* by Konrad Lorenz and *The Double Helix* by James Watson, and became equally interested in animal behavior and molecular biology. But it did not occur to me that these two subjects could be considered opposite ends of the spectrum in biology.

During the five years that I spent earning my bachelor's and master's degrees in zoology at what was then called Central College, in Bangalore, India, I continued to observe insects: I spent many hours watching, with great curiosity, the wasp colonies that were on every window sill, but could not identify the wasps. For my Ph.D., I applied to the Indian Institute of Science, arguably the best place in India to pursue scientific research. The Institute did not have a program in animal behavior, but it offered me the only slot available to study what was then called interdisciplinary molecular biology. For the next five years I greatly enjoyed studying the mechanisms of interaction between bacteria and bacteriophages.

On my first day at what was then the Microbiology and Pharmacology Laboratory, I was assigned a desk on which somebody had left a copy of India's premier science magazine, *Current Science,* dated 5 August 1974. Opening it casually, I saw an article entitled "Caste differentiation in the paper wasp *Ropalidia marginata* (Lep.)," by M. Gadgil and A. Mahabal.

After reading the paper, I suspected that the wasps I had watched in the Zoology Department of Central College were *Ropalidia marginata*. Gadgil was at the Centre for Theoretical Studies in the Indian Institute of Science, and I had met him briefly at the previous year's meeting of the Ethological Society of India. I sought him out to tell him that I knew of what I thought was a population of the paper wasps that he was studying. We went to Central College on his scooter early the next Sunday morning, and he confirmed that the wasps were indeed *Ropalidia marginata*, but disappointed me by saying that he no longer studied them. But then he said, "if you are interested, however, I can help you study them."

Thus I began to study *Ropalidia marginata*, by observing the colonies at Central College and also by keeping some colonies in the Microbiology and Pharmacology Laboratory. But I could pursue these studies only as a hobby since I had just begun the Ph.D. program in molecular biology. Most of my wasp watching had to be done on weekends. But even so, my interest in these insects led me to explore behavioral ecology and sociobiology and to read the works of W. D. Hamilton, E. O. Wilson, and other leaders in those fields. Thanks to Gadgil, and to the patience and understanding of my two supervisors, T. Ramakrishnan and K. P. Gopinathan, I was able to pursue my dual interests in molecular biology and animal behavior.

In 1979, when I had obtained my Ph.D. and had decided to make the study of animal behavior and ecology my profession, I met Mary Jane West-Eberhard when she came to Bangalore. She taught me how to individually mark the wasps and make behavioral observations, an experience which I cherish. The Centre for Theoretical Studies at the Indian Institute of Science offered me a research associateship for one year to analyze my data on *Ropalidia marginata* and write up the results. But Mary Jane West-Eberhard's visit had inspired me to begin new observations and my associateship was renewed repeatedly. I will always remain grateful to Madhav Gadgil, H. Sharat Chandra, E. C. G. Sudarshan, N. Mukunda, R. Rajaraman, and others at the Centre for Theoretical Studies for believing in me.

This book is the story of my adventures with *Ropalidia marginata* during the last twenty years, of my dream of understanding the evolution of that pinnacle of social life in animals, eusociality. I begin the book with a primer of eusociality, show why social insects should be studied, and point out the benefits of studying my genus of choice—*Ropalidia*—and my species of choice, *R. marginata.* In the second part, I discuss general aspects of the social biology of *R. marginata,* including nesting biology, behavioral caste differentiation, dominance behavior, and age polyethism. The third part focuses directly on the evolution of eusociality. Beginning with a brief account of the theoretical framework adopted, I explore the genetic, ecological, physiological, and demographic factors that influence the evolution of eusociality. I wrap up this part with a first-approximation unified model that integrates various factors that influence the inclusive fitness of workers and solitary nesters, and point out the factors that remain to be considered. In the fourth part, I end with a discussion of some general issues, perhaps the most important of which is the possibility of reversal of social evolution.

Much of the work described here is the result of collaboration with many gifted students and colleagues: Maulishree Agrahari, H. S. Arathi, Seetha Bhagavan, J. T. Bonner, Partiba Bose, Swarnalatha Chandran, K. Chandrashekara, Madhav Gadgil, A. P. Gore, N. V. Joshi, Sujatha Kardile, A. S. Mahabal, Rashmi Malpe, K. Muralidharan, Padmini Nair, Dhrubajyothi Naug, Sumana Rao, Sudha Premnath, M. S. Shaila, Mallikarjun Shakarad, A. Shanubhogue, Anindya Sinha, Thresiamma Varghese, Arun Venkataraman, and C. Vinutha. I thank them for their friendship and their collaboration. This is our collective story and I have the privilege of being the spokesman.

Several people provided technical help while the research described in this book was being conducted: Doddaiah, Augustin Dorairaj, S. Ganesh, G. Jayashree, Milind Kolatkar, Sheela Menon, N. Murugeshachar, Padmini Nair, Sai Prabha, M. Ramachandran, Preeti Roy, N. Roopa, U. N. Shyamala, S. Uma, D. Vasantha, V. Venkatesh, C. Vinutha, and Harry William. I thank all of them. In writing this book, I have been ably assisted by Swarnalatha Chandran, who word-processed most of the innumerable drafts, and by Milind Kolatkar, who meticulously prepared all the illustrations at times when most human beings should be, and were, asleep.

I owe a special debt of gratitude to my wife, Geetha, and my son, Vikram, for admitting *Ropalidia marginata* into our family—we have had colonies all over our home, our newly built house was christened *Ropalidia,* and of course we have never ceased to talk about these wasps. Geetha has, on more than one occasion, cared for and fed hundreds of wasps when I was away from home.

I am grateful to Michael Fisher, Ann Downer-Hazell, and Nancy Clemente of Harvard University Press for waiting with patience for the manuscript to be delivered and for sending encouraging e-mails from time to time. It has been a pleasure to work with them.

The following people read and commented on one or more chapters in the book or on one or more research papers on which this book is based: J. Altmann, S. A. Barnett, Sam Beshers, J. T. Bonner, R. M. Borges, S. Braude, M. D. Breed, M. V. Brian, H. S. Chandra, E. L. Charnov, D. Cherix, M. Collet, E. C. Cox, B. J. Crespi, R. H. Crozier, R. Dawkins, W. G. Eberhard, G. C. Eickwort, W. Fortelius, N. Franks, M. Gadgil, S. Gadgil, G. J. Gamboa, W. Getz, D. Gordon, A. P. Gore, D. Haig, W. D. Hamilton, J. Heinze, J. M. Herbers, B. Hölldobler, J. H. Hunt, Y. Itô, R. L. Jeanne, Amitabh Joshi, N. V. Joshi, L. Keller, W. Kirchner, B. König, M. Lehrer, E. G. Leigh, Jr., H. K. MacWilliams, R. W. Matthews, C. D. Michener, V. Nanjundiah, P. Nonacs, K. C. Noonan, L. Packer, P. Pamilo, C. Peeters, D. C. Queller, R. Rajasekharan, A. Stanley Rand, Photon Rao, Sumana Rao, H. K. Reeve, G. Robinson, K. G. Ross, D. W. Roubik, S. F. Sakagami, T. D. Seeley, K. Slessor, C. K. Starr, D. L. Stern, B. L. Thorne, K. Tsuji, S. Turillazzi, W. T. Wcislo, J. W. Wenzel, M. J. West-Eberhard, D. E. Wheeler, E. O. Wilson, D. Windsor, M. Winston, and D. Yanega. In the early stages of my work O. W. Richards and J. van der Vecht helped identify specimens and patiently answered many questions. I take this opportunity to thank all of them for their kindness.

I thank J. M. Carpenter and J. W. Wenzel for granting permission to reproduce certain illustrations. The following gave permission to reproduce illustrations for which they hold the copyright: Academic Press, Birkhauser Publishing Ltd., Blackwell Science Ltd., Cornell University Press, Indian Academy of Sciences, Indian National Science Academy, Kluwer Academic/ Plenum Publishers, Oxford University Press, Springer-Verlag GmbH & Co. KG, and the Royal Society of London.

While writing this book, I was first a Homi Bhabha Fellow, then a B. P. Pal National Environment Fellow, and finally a Schering Fellow at the Wissenschaftskolleg. I am happy to record my appreciation to the Homi Bhabha Fellowships Council, the Ministry of Environment and Forests, and the Schering Foundation and the Wissenschaftskolleg zu Berlin for their generosity. My research has been generously supported by the Department of Science and Technology, the Ministry of Environment and Forests, the Department of Biotechnology, the Council for Scientific and

Industrial Research, the Indian National Science Academy, the Jawaharlal Nehru Centre for Advanced Scientific Research, and the Centre for Theoretical Studies and the Centre for Ecological Sciences of the Indian Institute of Science. I am grateful to the authorities of all these organizations. I especially thank the successive directors of the Indian Institute of Science, S. Dhawan, S. Ramaseshan, C. N. R. Rao, G. Padmanaban, and Goverdhan Mehta, for creating and sustaining an inspiring academic atmosphere.

I hope to continue to study *Ropalidia marginata* and *Ropalidia cyathiformis,* and to continue to explore the evolutionary forces that influence the balance between cooperation and conflict in these remarkable wasps. I would therefore welcome queries and comments from readers and potential social insect researchers; I can be reached by e-mail at <ragh@ces.iisc.ernet.in>.

PART I

Introduction

1   A PRIMER OF EUSOCIALITY

$\mathscr{I}$N HIS inimitable style Edward O. Wilson (1990) remarked that

> despite the accessibility and ease of study of social insects, they have
> received relatively little attention in general textbooks of biology,
> ecology and behavior, perhaps because their life cycles and anatomy
> initially seem more complex, their diversity greater and the literature
> consequently more "technical." All this remoteness is due to the acci-
> dent of our mammalian origins. If the first highly intelligent, linguis-
> tic species had been a termite instead of an Old World primate, the
> reverse perspective would exist. Vertebrates would be noted chiefly for
> their gigantism, scarcity, and unfamiliar anatomy.

I believe that this same accident of our mammalian origins also makes the
study of social insects particularly fascinating—these insects offer us the
unique possibility of discovering how animals belonging to an altogether
different subkingdom have dealt with the challenges and opportunities of
social life.

1.1.   Levels of Social Organization

Before we can begin to assess the impact of sociality, we must deal with the
tremendous diversity of social insects, not only in their structure but in
their social systems. One not entirely satisfactory way to cope with their
social diversity is to distinguish a class of "truly" social species from all
others that exhibit "lower" or varying degrees of social development. Such
"truly" social insects, on which insect sociobiologists focus almost exclu-

sively (but perhaps not wisely), are called eusocial (a term coined by Batra, 1966), and are defined as those that exhibit three properties:

1. cooperative brood care;
2. differentiation of colony members into fertile reproductive castes (queens or kings) and sterile nonreproductive castes (workers);
3. an overlap of generations such that offspring assist their parents in brood care and other tasks involved in colony maintenance.
(Michener, 1969; Wilson, 1971)

This system of classification explicitly recognizes equally well defined stages that are in a sense "lower" than the eusocial. Omit the criterion of overlap of generations and we have the semisocial. Omit also the criterion of reproductive caste differentiation and we have the quasisocial. Omit all three criteria and we have the communal, if there are aggregates of individuals of the same generation assisting each other in caring for the young; the subsocial, if there are offspring assisting their parents to produce more offspring; and the solitary, if there are no aggregates of two or more individuals caring for the young (Table 1.1). The term parasocial is sometimes used as a composite category to include communal, quasisocial, and semisocial. It is customary to recognize two further subdivisions of the eusocial—the primitively eusocial and the highly eusocial. The most widely accepted criterion for separating the primitively and highly eusocial stages is the presence of morphologically differentiated reproductive and nonreproductive castes in the highly eusocial species and their absence in the primitively eusocial ones (but see Cowan, 1991, for the other criteria). Although this definition of eusocial has been widely accepted and applied (and will be used here), there has recently been a spurt of papers expressing dissatisfaction with, and suggesting modifications to, it (Furey, 1992; Tsuji, 1992; Gadagkar, 1994b; Kukuk, 1994; Sherman et al., 1995; Crespi and Yanega, 1995; Costa and Fitzgerald, 1996; Wcislo, 1997).

I think that the definition is too restrictive and hence not very helpful for the study of altruism, which in my opinion is the central problem of eusociality. It is too restrictive because it excludes semisocial and subsocial insects and, even more important, it excludes birds and mammals that breed cooperatively, not to mention social amoebae (see, e.g., Gadagkar and Bonner, 1994), all of which show no less altruism than primitively eusocial wasps such as *Ropalidia marginata*. Somewhat paradoxically, the

Table 1.1. Levels of social organization in insects (After Wilson, 1971; Michener, 1969, 1974; Starr, 1984; and Crozier and Pamilo, 1996.)

Level	Continued care of young	Cooperative brood care	Reproductive caste differentiation	Overlap of generations	Morphological caste differentiation
Solitary	—	—	—	—	—
Subsocial	+	—	—	—	—
Parasocial					
Communal	+	—	—	—	—
Quasisocial	+	+	—	—	—
Semisocial	+	+	+	—	—
Eusocial					
Primitive	+	+	+	+	—
Advanced	+	+	+	+	+

definition clubs together primitively eusocial species and honey bees, which have less in common with each other, and separates primitively eusocial species and cooperatively breeding birds, which have more in common with each other (Gadagkar, 1994b). Sherman and his colleagues (1995) seem to agree to deal simultaneously with primitively eusocial insects and cooperatively breeding birds and mammals and propose reproductive skew as the way to classify species on the "eusociality continuum." Crespi and Yanega (1995) propose narrowing the definition of eusociality to those species showing alloparental care and irreversible caste differentiation because they believe that such "eusocial" species represent an "evolutionary threshold." Finally Wcislo (1997) proposes abandoning attempts at categorization itself, for he believes that behavioral classifications are "blinders to natural variation." I refer readers to Costa and Fitzgerald (1996) for an intelligent and fair commentary on these recent rumblings.

In the rest of this chapter I will attempt to give a panoramic overview of taxa with eusocial species and highlight various nuances of social evolution that different taxa are best suited to demonstrate.

1.2. The Hymenoptera

1.2.1. Ants

All ants belong to a single family, the Formicidae, in the order Hymenoptera. The living ants are classified into 14 subfamilies (Baroni Urbani, Bolton, and Ward, 1992), 297 genera, and over 8800 species (Hölldobler and Wilson, 1990). Ants unknown to science almost certainly outnumber known species; some 350 genera and 20,000 species are conservatively estimated to exist (Hölldobler and Wilson, 1990). In many ways ants are the premier social insects, numerically more abundant and with more species than any other eusocial group. Their diversity in feeding and nesting habits is most impressive, as is their geographic distribution—they range from the Arctic tree line in the north to the tip of South Africa and Tasmania in the south (Wilson, 1971). Of more interest here is the diversity of their social systems, which, owing to the fact that all ants are eusocial, is relatively modest compared to that of bees and wasps, for instance. The morphologically primitive ponerine ants may lack queens and often consist of groups of workers, one or more of which functions as an inseminated egg layer, or gamergate. Some species that have developed into social parasites may lack workers altogether and depend entirely on workers of their host species. Both these unusual states are undoubtedly derived from the eusocial state, which is the ancestral condition of all ants (Hölldobler and Wilson, 1990).

In nonparasitic species with queenright colonies, new colonies may be founded by a single queen, groups of queens, or groups of queens and workers. In the founding stage, queens mate, shed their wings, and metabolize their wing muscles and fatbodies to produce eggs that will develop into workers as well as to produce trophic eggs and/or salivary secretions that will feed the first batch of larvae (claustral founding). Some founding queens forage to obtain part of the nourishment required for their larvae, and there are good reasons to believe that such partial claustral founding is the evolutionarily primitive state in ants (Hölldobler and Wilson, 1990). As the eggs in the first batch complete development, they take on the duties of foraging, brood care, nest building and maintenance. Individual colonies may contain anywhere from about ten to several million workers and from one (or zero in the case of queenless ponerines) to several hundred queens. As in all hymenopteran societies, males are usually incidental to colonial life, being present only occasionally and in relatively small numbers, from the day of their eclosion to the day of their fatal nuptial flight. Morphological differentiation between queens and workers, and between different subgroups of workers specializing in different tasks in the colony, can be spectacular, and such specialization, along with an age-dependent work profile, often facilitates an efficient division of labor. In addition, along with termites, ants exhibit some very impressive forms of chemical communication, a factor of undoubted significance in the elaboration of their social organization.

In their mammoth monograph Hölldobler and Wilson (1990) recognize eight prominent lifestyles among ants: specialized predators, army ants, fungus growers, harvesting ants, weaver ants, social parasites, symbionts with other arthropods, and symbionts with plants. Studying ants with these various lifestyles, myrmecologists have often focused their attention on such questions as the evolutionary origins and the proximate mechanisms permitting extreme dietary specialization; the mechanisms of coordination between the hundreds of thousands of individuals during army ant raids (Franks and Fletcher, 1983; Franks, 1989; Franks et al., 1991); the evolutionary origins of fungus cultivation by leaf cutter ants (Mueller, Rehner, and Schultz, 1998); the patterns of competition between coexisting harvester ants and their impacts on the plant species they exploit by eating the seeds and help by dispersing seeds; the evolution of cooperative nest building, especially of nests that are woven with larval silk fibers; and the origin and evolution of social parasitism.

If the foregoing discussion suggests that ants provide rich opportunities for the investigation of general principles of animal ecology and evolution, that impression is correct. If it suggests that ants do not provide similar opportunities for investigating the evolution of sociality itself, then that impression is not entirely correct. Although ants do not exhibit the fine intergradations between the solitary and the eusocial seen among bees and wasps, there is often a subtle balance between cooperation and conflict in ant societies that, subjected to an integrated theoretical and empirical analysis, can contribute substantially to answering questions about social evolution (Bourke and Franks, 1995; Crozier and Pamilo, 1996). In many genera of the subfamily Ponerinae, morphologically differentiated queens are not produced, and mated workers (who retain their spermathecae, on account of the morphological primitiveness of the subfamily) take on the role of queens (Peeters, 1997). I believe that the ponerines provide an altogether unique opportunity to understand social evolution, or perhaps, the beginnings of the loss of eusociality. In Chapter 18, I will return to the possibility of such reverse social evolution and to the possible significance of ponerines in this context.

1.2.2. Bees

Bees evolved from wasplike ancestors, some 100 million years ago, perhaps about the time of the origin and rise to dominance of angiosperms (Michener, 1974, 2000; Winston, 1987). Largely on account of their mutualistic relationships and coevolution with angiosperms, bees have diversified enormously and today constitute some 10 families, 700 genera, and 20,000 species. Although most bees are solitary or social parasites, eusociality has arisen more often in lineages of bees than in any other group of social insects (Wilson, 1971; see also Winston and Michener, 1977; Cameron, 1993). Bees offer excellent opportunities to investigate the origin and evolution of sociality. Eusociality has arisen at least eight times among bees, and significant levels of sociality that fall short of eusociality have arisen many more times. Extant species of bees show the finest series of intergradation of every conceivable level of sociality from the solitary to the highly eusocial. Bees also display the highest levels (among all social insects studied) of temporal and spatial variation in levels of sociality, both between and within species (Wilson, 1971; Michener, 1974). Eusociality is seen today in one or more species belonging to two families (Halictidae and Apidae), three subfamilies (Halictinae, Xylocopinae, and Apinae), and seven tribes (Halictini, Augochlorini, Ceratinini, Allodapini, Bombini, Meliponini, and Apini) (Table 1.2) (Roig-Alsina and Michener, 1993; Alexander and Michener, 1995).

Species in the tribe Halictini may be solitary, communal, cleptoparasitic, or eusocial; some 300 species have eusocial colonies. These are small primitively eusocial bees that usually nest in the soil and mass provision their brood in individual, isolated cells. Social organization, caste differentiation, mating biology, kin recognition, and regulation of worker activity by queens are just some examples of the phenomena that have been most profitably studied in these taxa (Knerer and Plateaux-Quénu, 1966a, 1966c, 1966b, 1967a, 1967b; Michener, 1974, 1990). The tribe Augochlorini is less well known, but similarly contains solitary, communal, cleptoparasitic, and eusocial species; only some 25 or so species are known to have primitively eusocial colonies (Danforth and Eickwort, 1997; see also Michener, 1974). The tribe Ceratina has mostly solitary species, but some 3–4 species may have eusocial nests. Ceratina bees nest in dead, dry, pithy stems. In the eusocial nests one individual is more queenlike and the remaining are more worker-like by hymenopteran standards. The genus *Ceratina* has been made famous by the experiments of Sakagami and Maeta (1982, 1987), who attempted, somewhat successfully, to artificially induce sociality in basically solitary Japanese species. The allodapine bees also nest in hollow or pithy stems. However, in contrast to all other kinds of bees, they do not make partitions between the cells. As a result, the brood and adults can have continued interactions with each other. Unlike the ceratine bees, allodapines feed their brood progressively, that is, at intervals during their growth.

The bumble bees, along with the stingless bees and honey bees, belong to the socially more advanced subfamily Apinae. There is a single genus, *Bombus*, with about 200 eusocial species, apart from a few social parasites. Bumble bees are primarily temperate zone bees, widely distributed in northern parts of America and Eurasia. Because of their geographical distribution, their highly eusocial nature, and their considerable economic importance as pollinators, bumble bees have received much attention. Nests are usually subterranean, burrows previously occupied by mice or other rodents being favorite nesting sites. Queens are always larger than workers. The latter do not mate but can develop their ovaries and lay haploid eggs toward the end of the annual colony cycle. Queens prevent or postpone this as long as possible by producing pheromones that inhibit workers' ovarian development. (For a review see Wilson, 1971; Michener, 1974; Röseler and van Honk, 1990).

The stingless bees belong to the tribe Meliponini. This is a large group distributed throughout the tropical regions of the world, containing some 23 genera and 370 species, all eusocial. In the New World tropics stingless bees are more numerous and dominant than they are in the Old World tropics. Their colonies are perennial and they initiate new nests by swarming. Their sociality can be as highly developed as that of true honey bees, and their nest architecture is even more elaborate and complex. They are major pollinators in the tropical regions and probably fill the niche occupied by bumble bees in the temperate regions of the world (for a review see Wilson, 1971; Michener, 1974; Roubik, 1989).

Finally, the tribe Apini, consists of some 9–10 species of true honey bees, all belonging to a single genus, *Apis*. The best-known species are *Apis mellifera*, the European bee that has been domesticated and spread almost throughout the world by man; its Asian counterpart, *Apis cerana*, which has been domesticated but has not spread very far; the primitive Asian dwarf bee *Apis florea;* and the large and aggressive Asian rock bee *Apis dorsata*. Because *Apis mellifera* provides an irresistible model system for diverse kinds of studies, that species must be regarded as unique not only among insects but perhaps among all animal and plant species.

Table 1.2. Taxonomic distribution of eusociality (see the text for bibliographic citations).

Phylum	Class	Order	Family	Subfamily	Tribe	Common name	Remarks
Arthropoda	Insecta	Hymenoptera	Formicidae	14 subfamilies		Ants	297 genera, 8800 species, all highly eusocial or queenless species and social parasites derived from the highly eusocial
			Vespidae	Stenogastrinae		Hover wasps	6 genera, 50 species, primitively eusocial
				Polistinae		Paper wasps	29 genera, 800 species, including primitively eusocial, highly eusocial, and social parasites
				Vespinae		Hornets and yellowjackets	4 genera, 60 species, highly eusocial and social parasites
			Sphecidae	Pemphridoninae			Single eusocial genus, Microstigmus, 22 species, primitively eusocial
			Halictidae	Halictinae	Halictini	Sweat bees	23 genera, may be communal, cleptoparasitic, or eusocial; some 300 species have eusocial colonies
					Augochlorini	Sweat bees	25 genera; only some 25 species have eusocial colonies
			Apidae	Xylocopinae	Ceratinini	Carpenter bees	2 genera, mostly solitary; some 3–4 species have eusocial colonies
					Allodapini	Allodapine bees	11 genera; some 200 species have eusocial colonies but only 2–40% of the nests of a given species may be eusocial

Phylum	Class	Order	Subfamily	Family/Tribe	Common name	Description
			Apinae	Bombini	Bumble bees	Single, eusocial genus, *Bombus*; 200 species, not counting parasitic species.
				Meliponini	Stingless bees	23 genera, about 370 species, all eusocial
				Apini	Honey bees	Single genus, *Apis*, 6–10 species recognized, all highly eusocial
		Isoptera		Hodotermitidae	Lower termites	16 species, highly eusocial
				Indotermitidae	Lower termites	6 species, highly eusocial
				Kalotermitidae	Lower termites	332 species, highly eusocial
				Mastotermitidae	Lower termites	1 species, highly eusocial
				Rhinotermitidae	Lower termites	204 species, highly eusocial
				Serritermitidae	Lower termites	1 species, highly eusocial
				Stylotermitidae	Lower termites	28 species, highly eusocial
				Termopsidae	Lower termites	16 species, highly eusocial
				Termitidae	Higher termites	1685 species, highly eusocial
		Hemiptera		Pemphigidae	Aphids	Primitively eusocial, containing sterile soldier castes
				Hormaphidae	Aphids	Primitively eusocial, containing sterile soldier castes
		Coleoptera		Curculionidae	Ambrosia beetles	Primitively eusocial
		Thysanoptera		Phlaethripidae	Thrips	Primitively eusocial, containing subfertile soldiers
	Crustacea	Decapoda		Alpheidae	Snapping shrimps	Primitively eusocial by hymenopteran standards
	Arachnida	Araneae		Theridiidae	Spiders	May be primitively eusocial
Chordata	Mammalia	Rodentia		Bathyergidae	Naked mole rat	Primitively eusocial by hymenopteran standards

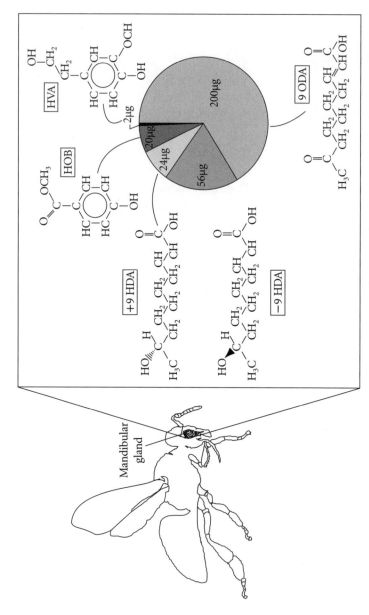

Figure 1.1. The queen mandibular pheromone, containing well-defined amounts of five of the most important components of the mandibular gland secretion, elicits most of the responses expected from workers. See the text for details. (Based on Winston and Slessor, 1992; redrawn from Gadagkar, 1997b.)

Honey bees live in populous colonies consisting of tens of thousands of sterile workers, a few hundred drones, and a single fertile queen. Communication between the queen and her workers is primarily mediated by pheromones (Free, 1987). The well-known effects of queen pheromones on workers include rapid detection of the presence or absence of the queen. A retinue of some 8 to 10 workers, the composition of which changes every few minutes, attends the queen; these workers feed and lick the queen and thereby acquire the queen pheromones and pass them on to other workers. The queen pheromones also inhibit the development of workers' ovaries and stimulate building and foraging activities. Considerable progress has been made in recent years in understanding the chemical composition of the queen pheromone and the biosynthetic pathways of its key components. Winston and Slessor (1992) have succeeded in identifying five of the most essential components of the queen pheromone, which together elicit most of the important behavioral responses observed in the workers (see also Plettner et al., 1997). One queen equivalent (the average amount present in a queen) of this so-called queen mandibular pheromone (QMP), consists of about 200 μg of 9-keto-(E)2-decenoic acid (9 ODA); about 80 μg of 9-hydroxy-(E)2-decenoic acid (9 HDA), of which about 56 μg is the ($-$) optical isomer and about 24 μg the ($+$) optical isomer; about 20 μg of methyl p-hydroxybenzoate (HOB); and about 2 μg of 4-hydroxy-3-methoxyphenylethanol (HVA) (Figure 1.1). If it appears at first sight that these chemical details are at best of superficial interest for understanding the evolution of eusociality, that impression is likely to be quite mistaken. I believe that the many details of the honey bee queen's pheromone biochemistry and physiology, and especially their biosynthetic pathways, will be of considerable help in resolving such fundamental questions concerning the evolution of eusociality as whether social insect colonies are evolutionarily stable or unstable, what the relative roles of individual and group selection are in molding sociality, what is the evolutionary antiquity of queens versus workers, and what are the possible genetic mechanisms for the origin of their altruistic phenotype (Gadagkar, 1996b, 1997a, 1997b; see also Chapter 17).

1.2.3. Wasps

In contrast to the ants, wasps constitute many families and, as with the bees, the majority are solitary or parasitic. With the exception of the genus

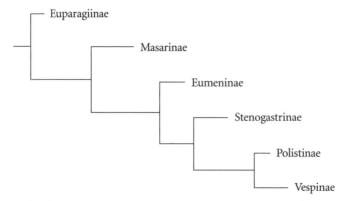

Figure 1.2. Cladogram for the subfamilies of Vespidae. (After Carpenter, 1982; redrawn from Carpenter, 1991.)

Microstigmus, which belongs to the family Sphecidae, all eusocial wasps belong to the family Vespidae. That family has six subfamilies: Euparagiinae, Masarinae, Eumeninae, Stenogastrinae, Polistinae, and Vespinae (Figure 1.2). Eusociality is restricted to the Stenogastrinae, Polistinae, and Vespinae, three subfamilies that appear to share a recent common ancestor not shared by the three solitary subfamilies (Figure 1.2). This conclusion was arrived at through cladistic analysis of morphological and behavioral data (Carpenter, 1982, 1991). Another cladistic analysis using nuclear 28S ribosomal DNA and mitochondrial 16S ribosomal DNA suggests, however, that Stenogastrinae, Polistinae, and Vespinae do not share a common ancestor. It suggests instead that Eumeninae is a sister group to Polistinae + Vespinae. Because the Eumeninae do not show eusociality, this means that eusociality must have arisen at least twice in the Vespidae. This separation of Stenogastrinae from Polistinae and Vespinae supports the views of such earlier experts as Vecht (1977) and Spradbery (1975). It appears therefore that vespid phylogeny is once again in flux (Schmitz and Moritz, 1998). Like bees, wasps have been favorite model systems with students of social evolution because they exhibit a series of intermediate stages between the solitary and the eusocial, offering rich possibilities for testing models for the origin and evolution of sociality.

The Stenogastrinae consists of six genera (Carpenter, 1988) and about 50 species of hover wasps that are somewhat poorly known, primarily on account of their restriction to the rainforests of the Indo-Pacific region.

The stenogastrine wasps appear to be strikingly different from polistine and vespine wasps in many ways—the adults are slender with long petioles; they build nests without pedicels, often mixing vegetable fibers with soil; and they have rather unusual methods of oviposition and brood rearing. The female wasp produces an abdominal secretion, makes it into a ball with her legs and mandibles, lays an egg on this ball while still holding it in her mouth, and then deposits the egg into an empty cell by attaching the sticky ball to the bottom of the cell. In terms of social organization, dominance relationships and division of labor, queen supersedure, intracolony genetic relatedness and behavioral flexibility among adult females, these wasps are not unusual compared with any other primitively eusocial species. It is interesting, however, that no species in the family has progressed any further in social evolution (Turillazzi, 1989, 1991). The use of the inevitably limiting abdominal secretion in rearing brood and the demonstrably inferior quality of these wasps' nesting material are thought to have prevented increase in colony size and hence higher social evolution (Turillazzi, 1991; Hansell, 1987).

The Polistinae is a large and diverse cosmopolitan group that is, however, concentrated and most diverse in the tropics, especially in the neotropics. Twenty-eight genera are recognized and about 800 species have been described. Recent cladistic analysis has transformed our understanding of the phylogeny of the Polistinae (Wenzel and Carpenter, 1994; Wenzel, 1998; see Figure 1.3). Nevertheless, further subdivision of the Polistinae based on phylogenetic considerations may be less rewarding to students of social evolution than a subdivision based on behavior and nest architecture. An elegant example of the latter has been provided by Jeanne (1980), who distinguishes two subgroups, the independent-founding and the swarm-founding Polistinae.

Independent founding is characteristic of five genera: *Polistes, Mischocyttarus, Belonogaster, Ropalidia* (not all species), and *Parapolybia* (for reviews see Gadagkar, 1991a; Reeve, 1991). Independent-founding polistines live in relatively small colonies (with rarely more than 100 adults) and construct small, simple, unenveloped combs that are normally suspended by a narrow pedicel (Figure 1.4). Queens initiate new colonies either singly or in small groups, but without the aid of workers (independently). On account of their small colonies, queens of independent-founding polistines often use overt physical dominance to control or influence their nestmates.

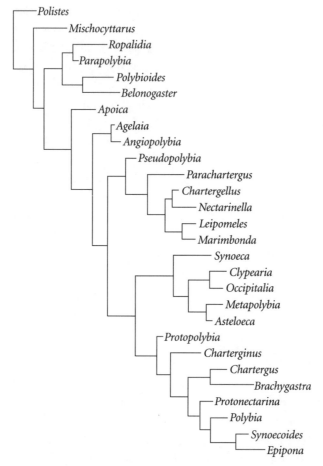

Figure 1.3. Cladogram for the genera of Polistinae based on adult and larval morphology and nest architecture. (Redrawn from Wenzel and Carpenter, 1994.)

Independent-founding species have a special problem with ant predation: direct physical resistance to marauding ants is difficult enough with small numbers of adult wasps, but when the nest and its brood have to be left completely unguarded while the single foundress is away foraging, it is impossible. Not surprisingly, independent-founding polistines have evolved a chemical defense against ants; a well-developed van der Vecht's gland on the sixth gastral sternum secretes ant-repellent chemicals, and an associated tuft of hairs acts as an applicator brush when the wasps periodically coat the pedicel and restrict the entry of ants (Jeanne, 1970; Jeanne,

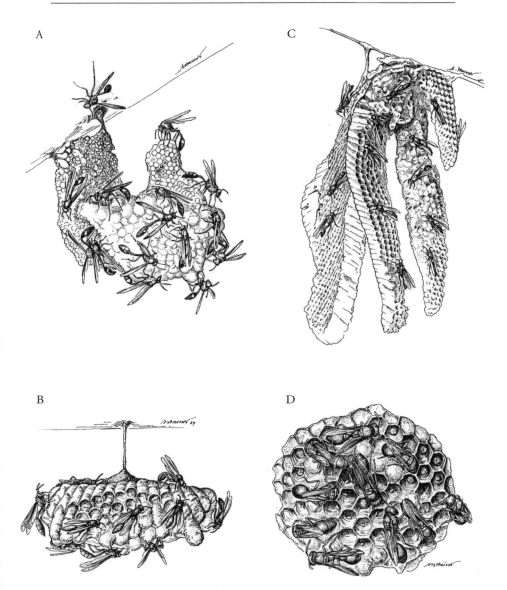

Figure 1.4. The mature nests of (A) *Belonogaster grisea,* (B) *Mischocyttarus drewseni,* (C) *Parapolybia varia,* and (D) *Ropalidia marginata.* (Reprinted from Gadagkar, 1991a, in *The Social Biology of Wasps,* ed. Kenneth G. Ross and Robert W. Mathews; copyright © 1991 by Cornell University; used by permission of the publisher, Cornell University Press.)

Downing, and Post, 1983; see also Kojima, 1982b, 1983a, 1992; Keeping, 1990). The ant-repellent chemicals in various independent-founding polistines have been identified as methyl palmitate, methyl stearate, methyl linoleate, butyl palmitate and related carboxylic acids (Post et al., 1984; Keeping, Crewe, and Kojima, 1995; Dani, Morgan, and Turillazzi, 1995).

It is easy to argue that independent-founding polistine wasps provide perfect model systems for investigating the plethora of forces that mold the evolution of social behavior (see Burian, 1996). Small colonies make it possible to mark and study every adult wasp, and the open combs hide nothing from the observer. The nests of many species are quite abundant and are often built in remarkably accessible places. A relatively primitive level of eusociality, characterized by a lack of morphological differentiation between queens and workers, an essentially behavioral (and therefore often reversible) mechanism by which queens suppress their workers, and the considerable flexibility in the social roles that the adult wasps are capable of adopting, all make independent-founding polistine wasps especially attractive model systems. In particular, these wasps provide an opportunity to compare and contrast the costs and benefits of social and solitary modes of existence because members of the same species routinely found single-foundress as well as multiple-foundress colonies (see Chapter 2).

Swarm founding is characteristic of some species of *Ropalidia* and the remaining 23 genera of the subfamily Polistinae (Jeanne, 1991b). In striking contrast to the independent-founding species, swarm-founding polistine wasps live in more populous colonies (often with 100 or more adults) and have correspondingly large nests that sometimes have several tiers of combs covered by an envelope. New colonies are always initiated by queens in the company of workers (that is, by swarms). Queens in swarm founders must find it impossible to subjugate or influence large numbers of nestmates by physical dominance and have, perhaps for that reason, evolved a pheromonal means of regulating worker reproduction. Swarm-founding species never have to leave their nest and brood unattended and usually have a substantial adult population to rid the nest of scouting ants. Swarm founders thus do not have a well-developed van der Vecht's gland. Instead, at least some of them have a well-developed Richard's gland on the fifth gastral sternum that is often used to lay an odor trail to guide the members of the swarm to a new nesting site (Jeanne, 1981; Jeanne, Downing, and Post, 1983; Downing, 1991a; Clarke et al., 1999).

The Vespinae is a more compact group, consisting of only 4 genera, *Vespa, Provespa, Dolichovespula,* and *Vespula,* and a total of no more than 60 species. For a study of the phylogenetic relationships and classification of the Vespinae, see Carpenter, 1987. *Vespa* and *Provespa,* commonly referred to as hornets, are largely restricted to Asia. *Vespa* (true hornets) contains 23 species and is endemic to eastern Asia, although some species extend into western Europe and the South Pacific Islands. *Provespa* is a poorly studied genus with only 3 species, all of which are nocturnal and endemic to the tropics of eastern Asia. *Dolichovespula* and *Vespula,* commonly referred to as yellowjackets, are distributed through out much of Europe, Asia, and North America. *Vespula* extends a bit farther south than *Dolichovespula,* both in Asia and in North America. Some species of *Vespula* have become established outside their normal range and have gained great notoriety. Among wasps, social parasitism is relatively common and relatively well studied in the Vespinae (Carpenter et al., 1993, and references therein; Matsuura and Yamane, 1990). Given all these characteristics, the Vespinae constitute excellent model systems for investigating social organization, division of labor, chemical communication, population dynamics, the evolution of swarming and polygyny, and of course the evolution of social parasitism.

At a time when hymenopteran eusociality was thought to be restricted to the families Apidae, Halictidae, Formicidae, and Vespidae, Matthews (1968a, 1968b) discovered eusociality in a sphecid wasp, *Microstigmus comes.* Sphecidae is a large family of primarily solitary nesters that nevertheless show repeated evolution of cooperative nesting, a trait of obvious significance for the evolution of eusociality. Although the sphecids have traditionally been called wasps, they really are phylogenetically quite distant from the Vespidae and indeed are more clearly related to the bees. *Microstigmus comes* is remarkably like an independent-founding polistine wasp. New nests may be initiated by a single mated female or by a group of foundresses, only one of which lays eggs while the remaining cooperate with the egg layer in nest building and nest defense, and in provisioning cells with collembolans. Eclosing daughters may stay on their natal nest, resulting in overlap of generations. The delicate nests, constructed with fibers from the underside of *Chryosophila* palm leaves and silk produced by the wasps, are suspended by a silk thread. The wasps are less than 4 mm long, so that observation of individually identified wasps is difficult. Ross and Matthews (1989b, 1989a) have used electrophoretic separation of iso-zyme markers to successfully overcome some of the difficulties of direct observation. These studies have confirmed the eusocial status of *Micro-stigmus comes.*

1.3.   Eusocial Insects outside the Hymenoptera

1.3.1.   Termites

Termites, constituting some 2291 fossil and living species, all belong to the order Isoptera and are all eusocial. Six to nine families are usually rec-ognized. Members of only one of them, the Termitidae, are considered "higher termites," while all others are considered "lower termites" (Table 1.2). Most termites, especially among the higher termites, are limited to the tropics. Although termites have reached levels of eusociality comparable to the highly eusocial Hymenoptera, there are fundamental differences be-tween termite eusociality and hymenopteran eusociality. In termites, the worker caste consists of both males and females; larvae contribute to col-ony labor; there are no dominance hierarchies within the colonies; there is no record of social parasitism; and the primary reproductive male, called the king, stays with the queen, helps her build the nest, and inseminates her intermittently. In the eusocial Hymenoptera, by contrast, the worker caste consists exclusively of females; larvae are usually incapable of contri-bution to colony labor; intracolony dominance hierarchies are common; social parasitism has evolved repeatedly: and the male inseminates the fe-male during the nuptial flight and dies soon after—there is thus no king to be seen in the colonies. Finally, an apparently nonsocial difference, but one that may have had a profound influence on social evolution, is that ter-mites feed almost exclusively on cellulose and need the help of symbiotic protozoans or bacteria to digest it.

Those studying the origin of termite eusociality, have quite naturally fo-cused on the lower termites, especially the two older living lineages of Isoptera, the Mastotermitidae and the Kalotermitidae. Two generalizations have pervaded the literature:

1. Termites are best described as social cockroaches because of the ex-tensive similarities between the most primitive termite family,

Mastotermitidae, and primitive blattoid cockroaches, especially the family Cryptocercidae.

2. Termites are completely dependent on their intestinal symbiotic protozoa and bacteria and necessarily have to acquire them from their nestmates not only at the beginning of their life but also after every molt, when they lose them along with the lining of the intestinal walls.

Recent work has begun to question both these statements. Cladistic analysis by Thorne and Carpenter (1992) suggests that such characters of Mastotermitidae as are reminiscent of Cryptocercidae may in fact be derived and that Cryptocercidae may therefore be a sister group to other cockroaches rather than to the termites. Thorne (1990) has produced some experimental evidence that transfer of symbionts can occur from *Cryptocercus* to the dampwood termite *Zootermopsis* (but see Nalepa, 1991).

Although termites do not exhibit significant intermediate stages in social evolution, what variability they do exhibit is perhaps the best hope for any empirical investigation of the question of the origin of termite eusociality. Several investigators have turned their attention to such variability (reviewed in Shellman-Reeve, 1997). Shellman-Reeve (1997) and Roisin (1993, 1994, 1999) makes the most convincing case yet that genetic, ecological, and other factors potentially involved in the evolution of termite sociality can be investigated in much the same way as in the study of eusocial Hymenoptera. That should go a long way to demystify the Isoptera for students of the Hymenoptera.

1.3.2. Aphids

In 1977 S. Aoki (1977a, 1977b) discovered a sterile soldier morph in the Japanese aphid genus *Colophina*. Since then soldiers have been found in about 50 species of aphids but always restricted to two closely related, gall-producing families, Pemphigidae and Hormaphididae. These soldiers are first or second instar larvae that defend their colonies by attacking predators, often in a self-sacrificing manner. The soldier morph may or may not be morphologically specialized and even when morphologically specialized for defense, it may or may not be obligatorily sterile. Several preadaptations that might have facilitated the evolution of altruistic soldiers among aphids can be identified (Stern and Foster, 1997). When first instar larvae function as dispersers, they are well sclerotized to avoid desiccation and have well-developed morphological adaptations for moving about and for breaking and entering galls made by conspecifics. First instar foundresses are morphologically and behaviorally adapted to fight with each other to gain and retain access to good gall-initiation sites. Where galls are formed, they provide an ideal environment for clonal reproduction, which results in high genetic homogeneity on the one hand and the opportunity to defend a valuable resource on the other. First instar soldiers not only defend but also perform housekeeping functions involving active cleaning of the gall and are equipped with morphological and behavioral adaptations for cleaning. All of these features of first instar larvae could well have served as preadaptations for their subsequently evolved defensive functions (Benton and Foster, 1992).

1.3.3. Thrips

Thrips belong to the insect order Thysanoptera, containing some 5000 species, mostly phytophagous but some predatory. A minority of the phytophagous species makes galls (Ananthakrishnan, 1973; Ananthakrishnan and Raman, 1989; Parker, Skinner, and Lewis, 1994). Crespi (1992) has demonstrated eusociality in 2 species of Australian gall thrips, *Oncothrips tepperi* Karny and *O. habrus* Mound. This is especially interesting because thrips are haplodiploid. Galls are initiated by single inseminated macropterous (fully winged) females in the spring. After fighting off other similar females for possession of a presumably valuable, young, growing phyllode tissue of *Acacia oswaldii* and *A. melvillei* respectively, the foundress oviposits inside the gall. Her offspring hatch, feed, develop, and eclose inside this gall. She produces four kinds of offspring: macropterous females (like her), macropterous males, micropterous (short-winged) females, and micropterous males. The term micropterous is somewhat distracting because the important feature of micropterous adults is their enlarged and armed forelegs, specialized for fighting. Sure enough, micropterous adults (both females and males) eclose earlier than macropterous females and males. Notice the analogy with the first batch of brood becoming workers and subsequent batches becoming future reproductives in the social Hymenoptera.

Crespi (1992) has convincingly demonstrated that in both species, micropterous adults attack and attempt to kill *Koptothrips* species (inquiline thrips that invade galls of other species, kill the gall formers, and breed inside), lepidopteran larvae, and *Iridomyrmex humilis* (now, *Linepithemia humilis*) ants, and do so more often than foundresses. He has also provided evidence that *Koptothrips* species constitute a real threat to the *Oncothrips* and that the micropterous offspring provide substantial protection for the foundresses. The micropterous adults are therefore termed "soldiers." Dissection of foundresses and micropterous adults has shown that although many soldiers had developing oocytes, their ovarian development was clearly inferior to that of the foundresses. Thus *O. tepperi* and *O. habrus* appear to satisfy all three criteria required to label them eusocial. There is overlap of generations; the morphological specialization and defensive behavior of the micropterous females and their consequences for the survival of the foundresses certainly implies cooperative brood care; and there is some level of reproductive caste differentiation or at least subfertility on the part of the micropterous adults. Notice however, that, in the thrips, unlike the Hymenoptera, and like the termites, the soldiers (workers) can be of either sex, at least in some species. The presence of the gall—a useful resource worth defending—and the need to defend it from invading *Koptothrips* appear to be the main selective forces responsible for the evolution of soldiers. The discovery of an independent haplodiploid taxon containing some (but not all) eusocial species is sure to enrich the study of social evolution in insects (Crespi and Mound, 1997; Gadagkar, 1993).

1.3.4.   An Ambrosia Beetle

By demonstrating eusociality in the ambrosia beetle *Austroplatypus imcompertus,* Kent and Simpson (1992) have brought a new order of insects into the eusocial fold. Even though the order Coleoptera is the largest insect order, and boasts of more species than any other animal group, and exhibits an impressive diversity of lifestyles, no example of eusociality was known from that order until 1992. *Austroplatypus compertus* lives in galleries in the heartwood of living eucalyptus trees in Australia. Extended parental care and cooperation, including helping behavior and division of labor, are widespread among bark beetles (Scolytidae) and ambrosia beetles (Platypodidae), but all species studied so far, except *A. compertus,* fall short of the full suite of characters needed for being classified as eusocial (Kirkendall, Kent, and Raffa, 1997). A systematic search for eusociality outside the Hymenoptera and Isoptera should continue to yield rich dividends.

1.4. Eusociality outside Class Insecta

1.4.1. A Coral-Reef Shrimp

If Kent and Simpson (1992) introduced a new insect order (Coleoptera) into the eusocial fold, Duffy (1996) brought in a new class of Arthropoda (Crustacea), when he demonstrated eusociality in the coral-reef, sponge-dwelling, snapping shrimp, *Synalpheus regalis.* This is a snapping shrimp that lives in colonies consisting of 3–319 (mean = 149) individuals, with a single adult reproducing female per colony. Genotype frequencies of polymorphic allozyme loci and other lines of evidence suggest that most colony members are offspring of the queen. Cooperative brood care is not related to food acquisition, since food is apparently usually unlimited, and if it is not, there is little that the shrimps can do to improve the situation. The role of the nonreproducing individuals is to defend the colony from intruders. Laboratory experiments using a heterospecific intruder demonstrate that large, nonreproducing colony members are significantly more likely to make contact with, attack, and kill intruders. The need to defend the spongy canals that constitute a valuable resource, and the possession of a powerful weapon (the major chela) with which to do so, are thought to be important preadaptations that may have favored the evolution of eusociality in this diploid group of animals (Duffy, 1996). Duffy (1998) and Duffy and Macdonald (1999) have recently added another two species of *Synalpheus* to the eusocial list (see also Duffy, Morrison, and Rios, 2000).

1.4.2. A Eusocial Spider?

Social behavior characterized by colonial nest building and sometimes by cooperative brood care has been recorded in a few dozen of the 34,000 described species of spiders. Reproductive caste differentiation has not been clearly established. However, Vollrath (1986) has claimed that *Anelosimus eximus,* a neotropical theridiid spider, is a "promising possibility" as a candidate for eusociality. His evidence for the existence of nonreproductive individuals is based on the absence of sperm in the receptacles of many females and the presence of far fewer egg sacs than the number of females present. Subsequent authors, however, have fought shy of claiming that any spider is eusocial. Their reluctance is best expressed in the words of Avilés (1997): "The term "eusociality" . . . should be restricted to those cases in which sterility or subfertility, rather than being a side product of competition, has been selected for as a socially adaptive trait . . . There is no evidence that such has been the case in any of the cooperative spiders."

1.4.3.   The Naked Mole Rat

Finally, Jarvis (1981) has added a mammal, the naked mole rat (*Heterocephalus glaber*), to the list of eusocial species. When asked in the mid-1970s why vertebrates had not attained eusociality, R. D. Alexander is said to have hypothesized a fictitious eusocial mammal: "a completely subterranean rodent that feeds on large tubers and lives in burrows inaccessible to most but not all predators, in a xeric tropical region with heavy clay soil" (Sherman, Jarvis, and Alexander, 1991). Jarvis (1981) first presented evidence for the existence of a remarkably similar beast—the naked mole rat that lives in colonies consisting of 25–295 (mean = 74) individuals in underground tunnels and feeds on tubers in the hot arid regions of Kenya, Somalia, and Ethiopia. Field and laboratory studies have since confirmed not only overlap of generations, cooperative brood care, and reproductive caste differentiation (only 1 female and 2–3 males in the colony breed), but also nonreproductive division of labor and polyethism (Tofts and Franks, 1992; Clarke and Faulkes, 1998), recruitment of nestmates to food sources (Judd and Sherman, 1996), nestmate discrimination (O'Riain and Jarvis, 1997), nepotism (Reeve and Sherman, 1991), inbreeding (Reeve et al., 1990), and queen activation of lazy workers (Reeve, 1992). Even more striking is the finding that long periods of inbreeding are interrupted by periodic outbreeding facilitated by the rare occurrence of a morphologically and behaviorally specialized dispersive morph (O'Riain, Jarvis, and Faulkes, 1996). Because of the evolutionary distance between eusocial arthropods and the naked mole rat, the discovery of eusociality in the latter has provided a uniquely powerful opportunity to understand the forces that have driven the evolution of eusociality (Alexander, Noonan, and Crespi, 1991; Sherman, Jarvis, and Alexander, 1991; Braude and Lacey, 1992; Cohn, 1992; Sherman, Jarvis, and Braude, 1992; Burda and Kawalika, 1993; Jarvis and Bennett, 1993; Jarvis et al., 1994).

Although eusociality was once thought to be restricted to the ants, bees, wasps, and termites, we have just seen examples of eusocial aphids, thrips, ambrosia beetles, marine shrimps, possibly a spider, and the naked mole rat. Perhaps this is just the beginning and we may yet see many more widespread examples of eusociality in the animal kingdom. As more attention is paid to various groups with examples of eusociality, especially when researchers with cladistic tools arrive on the scene, we may begin to discover examples of multiple origins (and losses) of eusociality in these groups. While this counting game (of numbers of eusocial species and numbers of independent origins and losses of eusociality) is welcome, it is important not to ignore levels of social organization that just fall short of eusociality—the latter may be even more rich with clues to the evolution of eusociality. For excellent reviews of these, often obscure, groups of social, but not necessarily eusocial groups, see Choe and Crespi (1997).

2   IN PRAISE OF *ROPALIDIA*

*R*OPALIDIA is a relatively large genus in the subfamily Polistinae, containing some 136 species distributed in tropical Africa, southern Asia, Australia, and Okinawa. The Indo-Australian species have been revised and described by Vecht (1941, 1962), the Australian and New Guinea species by Richards (1978), the Philippine species by Kojima (1982c, 1984) and Kojima and Tano (1985), the Nepalese species by Yamane and Yamane (1979), and the Indian species by Das and Gupta (1989). Although Vecht (1962) recognized three subgenera, Richards (1978) recognized six, and Das and Gupta (1989) recognized five, it is now the practice not to recognize any subgenera at all. Following a cladistic analysis, Kojima (1997) concluded that there is no basis for recognizing the current subgenera of *Ropalidia*. *Ropalidia* has often been considered a crucial genus in understanding social evolution in wasps (Wilson, 1971), because *Ropalidia* is the only polistine genus that contains both independent-founding and swarm-founding species. Recall from Chapter 1 that the differences between these kinds of species is considerable. Independent-founding species are characterized by small colonies; simple, small, open nests, often suspended by a narrow pedicel; relatively little queen-worker dimorphism; behavioral control of workers by the queens; a well-developed van der Vecht's gland on the sixth gastral sternum for defense against ants; and the ability of queens to found new nests independently of workers. Conversely, swarm-founding species are characterized by large colonies; complex, large, enveloped nests, often containing many tiers of combs; relatively greater queen-worker dimorphism; pheromonal control of workers by queens; a well-developed Richard's gland on the fifth gastral sternum that probably helps lay an odor trail during swarm founding; and the apparent inability of queens to found nests independently of workers. Studies of *Ropalidia* are therefore expected to throw light on the evolution of swarm founding from independent founding; of complex, multitiered, enveloped nests from simple ones; of chemical control of workers from physical control; of queen-worker dimorphism from monomorphism; and perhaps also of polygyny from monogyny.

Morphological, behavioral, and architectural diversity (in nest building) within the genus *Ropalidia* is truly impressive. For example, *Ropalidia montana* is a swarm-founding species in peninsular India, described by Carl (1934) and studied by Yamane, Kojima, and Yamane (1983) and by Jeanne and Hunt (1992). The large, enveloped nests measure from 25–42 cm in length and 25–33 cm in width, contain 60,000 to 145,000 cells arranged in 19–29 combs, some 9000 to 61,000 adult wasps, of which 0.1 to 1.4% are queens and about 1 to 47% are males (Yamane, Kojima, and Yamane, 1983; Jeanne and Hunt, 1992). Queen-worker dimorphism is quite distinct, with queens being significantly larger in body size than workers (Yamane, Kojima, and Yamane, 1983). Then there is the "wasp city" of *Ropalidia plebiana* described by Richards (1978), consisting of dense aggregations of up to several thousand combs, persisting for more than 10 years. Richards suspected that each comb is unlikely to be a separate colony and that there might be some form of cooperation among wasps belonging to neighboring combs. More recent observations by Itô and Higashi (1987) suggested that each comb is indeed an independent colony. But Richards' conjecture about cooperation seems to be borne out by the observations of Itô, Yamane, and Spradbery (1988), who found that when they accidentally disturbed one nest in the aggregation, wasps from nearby nests attacked them. Makino and his colleagues (1994) have added another fascinating detail to the unusual biology of *R. plebiana:* groups of foundresses occupy and reuse empty combs from the previous year, and as the brood develops they actually cut the combs into smaller units, each of which presumably becomes a separate colony. Nothing is known about queen-worker dimorphism in this species but as much as 93% of the females can be inseminated and 72% of the females can have well-developed ovaries (Itô and Higashi, 1987).

At the other extreme are the independent-founding species, and there is a great deal of diversity among these as well. *Ropalidia formosa* is a nearly solitary species from Madagascar (Wenzel, 1987). This species often nests in dense aggregations of hundreds of nests, but each comb is occupied usually by a solitary female. The species is also unusual in many other ways: it probably has the smallest mature nest size—the largest nest observed had 56 cells—and there can be as few as 3 cells at the first pupation and 8 cells at the first offspring eclosion. Males are produced very early— they may even be the second or third offspring produced on a nest. The mean number of foundresses per nest was estimated to be 1.03 (Wenzel, 1987). Another unusual species is *Ropalidia ignobilis*, which displays the most extreme queen-worker dimorphism recorded for any independent-founding polistine wasp (Wenzel, 1992). Although queens of one colony may be smaller than workers of another colony, queens are always larger than workers of their own colonies. Wenzel's (1992) study provides clear evidence that castes in the small-colony, independent-founding *R. ignobilis* resemble those of advanced, swarm-founding Polistinae. In both cases— the ultra-small, nearly solitary condition of *R. formosa* and the extreme queen-worker dimorphism of *R. ignobilis*—it is not clear whether the unusual condition is primitive or derived within the genus *Ropalidia*. Clearly, *Ropalidia* holds many a clue to the origin and evolution of sociality.

Ropalidia rufoplagiata represents yet another kind of extreme situation—perhaps unique among all known polistine wasps (Sinha et al., 1993). Thirty-three out of 46 females on a nest were observed to oviposit on 1–17 occasions each; 12 of the 33 egg layers were mated. No clearly dominant egg layer could be identified. More surprisingly still, most young individuals foraged while most older individuals oviposited and exhibited dominance behavior and egg cannibalism. West-Eberhard (1978a) postulated a polygynous, "rudimentary-caste–containing" stage as an intermediate stage between casteless group living and a stage with a regularly occurring worker caste (see Chapter 16). But her hypothesis has been criticized because no such species has apparently been found (Carpenter, 1991; but see Itô, 1993b). *R. rufoplagiata* may well be the missing link (Sinha et al., 1993).

The nests of *Ropalidia* also vary considerably, both among the simple, open combs of the independent-founding species and among the complex, enveloped combs of the swarm-founding species, providing ample opportunities to study the evolution of nest architecture (Jeanne, 1975; Kojima,

1982a; Kojima and Jeanne, 1986; Yamane and Itô, 1994). The construction of the envelope itself appears to have evolved independently at least three times in the genus *Ropalidia* (Spradbery and Kojima, 1989).

Among the independent-founding species the proportion of single-foundress nests is an important trait in the context of social evolution because the more the multiple-foundress nests the greater the proportion of individuals opting for the sterile, or nearly sterile, altruistic worker roles. The proportion of single-foundress nests is also highly variable in *Ropalidia,* being up to 80% in *Ropalidia gregaria spilocephala* (Itô et al., 1996), 45% in *Ropalidia fasciata* (Itô and Iwahashi, 1987), and about 35% in *Ropalidia marginata* (see also chapter 4 in Shakarad and Gadagkar, 1995). Like other independent-founding polistines, those of *Ropalidia* protect their nests against ants by rubbing the nest pedicel with an ant-repellent substance secreted by the van der Vecht's gland (Kojima, 1982b, 1983a). Not only has this substance apparently been lost in the swarm-founding species of *Ropalidia,* but even among the independent-founding species, its role in protecting the nests against predation by ants varies considerably. Some species build very large numbers of closely placed secondary pedicels, which are apparently not coated with ant repellent (Gadagkar, 1991a; see also Chapter 3). Here is an opportunity to study the evolution of chemical defense against ants as opposed to physical guarding by the adult wasps.

Ropalidia shares with three other old world polistine genera—*Parapolybia, Belonogaster,* and *Polybioides*—the behavior of peritrophic sac extraction and meconium ejection (see Chapter 1). This behavior has been retained even in the swarm-founding species of *Ropalidia,* in spite of imposing severe constraints on the evolution of nest architecture (Kojima and Jeanne, 1986). In some *Ropalidia* species variable proportions of the larvae appear to be able to eject the meconium without adult aid (*R. turneri,* 85%; *R. romendi,* 49%; *R. fasciata,* 100%), although adults routinely do it for them. In contrast, adults do not remove the meconium at all in *Polistes* and in *Belonogaster,* larvae die if the adults do not remove the meconium for them (Kojima, 1983b, 1996b; Kojima and Jeanne, 1986). Thus *Ropalidia* again appears to hold the key to understanding the evolution of this complex behavior of adults and larvae.

The genus *Ropalidia* is no less appropriate for studying the role of the sting in social evolution, given the obvious differences in defensive requirements of small, independent-founding species and the large swarm-founding species. Macalintal and Starr (1996) performed a comparative study of the morphology of the sting of 39 species of *Ropalidia*. They tested three perfectly reasonable hypotheses about the sting in the swarm founding species:

1. a larger sting, relative to body size as a whole;
2. a large furcula (a small sclerite at the base of the sting shaft that aids in the precision of stinging), relative to the size of the sting as a whole;
3. more highly developed barbs on the sting lancets (the barbs can prevent the stinger from being withdrawn), leading to a higher probability of sting autotomy.

In spite of Starr's favorite hypothesis that "the sting's the thing" among enabling mechanisms in the origin of eusociality (Starr, 1985), none of these hypotheses was supported. As Macalintal and Starr (1996) point out, perhaps one should look at other parts of the sting or the venom apparatus, or perhaps one should look for differences in the venom chemistry—whichever it is, *Ropalidia* holds the key.

There are two quite different ways to utilize the power of *Ropalidia* to unravel the mysteries of social evolution. One is to focus on the tremendous diversity within the genus and undertake explicitly comparative studies. The other is to concentrate on one or a small number of species and conduct detailed analyses of cooperation and conflict and assess the costs and benefits of social life, especially by contrasting single- and multiple-foundress associations. I have chosen the latter option, and to see why my species of choice, *Ropalidia marginata* in peninsular India, is exceptionally well suited for my purpose, the reader will have to wait until the end of Chapter 4 or even until the end of the book.

PART II

Social Biology

3 SOME METHODOLOGICAL NECESSITIES

$\mathscr{I}$N THIS chapter, I will briefly describe *Ropalidia marginata* and some of the methods that have been employed in this study. In doing so I will concentrate on those aspects of the methodology that were developed or modified specifically for study of *R. marginata*. Additional methodological details are given, where necessary, throughout the book.

3.1. *Ropalidia marginata*

The social wasp *Ropalidia marginata* (Figure 3.1) was first described by Fabricius (1793), who named it *Vespa ferruginea*. It was also described as *Epipona marginata* by Lepeletier (1836) and as *Icaria marginata* by Saussure (1862) and by Bingham (1897). The combination *Ropalidia marginata* was created by Vecht (1941) and has since been used by Vecht (1962), Das and Gupta (1983, 1989), and others. *R. marginata* has been recorded from India (from most of the states except the northeastern states), Sri Lanka, Pakistan, Myanmar (formerly Burma), Vietnam, Indonesia, Malaysia, Thailand, Philippines, Papua New Guinea, Australia, and the New Caledonia and Marianna islands in the Pacific ocean (Vecht, 1941, 1962; Das and Gupta, 1983, 1989).

My identification of *Ropalidia marginata* from India, initially confirmed by J. van der Vecht and O. W. Richards, is based on Vecht (1941, 1962) and Das and Gupta (1983, 1989). Males are relatively easy to identify on account of their yellowish clypeus and their 13 segmented antennae with tyloids and a long and curved ultimate segment (Figure 3.1). Although all specimens from India have been assigned to the subspecies *R. marginata marginata* (Vecht, 1962; Das and Gupta, 1983, 1989), in my experience

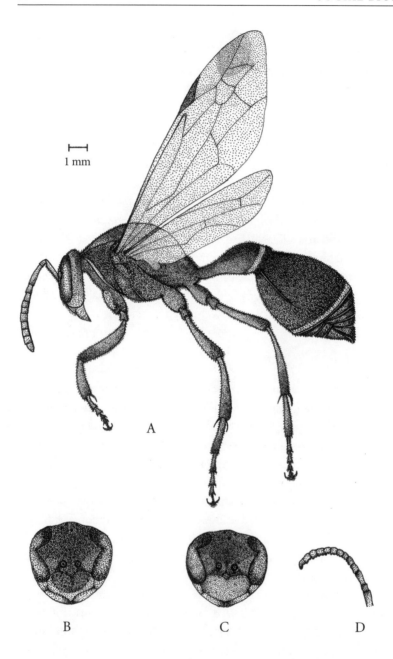

Figure 3.1. *Ropalidia marginata:* A, female in profile; B, head of the female, front view; C, head of the male, front view; D, male antenna, with the tyloids shown. (Drawing by Thresiamma Varghese.)

there is considerable variation, especially in the color pattern. I will therefore refer to all the material used in this study as *Ropalidia marginata* without assigning any subspecies status.

3.2. The Vespiary

R. marginata colonies are frequently built on eaves and windows of undisturbed buildings and other manmade structures, and occasionally on leaves and branches of some species of shrubs or trees in urban habitats. In many instances my students and I have studied colonies in their original location. For example, the initial studies of *R. marginata* described in Gadagkar (1980) and Gadagkar and Joshi (1983) and of *R. cyathiformis* described in Gadagkar and Joshi (1982a, 1984, 1985) were conducted in the original location of the nests. But this procedure often entails two kinds of difficulties. One is that the nests tend to be in relatively inaccessible locations—too high up or inside people's homes. On more than one occasion I have persuaded generous people to let me use their bathrooms and bedrooms for my observations, while they themselves moved to another part of their homes. The second and even more serious problem is frequent predation by *Vespa tropica*, whose workers almost systematically search for *Ropalidia* nests in the most likely places and prey upon the brood. On some occasions, a searching *V. tropica* worker has led me to colonies of *R. marginata*. But many a long-term experiment, especially with *Ropalidia cyathiformis* has been prematurely terminated by *V. tropica* predation.

Two fortuitous events taught me how to deal with these twin problems. On one occasion I had collected a small colony of *Ropalidia marginata*, the nest and its three adult wasps, in a small plastic box about 8 cm in diameter and about 10 cm high. Being too tired to measure and dissect the wasps, I carelessly left them in that box for the night. By the time I returned the next morning, the wasps had scraped enough pulp from their former nest, now lying on the floor of the box, and had constructed a new nest with several cells and a couple of eggs on the top corner of the box. The queen was sitting guard on the nest while the two workers were going back and forth between the old nest and the new nest and expanding the new nest. This incident convinced me that *R. marginata* is a hardy laboratory species that can withstand considerable disturbance and that will readily build or rebuild its nest in laboratory cages. Since then my students

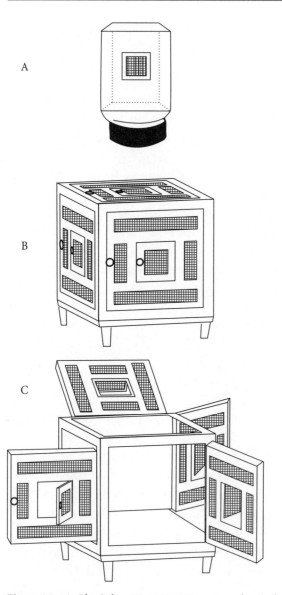

Figure 3.2. A: Plastic box 22 × 11 × 11 cm, used to isolate individual wasps, initiate single-foundress colonies, or to rear small colonies of *Ropalidia marginata*. B and C: Wood and wire-mesh cages 45 × 45 × 45 cm used for rearing large colonies of *R. marginata*.

and I have had hundreds of wasps build nests and establish active, apparently normal colonies in laboratory cages of various dimensions, ranging from plastic boxes with one or a few individuals to wood and wire-mesh cages 45 × 45 × 45 cm (Chandran and Gadagkar, 1990) with large number of wasps Figure 3.2). The wasps in these closed cages are provided with a piece of soft wood as a source of building material, final instar *Corcyra cephalonica* (Staint.)(Lepidoptera; Pyralidae) larvae, honey, and tap water.

The second fortuitous event occurred during a collection trip to the city of Mysore, about 150 km from Bangalore, where my laboratory is. The large sprawling campus of the Central Food Technology Research Institute has always been one of our favorite spots for *R. marginata* hunting. On this occasion my students and I came across a small pumphouse, no more than 4 × 4 × 6 m that seemed to be in disuse. The several large windows were open, but each window had a wire-mesh screen and the openings in the wire mesh were about 0.75 × 0.75 cm each. I have no idea why each open window was covered with such a wire mesh, but clearly the mesh had, from our point of view and from the point of view of the only living occupants of the room, one profound consequence: it did not permit the entry of *Vespa tropica*. The result was that we saw in that room, as we have never done before or since, some 5–6 extremely large nests of *R. marginata*. Each nest was the size of a large pancake, about 15 cm in diameter, and contained hundreds of adult wasps and thousands of cells with brood. This incident revealed most dramatically what *R. marginata* is capable of when freed from predation by *V. tropica*. And the message for us was clear: we could protect our observation nests from *V. tropica* by covering them with a similar wire-mesh screen. We began to experiment, first by covering naturally occurring nests with a wire-mesh hood and then by constructing cages for individual colonies with a wooden frame and wire-mesh walls. Finally, with the aid of a generous grant from the Department of Science and Technology of the government of India, we constructed the vespiary— a room measuring 9.3 × 6 × 4.8 m and covered on all four sides by wire mesh with openings 0.75 × 0.75 cm.

The vespiary has since permitted us to transplant a large number of nests, which do very well there, and to conduct experiments and observations uninterrupted by *V. tropica* attacks. It must be said, however, that such protection nearly breaks down during June through August every year, when the density of *V. tropica* and *V. affinis* is so high that some work-

ers do manage to crawl into the vespiary. In such situations we end up having to physically guard the vespiary against *Vespa*, much the way independent-founding polistines guard their nests against ants! To transplant a
nest into the vespiary or elsewhere, we collect all adults and the nest from
its natural location and bring them to the new location. Then we glue the
nest to a suitable substratum, gradually release all the adults onto the glued
nest, and, if necessary, enclose the newly released adults in a plastic box
around the nest for several hours. Later the adults continue to expand the
nest, forage in the new surroundings, feed their brood, and produce large
colonies, occasionally even approaching the size of the nests we first witnessed in Mysore.

Thus we now conduct our studies of *R. marginata* in three kinds of situations: natural undisturbed nests in their original locations; transplanted,
free-foraging nests in the vespiary; and transplanted nests in closed cages.
The free-foraging nests in the vespiary are natural in almost every respect
except for the protection from *Vespa* that the vespiary provides. We do not
feed the wasps—they forage for themselves from natural sources. Besides,
the adult wasps are free to leave their nests and initiate new nests or join
other nests, inside the vespiary or outside it. Indeed, wasps emigrating
from nests inside the vespiary as well as those coming from outside the
vespiary often initiate new nests inside the vespiary. In a 2-year study, we
recorded and followed, until the eclosion of the first adult offspring, 80
newly initiated nests in the vespiary (Shakarad and Gadagkar, 1995).

When the situation demands that the wasps of a colony be isolated from
other wasps or be prevented from leaving, we can also keep *R. marginata*
colonies in closed plastic boxes (for small colonies; Figure 3.2A) or in
wood or wire-mesh cages (for large colonies; Figure 3.2B and C). To do so
we either transplant a nest, as described above, into a closed cage, or release one or more wasps into a closed cage and wait for them to initiate a
new nest, which they readily do. As described above, such wasps are provided with food, water, and building material. Most colonies in closed
cages build apparently normal nests and maintain apparently normal social organization (Chandran and Gadagkar, 1990).

Unfortunately, we have not been so successful in transplanting and similarly protecting from predation nests of *R. cyathiformis*. For reasons that
we don't fully understand, *R. cyathiformis* is much more sensitive to any
kind of disturbance (see also Chapter 15). I have the impression that rather

than attempt to defend the nest and protect their brood against disturbance, these wasps flee and avoid any risk to themselves. This behavior
means that we cannot protect *R. cyathiformis* from predation and thus
cannot conduct long-term studies. The poor state of our knowledge of *R.
cyathiformis*, relative to our much better understanding of *R. marginata*, is
entirely attributable to this intriguing behavioral difference between the
two species. Surprising as it may seem, *R. cyathiformis* adults are extremely
sensitive even to a wire-mesh hood that I have often attempted to put
around their nests in their original locations. Although these wasps are
much smaller than *R. marginata* and hence should be able to fly in and out
of the wire mesh much more easily, most of them, sometimes including
the queen, leave the nest and sit just outside the mesh hood; they do not go
back to their nest until I remove the hood. If confined to a closed cage,
they seldom build a new nest or expand their own transplanted nest. They
are reluctant to accept *Corcyra cephalonica* larvae (even the much smaller
first or second instar larvae) or any other kind of food. We have now begun
to learn to hand feed them and hope that eventually we will learn to per-
suade *R. cyathiformis* to inhabit the vespiary along with *R. marginata*.

3.3. Nestmap and Census Records

Once a nest is chosen for present or future study, either in the vespiary or
outside, we maintain a certain amount of baseline data on it, including information about the growth of the nest, development of the brood, and
eclosion and death of the adult wasps. Information about the nest and
brood is maintained in so-called nestmap records. A drawing of the nest is
made and each cell is given a unique number; the map is updated about
every 2 days (Figure 3.3). A record is then kept of the contents of each cell,
daily or at least 3–4 times a week (Figure 3.4). Since the brood fortunately
does not move around, the cell number in which an egg, larva, or pupa is
present constitutes its unique identity. By noting down the cell number
from which each adult ecloses, we can follow every individual from the
time it was laid as an egg up until it becomes an adult and eventually dies
or disappears. By recording the identity of the wasp that laid the egg in
question, we can also infer the genetic relationships between every pair of
individuals in the colony (egg, larva, pupa, or adult)(for examples, see
Gadagkar et al., 1991a, 1993b).

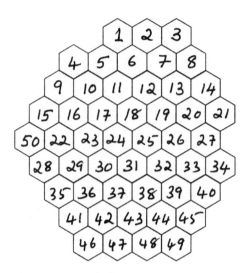

Figure 3.3. A typical nestmap.

An essential part of our methodology is the ability to mark every adult wasp for unique identification. This we do by placing small spots of colored paint on the thorax, abdomen, and, if necessary, on the wings of the wasps, using Testors enamel paints (from the Testors Corporation, Rockford, Illinois 61104, USA). Some details of our method of marking may be helpful. First, we do not handle the wasps while marking them. A little patience and skill permits putting the right sized spot of the right color on the required part of the body while the wasp is engaged in some activity. Since the paint we use is nontoxic and quick drying, this procedure causes almost no disturbance to the wasps. Each wasp is then assigned a code name based on the kind of paint marks it has received. By using one or two letter codes for each color and each part of the body and by always using various combinations of colors and body parts in a preset sequence, I initially developed a set of 120 codes: 10 colors + 1 blank = 11 × 11 = 121 combinations −1 (so that no wasp is entirely unmarked) = 120 unique codes. The advantage of the predetermined sequence in which the codes are used is that we already know the code names of wasps that are yet to eclose so that general-purpose coding sheets can be prepared and computer programs can be written. However, 120 codes soon became inadequate when we began to transplant several colonies into the vespiary and began to worry about possible movement of wasps, either deliberate or accidental, from one nest to another. Hence I developed a new system in

Ropalidia marginata, Colony T07
A portion of the nestmap record

DATE		18/4/87	20/4/87	21/4/87	23/4/87	25/4/87	27/4/87
TIME		2200	2100	2130	1700	2130	2230
CELL	1	L	L	L	L	L	L
	2	L	L	X	E	E	E
	3	X	X	—	—	—	—
	4	E	E	E	X	—	—
	5	P	P	P	P	E	E
	6	E	E	E	E	E	E
	7	L	P	P	X	E	E
	8	E	E	—	—	—	—
	9	E	E	E	X	E	E
	10	E	E	E	X	E	E
	49	E	E	E	E	—	—
	50	E	E	E	—	—	—
# EGGS		22	22	22	18	21	22
# LARVAE		20	19	13	11	9	10
# PUPAE		7	5	5	4	2	2
# EMPTY CELLS		1	4	8	11	8	5
#PARASITIZED CELLS		0	0	0	0	0	0
# TOTAL CELLS		50	50	48	44	40	39

E = EGG; L = LARVA; P = PUPA; X = EMPTY

Figure 3.4. A typical page from the laboratory notebook used for maintaining nest-map records.

which we have a practically infinite number of codes with the same 10 colors and 5 possible body parts to be marked (10 colors + 1 blank = $11^5 - 1$ = 161,050) (Table 3.1). Wasps are marked, usually on the day of their eclosion, and a census of all wasps present on the nest after dusk is taken about every other day (Figure 3.5). Almost always the wasps are back on the nest by dusk so that the census record gives information on how many and which wasps are still present in each colony. The disappearance of any

Table 3.1. Animal codes (new system).

1	R----	41	YB---	81	SG---
2	-R---	42	YL---	82	SD---
3	RR---	43	YO---	83	SW---
4	Y----	44	YS---	84	SX---
5	-Y---	45	YG---	85	GR---
6	YY---	46	YD---	86	GY---
7	B----	47	YW---	87	GB---
8	-B---	48	YX---	88	GL---
9	BB---	49	BR---	89	GO---
10	L----	50	BY---	90	GS---
11	-L---	51	BL---	91	GD---
12	LL---	52	BO---	92	GW---
13	O----	53	BS---	93	GX---
14	-O---	54	BG---	94	DR---
15	OO---	55	BD---	95	DY---
16	S----	56	BW---	96	DB---
17	-S---	57	BX---	97	DL---
18	SS---	58	LR---	98	DO---
19	G----	59	LY---	99	DS---
20	-G---	60	LB---	100	DG---
21	GG---	61	LO---	101	DW---
22	D----	62	LS---	102	DX---
23	-D---	63	LG---	103	WR---
24	DD---	64	LD---	104	WY---
25	W----	65	LW---	105	WB---
26	-W---	66	LX---	106	WL---
27	WW---	67	OR---	107	WO---
28	X----	68	OY---	108	WS---
29	-X---	69	OB---	109	WG---
30	XX---	70	OL---	110	WD---
31	RY---	71	OS---	111	WX---
32	RB---	72	OG---	112	XR---
33	RL---	73	OD---	113	XY---
34	RO---	74	OW---	114	XB---
35	RS---	75	OX---	115	XL---
36	RG---	76	SR---	116	XO---
37	RD---	77	SY---	117	XS---
38	RW---	78	SB---	118	XG---
39	RX---	79	SL---	119	XD---
40	YR---	80	SO---	120	XW---

Note: R = red, Y = yellow, B = dark blue, or L = light blue, O = orange, S = silver, G = gray, D = dark green, W = white, X = gold. 1st position = thorax, 2nd position = abdomen, 3rd position = left wing, 4th position = right wing, 5th position = for future use. Only the 1st set of 120 codes utilizing the thorax and abdomen are shown. When other parts of the body such as the wings are also used, the number of possible codes is a whopping 161,050.

Ropalidia marginata, Colony T07,
A portion of the Census Record

DATE	18/4/87	20/4/87	21/4/87	23/4/87	24/4/87	25/4/87
TIME	2240	2230	2120	2145	2010	2225
1. R T	✓	✓	✓	✓	✓	✓
2. R A	✓	X	X	X	X	X
3. R T A	✓	✓	X	X	X	X
4. Y T	✓	✓	X	X	X	X
5. Y A	✓	✓	X	X	X	X
6. Y T A	X	X	X	X	X	X
7. D B T	✓	✓	✓	✓	✓	✓
8. D B A	—	✓	✓	✓	✓	✓
9. D B T A	—	✓	✓	✓	✓	✓
10. L B T	—	✓	✓	✓	✓	✓
11. L B A	—	—	—	—	✓	✓
12. L B T A	—	—	—	—	—	✓
13. O T	—	—	—	—	—	✓
# FEMALES	6	8	5	5	6	8
# MALES	0	0	0	0	0	0
# ADULTS	6	8	5	5	6	8

✓ = Present; X = Absent; — = Not eclosed

Figure 3.5. A typical page from the laboratory notebook used for maintaining census records.

wasp during two or more consecutive census records is taken as death or disappearance of that wasp from that colony. The census records also provide very useful information about the ages of the adult wasps present on the nest at any given time.

3.4. Sampling and Recording Behavior

Most of the studies discussed in this book may be described as quantitative ethology, a discipline requiring many precautions to avoid the numerous pitfalls inherent in any attempt to "measure behavior." It was my good fortune to have been advised to read a paper entitled "Observational Study of Behavior: Sampling Methods" by Jeanne Altmann (1974) before I began my first quantitative ethological study of R. marginata. In devising our sampling methods, my students and I have paid strict attention to all the advice in Altmann's paper. I began my ethological study of R. marginata by many days of ad libitum sampling, to construct an ethogram and get qualitative information needed to design a quantitative ethological study. Indeed, we resort to ad libitum sampling from time to time under two circumstances—when we have a new observer and when we embark on a new kind of study. For the quantitative sampling of behavior we use a combination of instantaneous scans, recording all occurrences of chosen behaviors and, somewhat less frequently, focal animal sampling. Instantaneous scans consist of recording a snapshot of the behavioral state of each individual in the colony. From these records, we compute time-activity budgets, the proportions of time spent by different individuals in performing different behaviors. Such budgets are appropriate for behaviors that are of reasonable duration and that are reasonably common. Many other behaviors of obvious biological significance are rare and are usually of short duration. These are therefore more appropriately thought of as events rather than states. Information obtained from scans for such behaviors is insufficient and often inaccurate. Hence we record the number of times such rare behaviors are performed by different individuals during separate sessions. During these sessions any occurrence of any of the chosen behaviors by any individual is recorded. During focal animal sampling, a randomly chosen individual wasp is followed continuously for a fixed period of time and a continuous record of its changing behavioral state is recorded at intervals of every 10 s. By making the duration of each sampling session small and using a large number of such sessions with breaks in between, we greatly increase the accuracy of our recordings. We have used 5 min as the duration of each session with a 1-min break. The provision of 5 min for scans is notional and the actual time required for noting down the behavioral states of all the individuals varies from a minute or so to as much as 5 min, depending on the number of individuals present. Nevertheless, the behavioral state of each individual is noted at the instant we set eyes on that individual. The different sampling methods—instantaneous scans, recording of all occurrences of rare behaviors, and focal animal sampling—are randomly intermingled with each other. Observations are begun and terminated by time-contingent rules. Because the rates of different behaviors may vary depending on the time of day, we typically observe colonies uniformly throughout the day. In focal animal sampling, the individual to be focused on is chosen randomly and this choice is made beforehand. All of these procedures help to sample behavior as objectively as possible without introducing much observer bias, especially bias caused by the fact that some individuals and some behaviors may be more conspicuous than others.

I am sometimes asked whether such a "mechanical" approach to recording of behaviors does not prevent us from seeing unexpected, interesting things that may happen on the nest during our sampling sessions. The answer is no; it does not prevent us from "seeing" unexpected new things, because we train ourselves to pay attention to any unexpected, interesting behavior patterns and record them in writing on the back of our coding sheets or by speaking into a tape recorder. Such qualitative information then becomes the basis of subsequent objective, quantitative studies. Thus, while we "see" and record unexpected new things, we do not pretend to "measure" them the first time we saw them.

When I began these studies in 1980, it was still common practice, certainly in India, for ethologists to do "field work" for many months or years before commencing any significant analysis of their data. Moreover, during the field work it was common practice to make notes in long hand, almost as if keeping a diary. Then, as I witnessed so often, there was a long, painfully tedious, and monotonous period of "coding" the data for possible computer-aided analysis. This involved converting the notes into simple, repeatable number or alphabet codes so that a computer program could read the data and compute the required quantities.

To avoid this drudgery I decided to precode all my data. As mentioned above, my wasps have well-defined code names. With the aid of *ad libitum* sampling, I made as complete a list as possible of the various behaviors shown by the wasps. This list, which has some 79 entries, has simple, neomonic, two-letter codes for each behavior (Table 3.2). All new students in my laboratory undergo a training period, lasting at least a few weeks, during which they learn to recognize the behaviors of the wasps, memorize their code names, learn how to mark the wasps and recognize their code names, and receive instruction on how to keep nestmap and census records and use the different sampling methods.

Table 3.2. A partial ethogram for *Ropalidia marginata*.

No.	Code	Individual Behaviors
1	AN	Antennate nest
2	AS	Abdomen shake (up and down)
3	AW	Abdomen wag (sideways)
4	BB	Bring building material
5	BF	Bring food
6	BJ	Body jerk
7	BL	Bring liquid
8	BN	Build new cell
9	BP	Build pedicel
10	BW	Break down walls of old cells
11	BX	Return with nothing
12	C1	Initiate cannibalism of egg
13	C2	Initiate cannibalism of larva
14	C3	Initiate cannibalism of pupa
15	CB	Collect building material
16	CD	Colony defense (against ants, flies, etc.)
17	CE	Clean empty cell
18	CM	Mouth pupal cap
19	CP	Coat pedicel (with ant repellent)
20	EL	Egg laying
21	EO	Extend walls of existing cells
22	FA	Fan wings for cooling
23	FE	Feed (self)
24	FL	Feed larva
25	FO	Forage (= absent from nest)
26	FY	Fly around the nest
27	GR	Groom (self)
28	IP	Inspect pupa
29	MH	Make hole in pupal cap for inspection
30	MP	Mouth pedicel
31	MS	Mouth substrate around the pedicel
32	NB	Nibble back of nest
33	OP	Open cap of pupal cell to assist eclosion
34	PE	Inspect egg cell
35	PL	Inspect larval cell
36	RM	Remove meconium
37	RP	Remove dead pupa
38	SA	Sit with raised antennae
39	SI	Sit
40	SW	Sit with raised antennae and wings
41	WA	Walk
42	WJ	Wing jerk
43	AB*	Aggressively bite

Table 3.2 (continued)

No.	Code	Individual Behaviors
44	BI**	Being aggressively bitten
45	AE	Antennate another wasp
46	BE	Being antennated
47	AG	Allogroom
48	GG	Being groomed
49	AP	Approach another individual
50	PP	Being approached (without withdrawal)
51	WW	Withdraw when approached
52	AT*	Attack
53	BA**	Being attacked
54	BO*	Being offered regurgitated liquid
55	OF**	Submissively offer regurgitated liquid
56	CH*	Chase
57	BC**	Being chased
58	CR*	Crash land on another individual
59	BR**	Being crashed upon
60	HM*	Hold another individual in mouth
61	BH**	Being held in mouth
62	NI*	Nibble
63	NN**	Being nibbled
64	PK*	Peck
65	KK**	Being pecked
66	SB	Snatch building material
67	LB	Lose building material
68	SC	Solicit
69	SS	Being solicited
70	SF	Snatch food
71	LF	Lose food
72	SL	Snatch liquid
73	LL	Lose liquid
74	SO*	Sit on another individual
75	BS**	Being sat on by another individual
76	M1	Mutual antennation
77	AM	Aggressive mutual antennation
78	AA	Mutual approach and withdrawal
79	FF	Falling fight

Note: * = dominance behaviors, ** = subordinate behaviors.

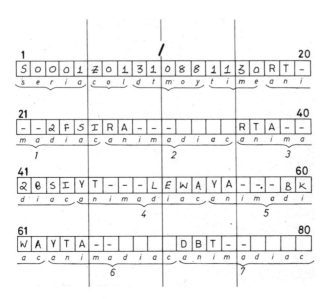

Figure 3.6. A typical page from the coding sheets that were used for recording scan data before the computer program for generating coding sheets was developed.

In addition to precoding the data, I also decided, from the very first study, to record the data in a form that could be directly transferred to a computer-readable medium—in those days punch cards! I designed simple, convenient coding sheets, an example of which is shown in Figure 3.6. Today we have come a long way. I now have a computer program that will instantly produce custom-made coding sheets for each observer, for each kind of experiment (Figure 3.7). It is easy to call up on the computer the relevant blank coding sheets, and data entry is rapid and painless. Data gathered from several hours of observation during the day can be conveniently transferred to the computer the same evening. Indeed, I have written a set of computer programs that will edit the electronic data files, help correct routinely occurring errors, compute time budgets and frequencies of rare behavior, and also generate dominance hierarchies. All this can be done in such a short period of time that the basic results of the day's "field work" can be examined that night before commencing the next day's work.

Now that commercial packages, notably the Observer from Noldus Innovation Technology, are available, all this may seem somewhat dated. Nevertheless, I have described the methods I developed in some detail be-

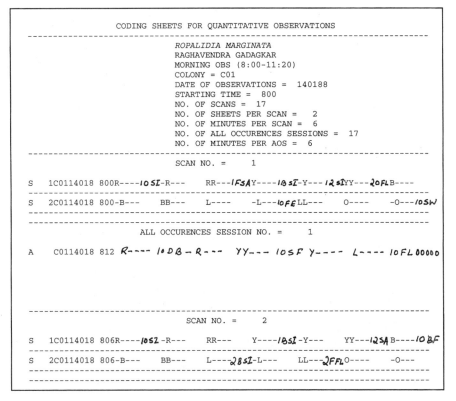

Figure 3.7. The opening page of the customized coding sheets generated by the computer program.

cause I believe that there are many beginners, especially in developing countries, who cannot afford to purchase the most recent update of commercial software. Lack of such equipment should not make them think that they cannot compete professionally with researchers who have more money. Indeed, I see no reason why a disproportionately large number of young people from poor, developing countries, with modest research budgets, should not practice quantitative ethology—a relatively inexpensive but an immensely satisfying enterprise.

4   NATURAL HISTORY

4.1.   Nesting Habits

*R*OPALIDIA MARGINATA is the most commonly encountered polistine wasp in rural and urban habitats in peninsular India. I have never seen it in any forested habitats and suspect that it prefers to nest in manmade habitats. It is not uncommon to find nests in gardens, hanging from the leaves of various species of shrubs and trees or sometimes on rocks. But by far the most preferred nesting sites are on relatively undisturbed buildings. Almost every window of the Zoology and Botany departments in the Central College in Bangalore has a flourishing nest. Some of the ecological information that I provide here comes from an extended study of that population. The first nests that I subjected to behavioral studies were located on several cypress trees in Cubbon Park in Bangalore. I and my students have frequently studied or collected large numbers of nests from various buildings on the campus of the Indian Institute of Science and the campus of the University of Agricultural Sciences, both in Bangalore, and the campus of the Central Food Technology Research Institute in Mysore. This of course is not a complete list of our collection sites—we have freely used nests from any place we could get them within about a radius of 500 km around Bangalore.

R. marginata (Figure 4.1) builds nests with simple, gymnodomous combs (open, without an envelope) (Richards and Richards, 1951), the construction of which begins with the laying down of the pedicel, which is usually 5–10 mm long and about 1–2 mm thick. The first cell is constructed at the tip of this pedicel and the subsequent cells are added either all around the first cell or only on one side so that in larger combs

Figure 4.1. A typical nest of *Ropalidia marginata* in Bangalore.

the initial pedicel may either end up being approximately in the center of the comb or at one end of it. As the nest grows in size the initial pedicel is enlarged in width and may grow up to 5–6 mm in diameter in large combs. In addition to enlarging the original pedicel, the wasps add at several points new, thinner pedicels (about 1–2 mm in diameter) that reinforce the attachment of the comb to the substratum. Most small combs (< 100 cells) have a single pedicel, while large combs (> 100 cells) often have multiple pedicels. A single comb is generally the rule, although I have once seen two combs within about 20 mm of each other with the adult wasps moving freely between them. There is some variation in nest architecture—the wasps clearly vary the shapes and sizes of their nests to suit the available nesting site (Gadagkar et al., 1982; Gadagkar, 1991a).

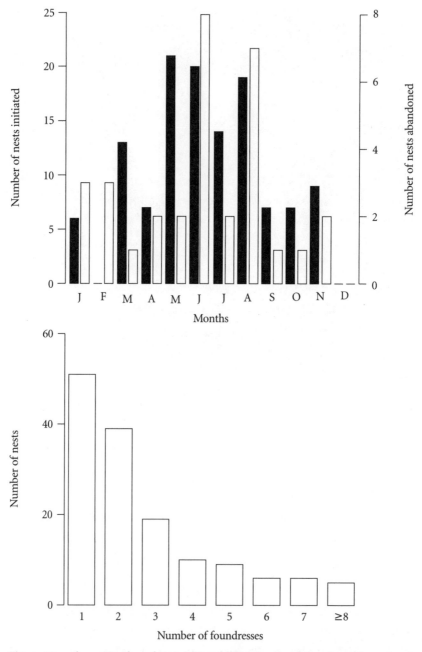

Figure 4.2. Above: Number of nests initiated (black bars) and abandoned (open bars) during different months of the year (Data from Gadagkar et al. 1982; Shakarad and Gadagkar, 1995.) Below: Frequency distribution of the number of nests with different numbers of foundresses. (Redrawn from Shakarad and Gadagkar, 1995.)

4.2. The Nesting Cycle

New nests are initiated more or less throughout the year, although initiation is more common during May to August and relatively rare during December to February (Figure 4.2). The large number of nests initiated during May to August is perhaps due to the warmer and more favorable temperatures during this part of the year, which result in the production of a large number of adults that can leave their natal nests and initiate new nests. Conversely, the initiation of very few nests during December to February is probably due to the low and unfavorable temperatures during this time of the year, which result in the production of very few adults and therefore a dearth of wasps to initiate new nests. Nests may also be abandoned any time of the year. Since many nests fail soon after initiation, the seasonal pattern of abandoning may mimic that of initiation to some extent.

As with many independent-founding polistines, new colonies of *R. marginata* may be founded either by a single female or by a group of females. A solitary foundress builds a nest, lays eggs, forages to feed her growing larvae, guards them from predators and parasites, and brings them to adulthood all by herself, until her daughters eclose and begin to stay back and assist her in rearing subsequent batches of brood. In multiple-foundress nests, only one female becomes the egg layer, or queen, and does little more than lay eggs, while the rest function as sterile workers and perform the tasks of nest building, maintenance, and brood care (Gadagkar, 1991a). Although the initial studies my colleagues and I conducted in Pune and Bangalore (Gadgil and Mahabal, 1974; Gadagkar, Gadgil, and Mahabal, 1982; Gadagkar et al., 1982) provided considerable information on the basic biology of this species, our understanding of colony founding and the natural history of young, pre-emergence colonies was very meager. This is not surprising because young, small colonies are easily overlooked on account of their small size. Some years ago we therefore undertook an extensive study, specifically focused on colony founding (Shakarad and Gadagkar, 1995), during which we monitored 145 naturally initiated pre-emergence nests over a period of 16 months. When a newly initiated nest was found, it was given a unique nest code and all the adult wasps on the nest were individually marked. Thereafter, records were maintained of the number of wasps attending each nest (census records) and the brood composition (nestmap records) at intervals of 1–2 days, at least until the first adult wasp eclosed.

Information about the number of wasps that cooperate in building and maintaining a nest is of obvious importance in the context of understanding the costs and benefits of social life. One of the aims of this study therefore was to construct a frequency distribution of the number of foundresses per colony, which as the reader will see below, is not as simple as it might appear. We found that the total number of marked wasps seen on any nest ranged from 1 to 31. However, the maximum number of wasps simultaneously seen on any nest on any given day (or night) ranged only from 1 to 25. We therefore estimated the number of foundresses per nest by averaging the number of wasps seen over all days of observation, as ranging from 0.38 to 21.54. We conclude therefore that newly initiated nests contain about 1–22 wasps (Figure 4.2). Of the 145 nests in this study, 23 nests contained only a single marked individual throughout the study period. Eight abandoned nests were adopted by a single wasp each. If all these 31 nests are classified as single-foundress nests, the percentage of single-foundress nests works out to 21.4%. However, for 20 additional nests, which contained more than one wasp on some of the days, the number of foundresses (averaged over all days of observation), when rounded off to the nearest integer, was 1, because the nests lost some of their foundresses. If we also treat these as single-foundress nests, as it is perhaps reasonable to do, then the percentage of single-foundress nests in the study population is 35.2%. The remaining 94 nests had a mean number of foundresses ranging from 1.5 to 21.5 and were considered multiple-foundress nests. Since the total number of wasps seen on all the 145 nests was 676, 4.6–7.5% of the founding population (depending on whether 31 or 51 is taken as the number of single-foundress nests) can be said to have nested solitarily, and the remaining 92.5–95.4% of the wasps can be said to have nested in groups. Because multiple-foundress nests have a single egg layer each, and every foundress on single-foundress nests is an egg layer by definition, there are at least 145 egg layers. In addition, we witnessed 13 cases of nest usurpation, 14 cases of queen turnover, and 9 cases of adoption of abandoned nests, leading to an estimate of 26.8% (181 out of 676), for the number of egg layers in the founding population. Thus some 73.2% of the population gave up reproduction and adopted altruistic worker roles. It is this apparently paradoxical behavior exhibited by such a large majority of the individuals in the population that will remain the major focus of attention throughout this book.

Immediately after nest initiation there is usually a rapid increase in the number of cells, all of which are filled with eggs. Foraging by the lone foundress or the cofoundresses (in multiple-foundress nests) is mostly to obtain building material during this so-called egg substage. When the eggs begin to hatch, the foundress(es) begin to forage for spiders, lepidopteran larvae, soft-bodied hemipterans, and so on, to feed the larvae. Mature larvae spin a cap of silk fibers on their cells and undergo metamorphosis. The period from the hatching of the first egg to the spinning of the first pupal cap is referred to as the larval substage, and the period from the spinning of the first pupal cap to the eclosion of the first adult is referred to as the pupal substage. The entire period from nest initiation to the eclosion of the first adult offspring is called the pre-emergence phase. Thereafter the nest is said to be in the post-emergence phase.

In many polistine wasp species, but not necessarily in *R. marginata* (Gadagkar, 1991a), the first offspring invariably become workers and take on the duties of foraging, nest building, feeding the larvae, and other tasks involved in nest maintenance and defense. This stage is referred to as the worker substage or the ergonomic substage. Later, males and nonworker females (referred to as reproductives, or gynes) are produced. This is called the reproductive substage. After the reproductives (male and female) eclose, there is a considerable amount of brood destruction during the so-called declining substage.

The pre-emergence phase and the post-emergence phase together constitute a colony cycle. If the nest is abandoned at the end of one such cycle, the colony cycle and the nesting cycle become equivalent and species that show this pattern are said to exhibit a determinate nesting cycle. The determinate nesting cycle may be seasonal or aseasonal; if it is seasonal, nests are initiated and abandoned at specific times of the year, and if it is aseasonal, nests are initiated and abandoned throughout the year (Gadagkar, 1991a; Jeanne, 1991b). Sometimes nesting cycles can be perennial but with strong seasonal variation in reproductive activity, foraging behavior, and brood rearing (Kojima, 1996a).

R. marginata differs from the typical temperate-zone polistine nesting cycle in important ways. The primary reason for the difference appears to be the tropical climate in which it lives. The absence of severe enough win-

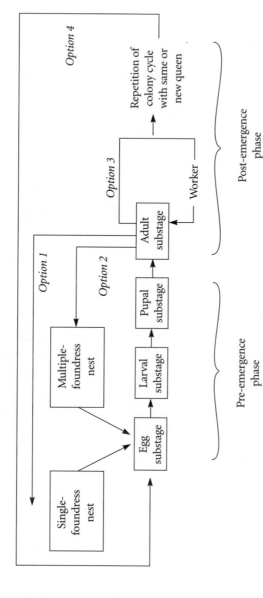

Figure 4.3. The perennial, indeterminate nesting cycle of *R. marginata*. For schematic convenience, the egg, larval, and pupal stages are shown as being distinct. In reality, there is considerable overlap between them, especially when several colony cycles are repeated on the same nest. Similarly, change of queens can take place at any time of the colony cycle. Note also that new colonies may be initiated at any time of the year and may also be abandoned at any time of the year and at any stage in the colony cycle. Female wasps have at least four different options. (Redrawn from Gadagkar, 1996a.)

ters to necessitate abandoning of the nests and hibernation of the adults, makes it possible for the nests to be active throughout the year. Thus although the nests go through all the stages of a typical determinate nesting cycle, often including the declining phase but without a distinct reproductive stage (see below), the nests are not necessarily abandoned at the end of one colony cycle. Instead, several colony cycles may repeat in the same nest, a pattern that has therefore been called an indeterminate nesting cycle (Chandrashekara et al., 1990; Gadagkar, 1991a; Jeanne, 1991b). As seen earlier, nests may be initiated and abandoned throughout the year, so that *R. marginata* colonies are indeterminate and aseasonal as well as perennial. Few species of polistine wasps studied so far seem to follow such a fascinatingly complex nesting cycle.

As might be expected, the demarcation between the egg, larval, pupal, and adult stages largely disappears during the subsequent colony cycles. Also, there is no clear demarcation between the worker substage and the reproductive substage. Indeed, the fates of wasps are largely undecided at eclosion. Most eclosing females appear to be capable of taking on both worker and queen roles, depending on the opportunities available. Thus female wasps eclosing virtually at any time of the year and at any phase in the colony cycle or nesting cycle have several options open to them (Figure 4.3). They can (1) leave their natal nests and initiate new single-foundress nests; (2) leave their natal nests and initiate or join other multiple-foundress nests along with their nestmates or even with wasps from other nests (see Chapter 9); (3) stay in their natal nest and spend their whole lives as sterile workers; or (4) stay and work for a while and later drive the original queen away and take over the natal nest as replacement queens.

The Shakarad and Gadagkar (1995) study of 145 pre-emergence colonies referred to above has shown that there are frequent deviations from the typical indeterminate colony cycle as well: a significant proportion of the nests exhibit complex and diverse fates (Figure 4.4). If a nest that is abandoned without producing a single adult offspring is said to have failed, then 48 nests out of 145 (33.1%) failed. Single-foundress nests may receive joiners at a very early stage and become equivalent to multiple-foundress nests, while multiple-foundress nests may similarly lose all but one foundress and become single-foundress nests. A nest may thus change from single- to multiple-foundress status and vice versa several times in its pre-emergence phase. Notice that the definitions of single- and multiple-

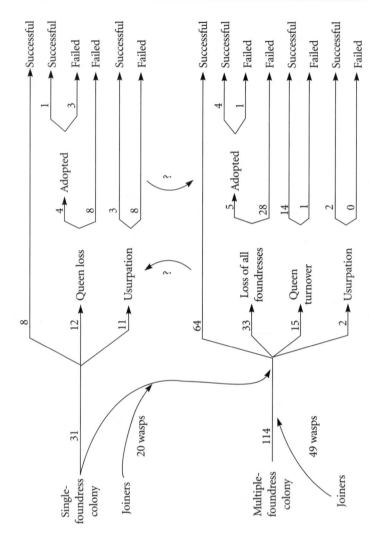

Figure 4.4. Diverse fates of pre-emergence nests. Numbers refer to numbers of nests unless otherwise indicated. Many multiple-foundress nests lost all but one foundress and became single-foundress nests and some of these later received joiners to revert back to the multiplefoundress state. The number of nests undergoing such transitions are not shown. (Modified from Shakarad and Gadagkar, 1995.)

foundress nests depend on whether one or more than one adult wasp are present during the pre-emergence phase of the colony cycle. Some nests may be usurped. When a joiner replaces the resident queen within a day or two of her joining, the nest is said to have been usurped. The original queen always disappears soon after the coming of the joiner. Of the 145 nests studied by Shakarad and Gadagkar, 13 (9%) were usurped. When a resident worker (not a joiner who has come in the preceding 1 or 2 days) replaces the existing queen, the nest is said to have experienced a queen turnover. As in the case of usurpation, the original queen always disappears. Of the 145 nests, 15 (10.3%) experienced queen turnovers in the pre-emergence phase. When a nest loses all its foundresses and then acquires a new wasp not originally belonging to it, that nest is said to have been adopted. Among the 45 nests that had lost all their foundresses in the Shakarad and Gadagkar study, 9 (20%) were adopted. Thus *R. marginata* colonies have a complex nesting cycle indeed.

Accurate estimates of the duration of egg, larval, and pupal developmental stages are of obvious importance in any attempt to quantify the costs and benefits of social life. Obtaining such estimates, however, is not easy. Our estimates (Table 4.1) show a great degree of variation both within and between study sites. Part of this variation may be genuine, but clearly part of it is attributable to inaccuracy in the methods of estimation. The observed variation in the duration of the egg stage, for example, is certainly at least partly due to egg cannibalism that goes undetected. Eggs may be eaten and replaced by new ones and several such consecutive replacements may occur before an egg successfully hatches. The variation in larval developmental times almost certainly reflects differences in food supply between nests and even between different larvae within a nest; the latter cases may be accidental or "deliberate" (see Chapter 12). A larva can complete development in as little as a week in Bangalore under laboratory conditions when the adults feeding the larva have a constant supply of food, but may sometimes take over 4 weeks under field conditions. The variation in pupal developmental times are the hardest to understand. The obvious hypothesis that a correlation between larval and pupal developmental times is the cause of this variation is not borne out because the correlation coefficients between larval and pupal development times are not significantly different from 0. This is true even in the large sample sizes from weekly observations in Pune (Gadagkar et al., 1982).

Table 4.1.   Brood developmental times. (Data from Gadagkar et al., 1982 and unpublished observations.)

	Developmental stage	Mean[a]	S.D.[a]	Sample size
Weekly observations in Pune	Egg	18	13	1221
	Larva	15	10	1052
	Pupa	16	8	1071
Daily[b] observations in Pune	Egg	12	8	64
	Larva	10	5	37
	Pupa	14	6	45
Weekly observations in Bangalore	Egg	27	15	43
	Larva	22	7	28
	Pupa	29	11	16
Daily[b] observations in Bangalore	Egg	18	4	91
	Larva	24	9	91
	Pupa	20	5	91

a. In days.
b. Or on alternate days.

Another variable of obvious importance is the lifespan of adult wasps. All male wasps eclosing on nests of *R. marginata* disappear within about 5–6 days of their eclosion and appear to lead a nomadic life, attempting to mate with female wasps who may be on foraging trips, and never return to their natal nests or to any other nest. Female wasps, by contrast, may remain on their natal nests for variable periods of time. A sample of data on 60 individually marked wasps on two naturally occurring nests in Bangalore illustrates the behavior of the females. The frequency distribution of residence times for these 60 female wasps is shown in Figure 4.5. This corresponds to a (mean ± S.D.) residence time of 27 ± 23 days and a range from 1 to 160 days. When a wasp disappears from one nest it may either have died (mortality) or initiated or joined another nest (emigration). In our records these two components cannot be distinguished directly. The (mean ± S.D.) age-specific day-to-day probability of remaining on the same nest (inset, Figure 4.5) has a value of 0.95 ± 0.04, which is nearly constant with age. This seems to suggest that mortality as opposed to emigration forms a large component of our estimates. The reasoning behind this is that mortality seems to occur largely during foraging trips because the wasps simply do not return to the nest at the end of the day. Perhaps they are lost or preyed upon. It is reasonable to assume that the probability of these events would be independent of the age of the individual and that the probability of emigration to found or join another nest would show at least some age dependence. Hence it may not be unreasonable to use the observed residence times to obtain approximate estimates of the lifespans of the wasps.

Given the perennial, indeterminate nesting cycle of *R. marginata*, it is not surprising that nests may last from a few days to several years and that colony sizes may vary from a few individuals and a few items of brood to hundreds of individuals and several hundred items of brood. This makes accurate determination of the distribution of colony lifespans or of nest sizes a difficult task, requiring long-term observation of identified nests over several years. The best data available on the distribution of nest lifespans and nest sizes in nature are shown in the bottom panel of Figure 4.5 and in Figure 4.6.

4.3. Enemies

Like any other organism, *R. marginata* has to contend with various enemies. Of minor importance are a strepsipteran that parasitizes adult females and brood parasites belonging to the dipteran family Tachinidae and the hymenopteran families Torymidae and Ichneumonidae (Belavadi and Govindan, 1981; Gadagkar, 1991a). My impression is that the adult wasps successfully ward off most of the parasites attempting to lay eggs in their nests. Of intermediate importance is predation of the brood by ants, against which the wasps have several lines of defense. A variety of ant species attempt to prey on the brood in *R. marginata* nests. The first line of defense against ant predation is the pedicel, which restricts the ants' routes of access. The second line of defense is the ant-repellent substance the workers smear on the pedicel. The prototype for these two lines of defense is the neotropical primitively eusocial wasp *Mischocyttarus drewseni* (Jeanne, 1970), where the pedicel is relatively much longer (than in *R. marginata*), making it more effective in restricting access to ants and also easier to keep coated with sufficient ant repellent. In *R. marginata* the pedicel is often so short that most ants can probably get onto the nests without crossing the pedicel. Besides, secondary pedicels are often constructed at several points to strengthen the attachment of the nest to the substrate. The very existence of secondary pedicels reduces the restriction of physical access to ants and may also reduce the efficacy of chemical re-

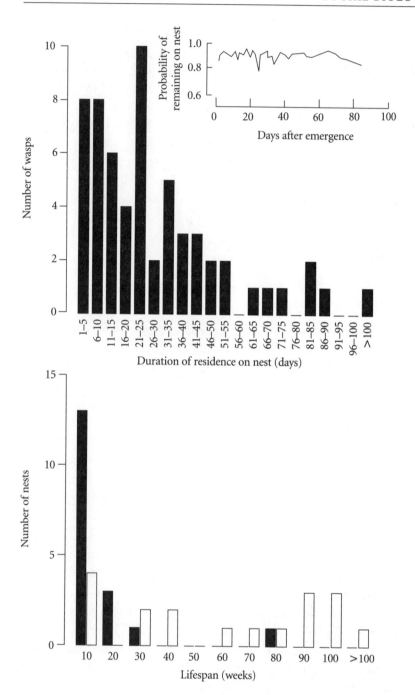

Figure 4.5 (facing page). Above: Frequency distribution of residence times on one nest of 60 female wasps of *R. marginata*. The age-specific day-to-day probability of remaining at the same nest (inset) of 0.95 ± 0.04 is nearly constant with age. Note that most of the points lie between 0.90 and 1.0. (Redrawn from Gadagkar et al., 1982.) Below: Frequency distribution of total lifespans (open bars) and minimum life spans (black bars) of *R. marginata* nests. Total lifespan is defined as the time between initiation and abandoning of a nest and is therefore known only for those nests for which both initiation and abandoning occurred during the period of study. Minimum lifespan is given only for those nests for which either the initiation or the abandoning alone is known. (Redrawn from Gadagkar et al., 1982.)

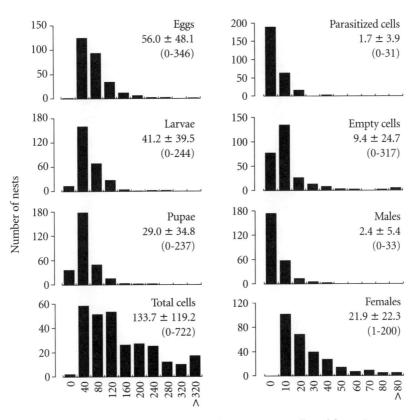

Figure 4.6: Size distribution of nests. (Data from 299 nests collected for various experiments in Bangalore between March 1982 to January 1998.) Mean ± S.D. and range of values are mentioned in each panel.

striction, since I have not seen the wasps rub their abdomens against the secondary pedicels. All these conditions probably diminish the importance of the first two lines of defense against ants (Gadagkar, 1991a).

The third line of defense is constant guarding of the nest by the adults, who become very agitated even when ants are moving nearby. This line of defense is of course unavailable to single-foundress nests, but larger nests of R. marginata appear to depend primarily on the third line of defense. The importance of the third line of defense underscores the idea that multiple-foundress associations may serve as an antipredator adaptation against ants. Unfortunately, we have no good quantitative data on the extent of predation by ants, but I suspect that few, if any, large nests are successfully attacked by ants.

Of decidedly major importance to R. marginata is predation by the tropical hornet Vespa tropica, against which the wasps have no defense whatsoever. V. tropica is undoubtedly the most serious natural enemy of R. marginata and plays a significant role in regulating its populations. As I have mentioned before, workers of the hornet species meticulously search for, locate, and consume almost the entire brood of colonies, small and large. Adult wasps are untouched and they invariably sit around the nest, completely helpless, until the hornet departs. They then return to the nest, inspect the cells with great agitation, and either cannibalize any remaining brood and abandon the nest or more frequently, continue to produce more brood on the same nest. In the latter situation the queen and some of the workers may stay while some of the workers may leave. There are other curious twists to the story: Hornets take only the brood and never prey on the adults of R. marginata. But the hornets readily take adult honey bee workers. While R. marginata make no effort to defend their nests against hornets, honey bees and some other wasps (e.g., Polistes olivaceous in southern India; personal observation) actively and sometimes successfully defend themselves against hornets.

What is R. marginata's most serious enemy—ant or hornet? More generally, should this distinction be given to the most devastating enemies seen today (hornets) or to those against whom the wasps have found it necessary to evolve effective defenses (ants)? A similar evolutionary conundrum about the relative importance of historical versus current predation concerns the function of the sting and the venom. The most striking feature of these wasps to any person who comes into casual contact with them is their unforgettable, rather painful sting. But as far as we can tell from current observations, there appears to be no profound role for the sting in the wasps' defense against parasites and predators. These are all interesting problems but are clearly beyond the scope of single species studies of the kind pursued here.

4.4. Reproductive Caste Differentiation

Let us now take up a more explicit consideration of the behavior and the interrelationships of the wasps on the nests. Much of the data in this section will be drawn from a study of 12 naturally occurring colonies in Bangalore and Mysore (Chandrashekara and Gadagkar, 1991a, 1991b, 1992). As might be expected from a eusocial species, the wasps of a colony exhibit reproductive caste differentiation into fertile queens and sterile workers. At any given time only one individual monopolizes all egg laying; the rest of the individuals, in all colonies studied, do not lay eggs while the queen is alive. The evidence for such monogyny is strong. Not only was this true for all 12 colonies in the Chandrashekara and Gadagkar study mentioned above, but during many hundreds of hours of observation by dozens of observers on a total of several hundred colonies, no one has ever seen more than one egg layer on any nest. The only exception was an experiment in which my students and I had deliberately knocked down a nest and removed the queen, to study the process of renesting by the previous occupants of the nest. Here we witnessed three of four cofoundresses (other than the queen) on a newly constructed nest, laying one egg each (see Chapter 6). However, this was clearly an abnormal situation and in any case was seen only in one of 12 such knocked-down nests. Moreover, all the three eggs laid by the cofoundresses were soon eaten by the queen and the cofoundresses then remained as sterile workers on that nest. Barring this exception, we have never seen any violation of monogyny even during times of extreme aggression or during periods of queen replacements. It is true that queens are often replaced by their nestmates, but this does not constitute simultaneous polygyny; it is instead called serial polygyny and has other interesting consequences (see Chapter 9).

Dissection of all the female wasps in a colony and examination of their ovaries confirms the behavioral observation of monogyny. Most colonies have three kinds of individuals: a queen with well-developed ovaries and at

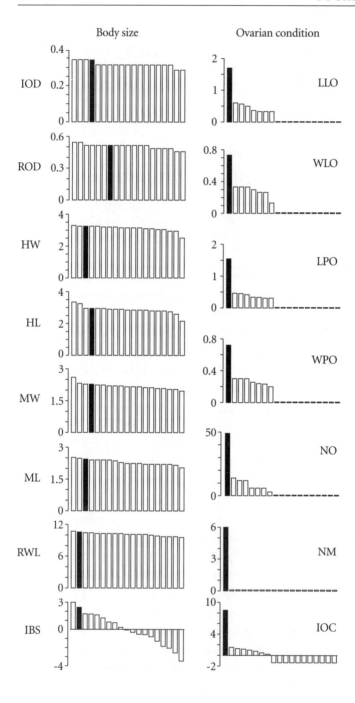

Figure 4.7 (facing page). Body size and ovarian condition in *R. marginata*. The individual values used to assess body size (left panel) and ovarian condition (right panel) are shown for all the workers (open bars) and the queen (black bars) in a typical colony. Left panel: IOD, interocellar distance; ROD, right ocello-ocular distance; HW, head width; HL, head length; MW, mesocutum width; ML, mesoscutum length; RWL, right wing length; and IBS, index of body size. Right panel: LLO, length of the largest oocytes; WLO, width of the largest oocyte; LPO, mean length of the proximal ooctyes; WPO, mean width of the proximal oocytes; NO, total number of oocytes; NM, number of mature eggs (proximal oocytes that, on microscopic examination, appeared to have a fully formed chorion, which gave them a characteristic pearly white appearance); and IOC, index of ovarian condition. For each panel, the values for different individuals are arranged in descending order. Each bar represents one individual. Note that in the body size measurements the queen is not the largest individual by any measure of body size (left panel) and that her body size relative to the worker varies from one measure to another. In the measures of ovarian condition (right panel) note that the queen always has the highest value for all measures of ovarian condition; several individuals have intermediate values; and a large number of individuals (all those other than the queen for the number of mature eggs) have a value of zero. All measurements except total number of oocytes and number of mature eggs are in mm. (Data from Chandrashekara and Gadagkar, 1991a.)

least one mature oocyte in each of the six ovarioles; a subset of individuals with partially developed ovaries containing several developing oocytes but usually no fully mature oocytes; and another subset of individuals with completely undeveloped ovaries containing threadlike ovarioles without any discernible oocytes. As might be expected from a primitively eusocial species, there is no morphological differentiation between the queen and the workers. Data on several measures of body size and a composite index of body size as well as several measures of ovarian development and a composite index of ovarian condition from a typical colony are shown in Figure 4.7. Notice that although the queen (with black bars) is among the larger individuals, by no measure of body size is she the largest individual in her colony. Notice also the very slight and gradual variation in body size among the members of a colony. On the other hand, the queen necessarily has the largest value for all measures of ovarian condition and is indeed qualitatively different from all her nestmates. Notice the very large variation in measures of ovarian development and the sharp drop in values between the queen and the intermediate individuals and again between the intermediate individuals and those with undeveloped ovaries.

The more powerful technique of multivariate statistical analysis helps

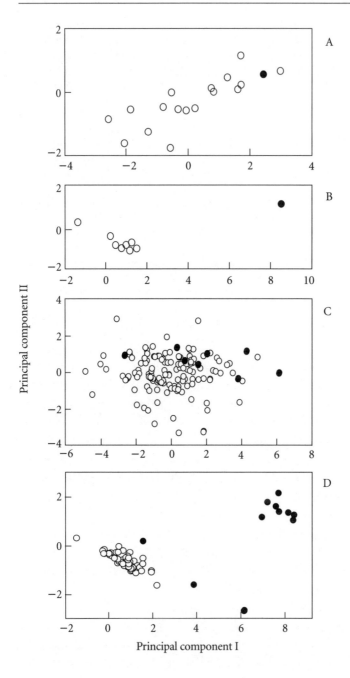

Figure 4.8 (facing page). A and B: Relative positions of queen and workers, with reference to their body size (A) and ovarian condition (B) in a typical colony. Data from Figure 4.7 are subjected to principal components analysis and the relative positions of the queens and workers are plotted in the space of the first two principal components. C and D: The relative positions of queens and workers from all the 11 colonies derived from a similar analysis. Note that in body size (C) queens and workers are not separated from each other, but that in their ovarian condition (D) queens and workers are well separated from each other. (Data from Chandrashekara and Gadagkar, 1991a.)

bring out more clearly the gradual variation in body size and the sharp variation in ovarian development. In Figure 4.8 the various measures of body size and ovarian development are subjected to principal components analysis and the relative positions of the members of a colony are plotted in the space of the first two principal components. Now we see clearly that the variation in body size is gradual and that the queen, though among the largest individuals in her colony, is not distinguished in any qualitative way from her nestmates (Figure 4.8A). The situation with ovarian development is quite different. The sharp delineation between the individuals with completely undeveloped ovaries, those with intermediate levels of ovarian development, and the queen is obvious (Figure 4.8B). Figures 4.7C and 4.7D depict the results of a similar analysis except that data from 11 different colonies are analyzed simultaneously. When variation across colonies is thus taken into consideration, a new result comes to the fore: while most workers in most colonies have poorly developed ovaries compared to the queens of most colonies (Figure 4.8D), many workers in many colonies are larger in body size compared to the queens of many other colonies (Figure 4.8C). Even if the queen of a colony is sometimes larger in body size than all her nestmates, the fact that workers of some colonies can be larger in body size than queens of other colonies is striking proof of the lack of morphological caste differentiation between queens and workers. If queens and workers are morphologically distinct, the process of caste differentiation must necessarily take place in one of the pre-imaginal stages because no morphological changes are possible in the adult stage. But in the absence of morphological differentiation between queens and workers the process of caste differentiation can also take place in the adult stage. If the wasps indeed get differentiated into queen and worker castes only in the adult stage, this makes it possible for them to utilize the multiple behavioral options provided by the indeterminate, perennial nesting cycle characteristic of the species.

As I have already mentioned, males eclosing on the nests stay for a few days and leave, never to return to the nest. Mating thus takes place away from the nest, and it is therefore difficult to observe and undertake any systematic study of mating biology; most of researchers' attention is focused on the nest, where the rest of the action is. Fortunately however, it is easy to determine whether a given female wasp is mated, because the females store sperm received from the males in a little gland called the spermatheca, and examination of the spermatheca under a microscope readily reveals the presence or absence of sperm. Of the 12 nests studied by Chandrashekara and Gadagkar, the spermathecae of all the wasps could be examined in 11 colonies. In 6 colonies the queen was mated and so was one additional individual in each colony, and these mated workers had ovaries with intermediate levels of development. Somewhat surprisingly, in the remaining 5 colonies, the queen herself was unmated although 2 of these colonies with unmated queens each had one mated worker with intermediate levels of ovarian development. This seemed somewhat unusual, and so we wondered whether these colonies with unmated queens were in some way abnormal. But closer examination revealed no apparent difference between the 5 colonies with unmated queens and the 6 colonies with mated queens. Using a total of 41 different variables that included measures of nest size, colony size, body size, and ovarian condition, as well as behavior of the adult wasps, we compared the two types of colonies using logistic regression analysis (treating the mating status of the queens as a binary variable) as well as analysis of variance (ANOVA). In the logistic regression analysis we did not find a single regression coefficient that was significantly different from zero, and in the ANOVA we did not find any significant added component of variance between colonies with mated and unmated queens, for any of the 41 variables.

It therefore appears that mating not only is unnecessary for an individual to fully develop its ovaries but also is unnecessary for an individual to assume the role of the queen, prevent all other individuals from laying eggs, and maintain normal social organization. But are unmated queens doomed to produce only male offspring? Probably not. Queens in *R. marginata* seldom leave their nests, but on more than one occasion my students and I have seen replacement queens leave their nests for several hours in the first few days after taking over the role of the queen. Such replacement queens have then been seen to produce both male and female offspring, which suggests that they can mate after taking over the role of the queen, if they have not done so earlier (Chandrashekara and Gadagkar, 1991b). However, the fact that unmated individuals can attain the position of the sole egg layer in colonies with apparently normal social organization suggests that male production by unmated females may be a significant avenue for gaining fitness in *R. marginata*.

4.5. *Ropalidia marginata* as a Model System

Several features of the biology of *Ropalidia marginata* discussed above point to its utility as a model system for understanding the evolution of eusociality. The nests are open (without an envelope, of the kind present in polybiine and vespine wasps, for example), and it is thus possible to observe all behaviors performed by all the wasps, including adult-adult as well as adult-larval interactions. The colonies are relatively small, and so it is easy to individually mark all wasps with unique spots of colored paints and to investigate the lifetime behavioral profiles of known individuals. The species is primitively eusocial and the absence of morphological caste differentiation that this entails makes it possible, at least in principle, for any or most adult wasps to take on either the role of queen or that of a worker, depending on the opportunities available. In a species where adults eclose as individuals already differentiated into queens and workers, any investigation of why some individuals accept subordinate roles as sterile workers must focus on pre-imaginal and perhaps largely physiological processes. In a primitively eusocial species such as *R. marginata*, by contrast, the most relevant investigations must focus largely on behavioral differences between different individuals before and after they begin to exhibit the queen and worker behavioral repertoires.

As we shall see later (Chapter 12), both pre-imaginal and post-imaginal processes appear to be involved in caste differentiation in *R. marginata*, which makes it an even more interesting model system. As we saw earlier, adult female wasps have multiple options open to them, such as leaving to found single-foundress nests, leaving to join multiple-foundress nests, staying in the natal nest as workers, and overthrowing the queen to take over the colony as a replacement queen. It must be mentioned that single-

foundress nests can produce female offspring who can mate and reproduce, so that an essentially solitary life cycle, though rarely seen, is by no
means impossible. The availability of these options permits the wasps to
follow complex behavioral strategies to maximize their fitness, thus setting
the stage for interaction of a variety of genetic, ecological, physiological,
demographic, and other factors that promote social evolution. The perennial, indeterminate nesting cycle of R. marginata makes the nests potentially immortal, providing ample opportunities for wasps eclosing at any
time of the year to exercise the various options potentially available to
them.

It is useful to contrast this situation with the annual nesting cycle of a
typical temperate-zone polistine wasp (West-Eberhard, 1969; Reeve, 1991):
New nests are initiated by overwintered females at the onset of spring; the
first one or two broods necessarily become workers (if only because of
the absence of any males during that time of the year); gynes (future
reproductives) and males are produced in the fall; and after that the nest is
abandoned, the queens, males, and all workers die, and the mated gynes
overwinter, to start the cycle all over again during the next spring. This
kind of cycle automatically limits the options available to wasps eclosing at
different times of the year. Wasps eclosing in early summer can seldom
found new colonies and cannot effectively challenge the queen. Wasps
eclosing in early fall can potentially do both, but it is too late in the year to
do either. In the case of R. marginata however, it is never too late in the
year to leave and start a new nest or to overthrow the queen and take over
the existing nest. The small, open nests, the primitively eusocial status, the
availability of multiple behavioral options, including solitary and colonial
nest founding, and the perennial, indeterminate nesting cycles combine to
make R. marginata a fitting subject for the investigation of the evolutionary forces responsible for promoting social life in insects.

5 BEHAVIORAL CASTE DIFFERENTIATION

5.1. Quantitative Ethology

*I*f the wasps that constitute a colony are all morphologically un-differentiated, how does the colony function as a social unit? How does *R. marginata* obtain the benefits of caste differentiation into queens and workers? How is labor divided between the members of a colony? How do workers achieve an appropriate balance between what is good for the colony as a whole and what might be best from their individual fitness point of view? These questions, which have been repeatedly asked, and answered with a great deal of success in the highly eusocial insects such as the honey bees (Winston, 1987; Seeley, 1995) and the ants (Oster and Wilson, 1978; Hölldobler and Wilson, 1990; Bourke and Franks, 1995), need to be asked separately, and addressed with different techniques, for primitively euso-cial species. And perhaps we should not expect too much help from the knowledge gained from studying highly eusocial species; at any rate we should not be blinded by what we have learned from studying the highly eusocial species—they are quite different from the primitively eusocial: they have pre-imaginal caste determination and morphological caste dif-ferentiation between queens and workers and often between subgroups of workers specializing in different tasks as well. A primitively eusocial species such as *R. marginata* has none of these features. However, morphologically similar individuals can be behaviorally diverse and, given that they have multiple options open to them, we should expect the wasps in a colony to pursue diverse behavioral strategies.

The key to answering the questions raised above must therefore lie in understanding interindividual differences in behavior. The first step in making this possible is to mark all the wasps in a colony so that they are uniquely identifiable. One only has to observe an unmarked colony of wasps and then observe after the wasps are individually marked to see the profound difference individual identification can make. Indeed, there is so much interindividual variability in *R. marginata* that long-term observations of marked wasps brings out the strikingly unique personalities of different wasps (Gadagkar, 1997d). Perhaps the most notable early proponent of this approach for studying primitively eusocial wasps was Phil Rau (Rau and Rau, 1918; Rau, 1933), and two studies that may be said to mark the beginning of the "modern" (post Hamilton, 1964) version of this approach are those by West-Eberhard (1969) and Jeanne (1972). In spite of the influential writings of all these pioneers (which I strongly recommend to all beginners), it is easy to underestimate the behavioral sophistication and especially the extent and significance of interindividual behavioral variability when humans deal with insects. I have therefore not approached my study of the behavior of the wasps any differently from the way I would have approached a study of a troop of chimpanzees. There is no denying, however, that such an approach carries with it an extra responsibility to avoid anthropomorphism and to be as objective as possible in observing and interpreting data. This responsibility I have attempted to discharge by doing quantitative, rather than qualitative ethology, and by being meticulous about using rigorous sampling methods and observation schedules. It also becomes evident from the observation of colonies of marked individuals that the wasps are not strikingly qualitatively different from each other—most individuals appear to be capable of, and seem to perform most behaviors, at least on some occasions. And this is not surprising, because in single-foundress colonies the lone foundress necessarily has to perform by herself all the tasks required to build a nest and bring her brood to adulthood. Interindividual variability then must be quantitative rather than qualitative—hence quantitative ethology.

5.2.   Time-Activity Budgets

One of my main goals has thus been to objectively measure quantitative behavioral variability among the members of a colony, discern any existing patterns in such variability, and attempt a biological interpretation of the discerned pattern. It seems reasonable to assume that such an exercise should also yield insights into the selective forces that favor worker behavior among individual wasps who are also capable of solitary nest founding and garnering the fitness benefits of direct reproduction. In any effort to objectively investigate quantitative behavioral variability, an important question is which behaviors to focus on. A perfectly reasonable approach would be to focus on those behaviors that appear to the human observer to be crucial for the well-being of the colony and the efficient rearing of brood—and this is what most previous studies had done and many studies continue to do. Such an approach would lead one to focus on behaviors involved in building the nest, obtaining food, feeding other adult wasps and the larvae, dominance and subordinate behaviors, and so on. Significant progress has been made through this approach by many students of other primitively eusocial wasps (West-Eberhard, 1969; Marino Piccioli, and Pardi, 1970, 1978; Pardi and Marino Piccioli, 1970, 1981; Jeanne, 1972; Litte, 1977, 1979, 1981; Kasuya, 1983; Strassmann, Meyer, and Matlock, 1984; and Keeping, 1992).

Without by any means abandoning, much less disparaging, such an approach, I decided to begin with a rather different approach. The question I asked myself was: Can one make an objective selection of an appropriate subset of behaviors without using criteria that seem "important" from the human point of view? To see if this was possible, I first made, in plain English, as complete a list as possible, of the various behaviors that adult wasps exhibit; I could recognize almost 80 distinct behaviors (Table 3.2). Next, I determined the proportion of time each wasp spent in each of these behaviors (see Chapter 3 for methods). It turned out, somewhat to my surprise, that all wasps spend about 95% of their time performing six behaviors: (1) sitting and grooming, (2) raising antennae (when the wasps are more alert), (3) raising wings (a sign of alarm), (4) walking, (5) inspecting cells, and (6) being absent from the nest.

Some of these behaviors may appear to be trivial, especially when compared with behaviors such as building the nest, feeding the larvae, dominance behavior, and so on. However, if the wasps spend such an overwhelming proportion of their time performing these behaviors, they must be of some significance. Let us therefore pursue this line of reasoning a little further. The time-activity budgets of wasps drawn from two nests with respect to the six behaviors listed above is shown in Figure 5.1. Notice that although each wasp spends about 95% of her time in the six behaviors put

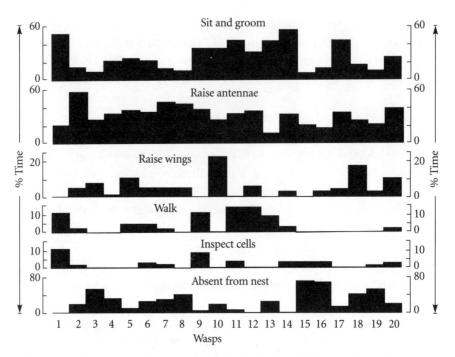

Figure 5.1. Time-activity budgets of 20 individually identified wasps drawn from two colonies of *R. marginata*. All wasps spend 85–100% (mean ± S.D. = 95.9 ± 0.4) of their time in the six behaviors shown. Note, however, that how the wasps allocate their time among these six behaviors is highly variable. (Redrawn from Gadagkar, 1985a.)

together, the manner in which different wasps allocate their time among these six behaviors is highly variable. For example, wasp 1 spent a large proportion of her time sitting and grooming and no time being absent from the nest. Wasp 2 spent a very high proportion of her time sitting with raised antennae; wasp 10, sitting with raised wings; wasps 11 and 12, walking; and wasps 15 and 16, being absent from the nest. What, if any, is the significance of such variability? Is the variability random or is there some underlying pattern? Is the variation continuous or are there subgroups of wasps that show less intragroup variability and high intergroup variability? If so, these subgroups could be thought of as forerunners of the morphologically differentiated subcastes seen among the workers of highly social species. A powerful way to detect any underlying pattern in such variability is to use multivariate statistical analysis. Principal components analysis is ideally suited for this purpose—it takes several intercorrelated input variables and, by making linear combinations of them, generates an equal number of uncorrelated output variables such that a small number of the output variables often accounts for a large fraction of the variance (Anderberg, 1973; Frey and Pimentel, 1978).

5.3. Cluster Analysis

It was my good fortune that when I realized that multivariate statistics is what was needed to take the next step in my study of R. *marginata*, my former roommate, and by then colleague, Niranjan V. Joshi, had just developed a computer program to use principal components analysis to analyze variability in the amounts of rainfall received in different parts of India and to discern some patterns in the seemingly chaotic variation (Gadgil and Joshi, 1983). Clearly, my problem was a parallel one—to make sense of the seemingly random variation in the time-activity budgets of the wasps in a colony. Thus began a very fruitful collaboration with Joshi which led to a number of insights that might not have been possible without the use of multivariate statistics. We subjected the proportions of time spent by the 20 wasps in each of the six behaviors to principal components analysis, the results of which are shown in Table 5.1. As expected, the first two principal components account for an overwhelming proportion of the variance in the data. The first principal component accounts for 72.3% of the total variance, with "absent from nest" as its dominant eigenvector ($= 0.8289$). The second principal component, whose dominant term is "raise antennae" (weightage $= 0.8219$), accounts for 20.2% of the total variance. Since the first two principal components together account for 92.5% of the total variance, it is entirely appropriate to represent each wasp as a point in the coordinate space of the associated amplitudes of the first two principal components. In other words, the method of principal components analysis allows reduction from six dimensions to two dimensions without loss of very much information. When plotted in this fashion, the points (wasps) fall into three apparently distinct clusters (Figure 5.2). The distinctness of the three clusters was confirmed by the method of nearest centroid—the distance between any point and the centroid of the cluster to which it belongs is less than its distance from the centroid of any other cluster. Individual 13 alone does not fall into any of the three clusters, because I believe, most of the data on this individual were collected when the nest was in the process of being abandoned owing to to predation by V. *tropica*. It must be emphasized here that the three clusters emerged as a result of an

Table 5.1. Eigenvectors of principal components, eigenvalues, percentage of variance, and cumulative percentage of variance. (From Gadagkar and Joshi, 1983.)

	Principal components			
Behavior	1	2	3	4
Sit and groom	−0.5305	−0.5025	0.0118	0.5835
Raise antennae	−0.1056	0.8219	−0.1701	−0.4060
Raise wings	0.0152	0.1012	0.9052	0.1252
Walk	−0.1343	−0.0690	−0.2960	0.6004
Inspect cells	−0.0457	−0.0524	−0.2357	0.1169
Absent from nest	0.8289	−0.2328	−0.0917	−0.3238
Eigenvalue	1.27×10^4	3.54×10^3	7.33×10^2	4.34×10^2
Variance (%)	72.30	20.20	4.18	2.48
Cumulative variance (%)	72.30	92.50	96.68	99.16

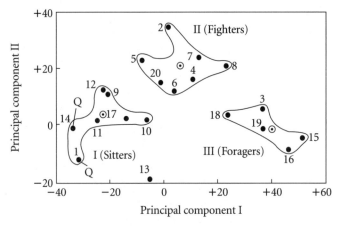

Figure 5.2. Behavioral castes of *R. marginata*. Twenty wasps are shown as points in the coordinate space of the amplitudes associated with the first two principal components. The points fall into three clusters (or castes) by the criterion of nearest centroid. Circled dot = centroid. Q = queen. (Redrawn from Gadagkar and Joshi, 1983.)

objective analysis of the data inasmuch as no a priori assumptions were made regarding the criteria to be used for classification or the number of clusters required.

Our confidence in these clusters was boosted when an independent method of classification generated identical clusters. Here we reduced the dimensionality of the data from six to just one, by computing the Pearson product moment correlation between all pairs of wasps using the time spent in each of the six behaviors. Such a correlation coefficient can be used as an index of similarity between individuals because wasps that are highly correlated in the manner in which they allocate their time among different behaviors can legitimately be said to be similar to each other. The question we asked again is whether there is a pattern in the levels of similarity and dissimilarity between different pairs of wasps. A one-dimensional index of similarity such as the Pearson product moment correlation can be conveniently subjected to cluster analysis by any one of a number of algorithms of hierarchical cluster analysis (De Ghett, 1978) (Figure 5.3). After such an analysis we found that, again, individual 13 was

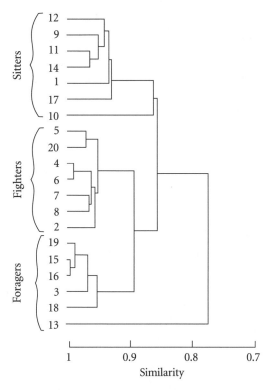

Figure 5.3. Hierarchical cluster analysis of the same 20 adults of *R. marginata* that were used in the principal components analysis shown in Figure 5.2. The similarity between individuals is the Pearson product moment correlation calculated on the basis of the percentage of time spent by the 20 wasps in the same six behaviors used for the principal components analysis in Table 5.1 and Figure 5.2. A single linkage algorithm was used for clustering. (Redrawn from Gadagkar and Joshi, 1983.)

well separated from all the others, and among the remaining 19 individuals, one could recognize three clusters with exactly the same composition
as the clusters obtained from the principal components analysis. This complete concurrence of the two methods is clearly an indication of the robustness of the clusters and gave me confidence to continue this line of
analysis.

5.4.   Sitters, Fighters, and Foragers

What is the biological significance of these clusters? This question can be
answered on two levels: first, with regard to the immediate consequence to
division of labor in the colony, which I will call the ergonomic significance
of behavioral caste differentiation; and second, with regard to the selective
forces that mold the behavior of wasps in a manner that yields these clusters, which I will call the evolutionary significance of behavioral caste differentiation. Let us approach this question at both levels, the ergonomic
and the evolutionary. A closer look at the results of the principal components analysis provides some insights. The dominant eigenvector in the
first principal component is associated with "absent from nest," and clusters I and III thus represent two extremes for this behavior, while cluster II
is intermediate. Similarly, the dominant eigenvector in the second principal component is associated with "raise antennae" and the members of
cluster II are thus different from those of clusters I and III in the time
spent with raised antennae. This line of reasoning suggests that one should
go back to the raw data and examine the behavioral profiles of the three
clusters (Figure 5.4A). Although it is not necessary for any one behavior
alone to show significant differences between the clusters (because the
clusters have been obtained by the consideration of six behaviors simultaneously), it is obvious from Figure 5.4A that the time spent in "sit and
groom," "raise antennae," and "absent from nest" are the most distinguishing attributes of clusters I, II, and III respectively. Thus I had no difficulty
in naming the wasps in cluster I sitters, but naming those in clusters II and
III was not as straightforward.

   At this point I decided to look at other behaviors that appear important to human observers but that the wasps do not spend large amounts of
time performing. Although I had decided to focus on behaviors the wasps
spent most of their time performing, I had of course recorded other behaviors, although by a different, more appropriate sampling method (see

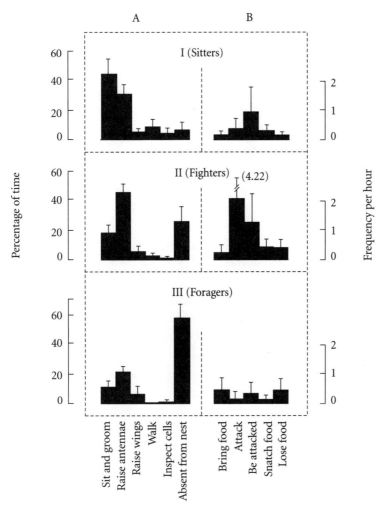

Figure 5.4. Mean behavioral profiles of the clusters obtained by principal components analysis and hierarchical cluster analysis shown in Figures 5.2 and 5.3. A: Mean percentages of time spent in each of the six activities used in obtaining the clusters. B: Mean frequencies per hour of the five behaviors not used to obtain the clusters. (Redrawn from Gadagkar and Joshi, 1983.)

Chapter 3). Not being included in the set of behaviors that the wasps spent 95% of their time in, these behaviors were naturally of short duration and were better measured by the frequency per hour with which they were performed. Some behaviors of this kind for which a reasonable amount of data existed were: (1) bring food, (2) dominance behavior (attack; see

Chapter 6), (3) subordinate behavior (be attacked), (4) snatch food (from incoming foragers), and (5) lose food (to other wasps on the nest).

These five relatively rare behaviors (which accounted for a very small proportion of the time of the wasps) can be used to construct alternative behavioral profiles of the three clusters obtained through principal components analysis of the proportion of time spent by the wasps in the six more common behaviors (Figure 5.4B). A high frequency of dominance behavior now emerges as a very conspicuous attribute of cluster II. Moreover, there is a significant positive correlation between the time spent by an individual with raised antennae (the conspicuous feature of cluster II in Figure 5.4A) and her frequency of attacking other individuals ($p < 0.01$). It seemed reasonable therefore to label the wasps in cluster II fighters.

Using a similar approach in an attempt to find a suitable name for the wasps in cluster III, I found that, relatively speaking, the most distinguishing feature of cluster III in Figure 5.4B is the frequency with which wasps brought food to the nest. Taken together with the fact that "absent from nest" is the most distinguishing feature of this cluster in Figure 5.4A, it seemed reasonable to call the wasps in cluster III foragers. Thus the adult wasps in *R. marginata* colonies were classified into three behavioral groups that can be called sitters, fighters, and foragers. It is worth recalling here that there are no morphological differences between the individuals in a colony, not even between the queens and workers, let alone between individual workers, who may specialize in different tasks. I believe that the kind of behavioral differentiation into sitters, fighters, and foragers I have demonstrated constitutes a rudimentary form of caste differentiation and hence I refer to it as behavioral caste differentiation.

It is important to note that the queens of both colonies (wasps 1 and 14) are sitters. The perceptive reader may have noticed that in this discussion of behavioral caste differentiation, I have not so far explicitly mentioned queens and workers. I did not analyze data on queens and workers separately, and I did not use the rate of egg laying as one of the behaviors while delineating the behavioral castes. Indeed, I deliberately ignored the fact that one individual was the queen and the rest were workers, during the observations and during all the data analysis presented so far, because an excessive emphasis on queen-worker dichotomy is a legacy from studies on highly eusocial species that is best avoided in studies of primitively eusocial species lacking morphological caste differentiation. Instead of making any a priori assumptions about queen-worker dichotomy in behavior, I found it more satisfying to establish the existence of behavioral caste differentiation, identify the sitters, fighters, and foragers, and then locate and attempt to interpret the position of the queen in this system of behavioral caste differentiation. Put another way, I do not think that queen-worker dichotomy, if it exists, is unimportant, but I do believe that it should not be taken as a given.

Now we are in a much better position to explore the biological significance of behavioral caste differentiation, both ergonomic and evolutionary. Sitters are those that spend much more time sitting and grooming than do other wasps. They do little or no foraging and seldom indulge in dominance behavior, either with other sitters or with any other wasps (see below). In the sitters group there are, in addition to the queens, other non–egg-laying members. The queens may be sitters because that may be the best strategy for conserving energy and maximizing egg laying. The non–egg-laying sitters may have some chance of reproducing in the future and will therefore emulate the behavior of the queens and thus maximize their chances of ascending to the status of the queen, in this or in another colony. Non–egg-laying sitters may thus be an example of what West-Eberhard (1978a) has called "hopeful queens." If this interpretation is correct, the sitters contribute little to the division of labor in the colony, though of course we must remember that at least some of the non–egg-laying sitters may be young individuals, yet to be recruited into the work force of the colony.

For understanding division of nonreproductive labor, we must perhaps turn to the fighters and foragers. Fighters spend a large proportion of their time with raised antennae, and show high levels of dominance behavior. Sitting with raised antennae probably serves the function of guarding the nest and its brood against parasites and predators. This interpretation is supported by the fact that wasps remain in this position for extended periods of time if the nest is disturbed either by human observers or by tachinid flies that parasitize their brood. Thus fighters may be like the soldiers of the highly eusocial species. What then is the function of dominance behavior, especially that shown by the workers rather than by the queens? One way to begin to understand the function of fighting by the fighters is to see who they fight with. Fighters show the highest frequency of dominance behavior toward other fighters and a lower frequency to-

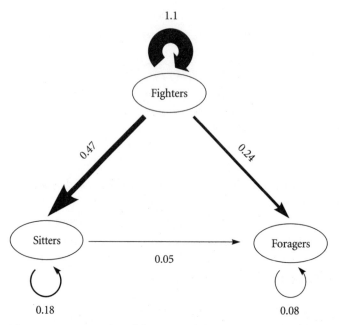

Figure 5.5. Frequencies of dominance behavior (computed as frequencies/individual/hour) within and between the behavioral castes. (Redrawn from Gadagkar and Joshi, 1983.)

ward sitters and foragers (Figure 5.5). It is conceivable that dominance shown by the fighters induces the other members of the colony to work. I have often observed wasps leave the nest as a result of repeated attacks from nestmates and later return with food or pulp. Further support for this idea comes from the observation that fighters also snatch food from other individuals to a large extent; there is a significant positive correlation between the frequency of dominance behavior and that of snatching food ($p < 0.01$). In other words, fighters may also function as "enforcers," coercing other wasps to go out and forage. But why should fighters fight more often with other fighters? One reason may be that fighters may have some chance of becoming future queens. Using high levels of aggression, especially toward other fighters, to stay on top of the social hierarchy may be another way of maximizing their chances of becoming queens in the future.

Foragers appear to constitute the principal worker force of a colony. They show the lowest frequencies of sitting and grooming, dominance behavior, and subordinate behavior and the highest frequencies of being absent from the nest and of bringing food. Therefore, they do not seem to be much involved in reproductive competition with their nestmates and may have the least chance of becoming future queens. In summary, sitters and fighters may both be potential queens and may be pursuing alternative strategies of maximizing their chances of becoming future queens, while foragers may be individuals that have little or no chance of becoming future queens and may thus be selected to work for the welfare of the colony and acquire indirect fitness.

When I completed this study, demonstrating behavioral caste differentiation using multivariate statistical analysis, I believed that it was the first study of its kind. But that was not so. One day Charles Michener pointed out to me, most politely, in the library of the Smithsonian Tropical Research Institute in Panama that Brothers and Michener (1974) had already conducted a very similar study on the primitively eusocial bee *Lasioglossum zephyrum* and had obtained results remarkably similar to those of my study. Subjecting quantitative behavioral data to principal components analysis, followed by cluster analysis, they had identified three clusters, which they had called, queens, guards, and workers. More recently, of course, several investigators have found it useful to employ multivariate statistical techniques to discern clusters of functionally similar individuals in social insect colonies that contain morphologically identical or similar individuals. This approach has often proved useful in the investigation of interindividual variability, division of labor, and social organization of insect societies (see, e.g. Fresneau, Perez, and Jaisson, 1982; Fresneau and Dupuy, 1988; Tsuji, 1988; Corbora, Lachaud, and Fresneau, 1989; Pratte, 1989; Robson et al., 2000).

5.5. Correlates of the Behavioral Castes

This was the state of my understanding when a graduate student, K. Chandrashekara, came along and offered to help me examine behavioral caste differentiation in greater detail. First we examined 12 additional colonies by the same methods of observation and statistical analysis. Next we demonstrated the existence of behavioral caste differentiation into sitters, fighters, and foragers in all these colonies (Chandrashekara and Gadagkar, 1991a). Third, we confirmed that the queens in 11 of these 12 colonies were sitters and that in only one case was the queen a fighter. With the large body of data from this study, we were able to make some more progress in understanding the biological significance of behavioral caste differentiation. In a slightly more sophisticated way we were now able to ask: In what ways are sitters, fighters, and foragers different from each other? We attempted to answer this question by using the method of logistic regression analysis.

We tested twelve variables—"sit and groom," "raise antennae," "absent from nest," "attack," "bring food," "snatch food," "feed larvae," "extend walls of cells," "index of ovarian condition," "index of body size," "dry weight (mg)," and "fat content"—for potential differences between sitters, fighters, and foragers. Each of these variables was separately modeled to test its influence on the probability of an individual's being a sitter, fighter, or forager (for further details, see Shanubhogue and Gore, 1987; Gadagkar et al., 1988, 1990; Cox and Snell, 1989; Chandrashekara and Gadagkar, 1990a, 1991a).

The coefficient of regression associated with "sit and groom" was significantly greater than zero ($p < 0.001$) in a comparison of sitters with either fighters or foragers. This can be interpreted to mean that sitters spend significantly more time in sitting and grooming than either fighters or foragers. Interpreting other coefficients similarly, we found that fighters spent significantly more time in "raise antennae" than either sitters or foragers did ($p < 0.001$), and foragers spent significantly more time in being absent from the nest than did either sitters or fighters ($p < 0.001$). These results were of course expected on the basis of the mean behavioral profiles of the castes, which were used to name them in the first place. However, we included these variables only as internal controls to establish the correct interpretation of the results of logistic regression analysis and to justify the use of this method to identify other correlates of the behavioral castes.

By examining other variables that were not used in deriving the clusters, we then attempted to establish the patterns of task allocation between the castes. Foragers brought food significantly more often than did either sitters or fighters ($p < 0.05$), and fighters did so significantly more often than did sitters ($p < 0.01$). Fighters showed dominance behavior significantly more often than did either sitters or foragers ($p < 0.05$), and sitters did so significantly more often than foragers ($p < 0.01$). Sitters and fighters snatched food, fed larvae, and extended walls of cells significantly more often than did foragers ($p < 0.05$) and were indistinguishable from each other by any of these three variables. Sitters had significantly better developed ovaries than either fighters or foragers ($p < 0.05$), and fighters had significantly better developed ovaries than foragers ($p < 0.05$). This result could have been due to the inclusion of queens in the sitter caste in 11 out of 12 colonies (recall that the queen was a sitter in 11 out of 12 colonies). When queens of all 12 colonies were excluded from the data set, we found that sitters and fighters still had significantly better developed ovaries than did foragers ($p < 0.05$), but sitters and fighters were now indistinguishable on the basis of their ovaries. The index of body size, dry weight, and fat content did not differ significantly between the behavioral castes. We also have more direct but preliminary evidence suggesting that body size is unlikely to be a determinant of dominance or egg-laying ability (Nair, Bose, and Gadagkar, 1990).

These results suggest that division of labor and social organization are closely linked to behavioral caste differentiation. The extranidal task of foraging is performed primarily by the foragers, while the intranidal tasks of feeding larvae and nest building are shared between sitters and fighters. Because both sitters and fighters have better developed ovaries than foragers, but are indistinguishable from each other by their ovaries, we can predict that new queens to replace old queens should be more likely to be drawn from either sitters or fighters, rather than from the foragers. The lack of any systematic differences between the behavioral castes in body size and fat reserves suggests that *R. marginata* gains the benefits of behavioral specialization among the workers without paying the cost of the inflexibility associated with morphological or physiological specialization.

5.6. An Experimental Approach

Correlation studies can only take us so far, and further progress in understanding the evolutionary significance of behavioral caste differentiation must come from different approaches. A powerful approach, especially to test the predictions made from correlational studies, is of course the experimental approach. Although not always feasible in evolutionary studies, the experimental approach is quite feasible in the present context. It is tedious but relatively simple to study several colonies, identify the sitters, fighters, and foragers and then remove the existing queen. Fortunately, in

R. marginata such experimental removal of the queen results in one of the workers taking over the role of the queen, and this often happens within a day or two of the removal of the original queen. One can therefore experimentally determine the probabilities with which sitters, fighters, and foragers become queens and test the prediction we have made by examining the correlates of the behavioral castes. With this aim in mind, we (Chandrashekara and Gadagkar, 1992) undertook an independent study of an additional 12 colonies. As before, the colonies were observed over a 2-week period, time-activity budgets of individually marked wasps were constructed, and these data were subjected to principal components analysis to identify the sitters, fighters, and foragers by the same criteria described before. After that, the queens of each of these colonies were removed, and after 24 to 48 hours, the same colonies were observed again for another 2 weeks. Removal of the queen, however, resulted in a disruption in the normal behavior patterns so that such a behavioral differentiation could not be clearly discerned in the post–queen-removal periods and even in some pre–queen-removal periods when the same colony was subjected to repeated queen-removal experiments.

In 9 out of 12 pre–queen-removal periods, however, the pattern was similar to that seen in undisturbed colonies. In all of these, the queen, as expected, was in the sitter group. After identifying the individual who takes over as the new queen during the post–queen-removal observations, we went back to the pre–queen-removal principal components plots and located the "wasp who would be queen." We refer to such individuals, during the period before they become queens, as potential queens. What then is the behavioral caste of the potential queens? Of the 9 potential queens we could thus identify (one in each colony), the potential queen was a sitter in six cases, a fighter in two cases, and a forager in one case. Roughly speaking, the number of potential queens that were sitters is higher than that expected by chance (there were 77 sitters out of a total of 184 wasps in the nine colonies), but the numbers of potential queens that were derived from the fighter and forager groups were less than expected by chance (there were 72 fighters and 35 foragers respectively, out of the total of 184 individuals in the nine colonies). In spite of 10 months of tedious work, the sample size of 9 potential queens is obviously too meager to permit any rigorous statistical analysis. Nevertheless, the fact that it was usually the sitters and not the foragers that become future queens is consistent with the interpretation I have made so far. But the significance of the fighters is not so clear, and perhaps will not be until we look more closely at the significance of fighting itself (see Chapter 6).

5.7. A Comparative Approach

Because of the relative difficulty of conducting appropriate experiments, and in sufficient numbers to yield large enough sample sizes, evolutionary biologists often turn to the comparative method and rely on nature's experiments. My natural inclination would have been to invest all my time and resources into delving deeper into the social biology of *R. marginata*. But the lure of the comparative approach and the easy availability of an equally fascinating, sympatric, congeneric social wasp, *Ropalidia cyathiformis*, made me temporarily stray away from *R. marginata* (Gadagkar and Joshi, 1982b). My hope was to find a species that was taxonomically closely related but behaviorally contrasting, so that some of our predictions, especially those regarding the evolutionary significance of the behavioral caste differentiation, could be put to yet another kind of test.

Thus I located a few colonies of *R. cyathiformis* (I did not have to go very far—they were right outside my office window, in the Centre for Theoretical Studies, Indian Institute of Science, Bangalore), marked the wasps for individual identification, and began observations. The first, and not so surprising, result was that more or less the same behaviors were shown by this species too. The dominance behaviors were of greater variety and complexity (see Chapter 6) but the behaviors used above for *R. marginata* were all present. Hence I constructed time-activity budgets for some 27 wasps, drawn from four colonies. The second, and somewhat more surprising result, was that the very same six behaviors—"sit and groom," "raise antennae," "raise wings," "walk," "inspect cells," and "absent from nest—accounted for about 95% of the time of each of the wasps. As in the case of *R. marginata*, there was a great deal of interindividual variability, perhaps even more than that seen in *R. marginata*, in the manner in which different wasps allocated their time among the same six behaviors. The third result was that, when subjected to principal components analysis, the first two principal components accounted for a very large fraction of the variance, as in the case of *R. marginata*. The fourth result was that when we plotted in the coordinate space of the first two principal components, we could identify three clusters of wasps that could, by the same kinds of logic used for *R. marginata,* be legitimately labeled sitters, fighters, and foragers (Gadagkar and Joshi, 1984). The similarity with *R. marginata* prevailed to the extent that hierarchical cluster analysis of the Pearson product moment correlation between pairs of wasps yielded three clusters with exactly the same composition generated by the principal components analysis.

I began to wonder what use the comparison would be if *R. cyathiformis* was indistinguishable from *R. marginata* in every respect. But such pessimism was short lived, since it turned out that there was one profound difference between the two species—the queens of *R. cyathiformis* were fighters, not sitters. It is hard to imagine a better situation for a comparative study aimed at predicting the evolutionary significance of behavioral caste differentiation, especially the significance of the rather surprising result that *R. marginata* queens were relatively docile sitters and that they let their workers do all the fighting. If *R. cyathiformis* was too different from *R. marginata,* then any differences we might discover with regard to behavioral caste differentiation might be attributable to these other differences, thus vitiating our comparison. The present situation was tailor-made for a comparative study: the two species were virtually indistinguishable in every other respect (in the context of quantitative ethology of course) and differed only with respect to the behavioral caste of the queen.

The foregoing analysis, as well as our observations on a number of additional colonies of both species (see also Chapter 6), show clearly that *R. marginata* queens are docile sitters who are not at the top of the behavioral dominance hierarchies of their colonies. Indeed, they are particularly inactive individuals who seldom initiate behavioral interactions with their nestmates. By contrast, *R. cyathiformis* queens are aggressive fighters who are always at the top of the dominance hierarchies of their colonies. They are usually the most active individuals in their colonies, frequently initiating behavioral interactions with their nestmates. If policing of the workers is required, they do most of it and do not leave it to their nestmates. The contrast between queens of the two species is especially dramatic in the early mornings: An *R. cyathiformis* queen is usually the first wasp on her colony to become active and move about the nest, spurring the rest of the wasps into activity. In contrast, an *R. marginata* queen usually remains inactive and stays sitting behind the nest well after some of the other individuals have become active and have even made one or two foraging trips.

Now why do queens of *R. marginata* and *R. cyathiformis* differ in this way? In the case of *R. marginata* queens, I have argued above that they were sitters because sitting may be the best strategy to conserve energy and maximize egg laying. Perhaps *R. cyathiformis* queens cannot afford to remain inactive and conserve energy—that may not be the route to maximizing egg laying in this species. In other words, an *R. marginata* queen may be a sitter because she faces relatively little reproductive threat from her nestmates, while an *R. cyathiformis* queen is a fighter because she faces relatively greater reproductive threat from her nestmates. This conclusion is supported by several facts. Although *R. marginata* queens can be replaced by their nestmates, their nestmates never lay eggs while the queen is still alive, as we saw in Chapter 4. *R. cyathiformis* colonies, however, sometimes have multiple egg layers—some sitters and even foragers also occasionally lay eggs while the queen (the principal egg layer) is still alive and is laying most of the eggs. The contrast between an *R. marginata* queen and an *R. cyathiformis* queen persists even when we compare monogynous colonies (with only one egg layer) of both species. Such a comparison is possible because polygyny (the presence of multiple egg layers) is not an invariant feature of *R. cyathiformis*. Another rather striking fact is that in single-foundress colonies of *R. cyathiformis,* the lone foundress (who of course is also the egg layer) is a sitter and not a fighter. In short, *R. cyathiformis* queens can also be sitters when they face no reproductive competition. Recall that even solitary foundresses can be classified as fighters because the criteria for being a fighter are based on behaviors other than fighting—that they have no one to fight with is not therefore a problem. I find it fascinating that, in spite of trying very hard, an *R. cyathiformis* queen is not entirely successful at suppressing egg laying by her nestmates, and while hardly making any apparent effort, an *R. marginata* queen is entirely successful at suppressing egg laying by her nestmates. We will see more of this, especially the significance of the docility of *R. marginata* queens, in Chapter 6, when we focus more directly on dominance behavior.

We saw that in *R. marginata* the queens are sitters, and the non–egg-laying sitters are most likely to be potential queens. The foragers constitute the principal work force of the colony. But the significance of the fighters was not entirely clear. We saw that in *R. cyathiformis* the queens are fighters, the non–egg-laying fighters are probably potential queens, and the foragers are again, almost certainly, the principal work force of the colony. Here the significance of the sitters is not so clear. To test the possibility that non–egg-laying fighters are in fact potential queens and in an attempt to discern the role of sitters, I conducted queen-removal experiments similar to those described for *R. marginata*, in section 5.5. These experiments were much harder to perform than those with *R. marginata*, because if disturbed *R. cyathiformis* much more readily abandon their nests than do *R. marginata*, which would attack and sting me instead of abandoning their nests. My students and I can also easily transplant nests of *R. marginata* into the vespiary and protect them against predation by the hornet *Vespa tropica*. But *R. cyathiformis* will not tolerate such transplantation. Thus my experiments with *R. cyathiformis* have often been prematurely terminated by predation. In the six successfully completed experiments (out of the many more that I began), not only was the queen always a fighter but the potential queen was also always one of the fighters (Gadagkar, 1987). This supports the idea that non–egg-laying fighters in *R. cyathiformis* are potential queens. Recall that this conclusion is based on observing a colony; identifying the sitters, fighters, and foragers; removing the queen; identifying the new queen; and then using the prequeen-removal principal components plots to determine the behavioral caste of the potential queen. Because I sometimes conducted successive queenremoval experiments on the same nest, such a procedure permitted me, in four out of the six experiments, to identify the behavioral castes of individuals who became queens after two successive queens were removed. In other words, queens were replaced by potential queens who were in turn replaced by the prepotential queen. In all four cases, the prepotential queen was a sitter. When a queen is removed and a fighter becomes the next queen, one of the sitters appears to use this opportunity to change her behavioral caste and become a fighter and await the next opportunity to become a queen. This supports the idea that just as fighters in *R. marginata* have at least a small chance of becoming future queens, so do sitters in *R. cyathiformis*.

Incidentally, the study of *R. cyathiformis* provided some further evidence of the robustness of the behavioral castes. My colleagues and I asked a question of obvious interest—whether the behavioral caste of an individual keeps changing with time or whether it remains constant.All the behavioral data available throughout the period of observation were pooled to give an average time-activity budget for each wasp, and these average time-activity budgets were used to obtain the clusters. To understand the effect, if any, of the age of the individual on its behavioral caste, it would be necessary to construct separate time-activity budgets for each wasp at different times in its lifespan. Unfortunately, data are often insufficient to permit this. However, for nine individuals time-activity budgets were prepared separately for each two weeks of their lives. With these time budgets their corresponding positions in the coordinate space of the principal components were computed. The fortnightly behavioral profiles were rather close to each other and to their corresponding mean profiles. This finding suggests that most animals did not drastically change their behavioral profiles with age. It is true, however, that in the first week of their lives, all wasps do little other than sit and groom. Very young individuals would thus get classified as sitters, but after this first week the behavioral caste of an individual seems stable at least for some time (Gadagkar and Joshi, 1984).

Another indication of the stability of behavioral caste differentiation comes from the fortuitous fact that during the course of my observation some of the wasps left their original nest and went on to found new nests nearby. Since I treated data on each wasp on each nest separately, I could confirm that each of the five individuals who left their old nests and founded a single new nest fell into the same cluster as before, even after they switched nests. Even more significant, this is also true for the individual that was not an egg layer in the old nest but became the sole egg layer on the new nest. All this does not of course mean that the behavioral castes cannot ever change in response to such drastic perturbations as the death of the queen or of a significant fraction of the foragers. Indeed, we saw above that when queens were experimentally removed, some sitters changed their behavioral caste and became fighters.

In summary, morphologically identical wasps in *R. marginata* and *R. cyathiformis* colonies can be classified into three behavioral castes, labeled sitters, fighters, and foragers, after their distinguishing features. As a result of assessment of the behavioral correlates of the castes, experimental removal of the queens to determine the behavioral caste of potential queens, and a comparative study of two species whose queens belong to different behavioral castes, we now have some understanding of the ergonomic and evolutionary significance of behavioral caste differentiation. In *R. marginata*, the queens are sitters, and the remaining sitters are potential queens, while the fighters also have some, though perhaps a somewhat lower, chance of becoming future queens and the foragers have the least chance, if any, of doing so. Foragers largely perform the risky extranidal tasks associated with foraging, while intranidal tasks are shared between the sitters and fighters. In *R. cyathiformis* there is a somewhat similar situation, but the queens are fighters and the remaining fighters are potential queens. The sitters also have some, but perhaps a lower, chance of becoming future queens, and the foragers again have the least chance, if any, of doing so. The risky extranidal tasks associated with foraging are largely performed by the foragers, and it appears that intranidal tasks are shared between the sitters and fighters, although my evidence for this is weaker than that for *R. marginata*.

I suggest that this system of behavioral caste differentiation has evolved in response to the predicament that the wasps find themselves in—most individuals have a finite probability of becoming queens in their lifetime and gaining direct fitness, and most individuals end up dying as sterile workers, with only indirect fitness to their credit. Wasps must then be selected to attempt to maximize their chances of becoming queens and gaining direct fitness. Such selection pressure may, however, work to the detriment of the colony as a whole and thus reduce the wasps' indirect fitness. Hence the wasps must also be selected to work toward the welfare of their colonies and to maximize their share of indirect fitness, just in case they die without any direct fitness—net inclusive fitness (direct component + indirect component) is what is finally reckoned by natural selection. It is not surprising therefore that queens do whatever it takes to maximize their rates of egg laying—become sitters in *R. marginata*, where they face relatively little reproductive threat from their nestmates, and become fighters in *R. cyathiformis*, where they face greater reproductive threat from their nestmates. Individuals who emulate the behavior of the queens—non–egg-laying sitters in *R. marginata* and non–egg-laying fighters in *R. cyathiformis*—have the highest chance of becoming future queens. Thus both queens and potential queens expend more effort to gain direct fitness and somewhat neglect indirect fitness. Fighters in *R. marginata* and sitters in *R. cyathiformis* have relatively lower probabilities of becoming future queens, but that probability is not zero. Not surprisingly, they perform mainly the less risky intranidal tasks. Finally, foragers in both species have the least chance of becoming future queens and, not surprisingly, they perform most of the risky extranidal tasks. Behavioral caste differentiation thus appears to permit the wasps to strike a fine balance between cooperation and conflict (Chandrashekara and Gadagkar, 1990b; Gadagkar, 1997d).

6 DOMINANCE BEHAVIOR AND REGULATION OF WORKER ACTIVITY

*S*EVERAL questions arise from what we have learned in Chapter 5 about behavioral caste differentiation and its possible ergonomic and evolutionary significance in *R. marginata*, and especially about the position of the queen in this system. Here I will attempt to address four such questions.

1. If the queen is indeed a meek sitter, how does she establish and maintain her complete reproductive monopoly in the colony, which we know she does with great success?
2. During the frequent queen replacements inevitable in the perennial, indeterminate, nesting cycle of *R. marginata*, how is the queen's successor chosen?
3. What is the significance of the dominance-subordinate behaviors, shown largely by the workers?
4. What wasp, if any, regulates worker activity in *R. marginata* colonies, with their docile queens and aggressive workers?

Before I attempt to answer these questions, let me examine the nature of dominance behavior a little more carefully. The study of 12 colonies by Chandrashekara and Gadagkar (1991a), used to assess the correlates of behavioral castes, is a good starting point.

6.1. The Nature of Dominance Behaviors

A variety of dominance behaviors are shown by the wasps in a colony. One wasp, dominant by definition, may attack, peck, chase, or nibble another, subordinate by definition. The frequencies of all these behaviors are pooled to obtain the frequency of dominance behavior for each individual. Similarly, the sum of the rates at which each individual is attacked, pecked, chased, or nibbled is her frequency of subordinate behavior. One member of a pair of individuals was nearly always (in 159 out of 163 interactions observed in 12 colonies) dominant over the other in all interactions between them, and thus their dominance-subordinate status was unambiguous. In other words, for any given pair of individuals that showed dominance-subordinate interactions between themselves, we can nearly always say unambiguously which wasp is dominant over the other. But how do we make a similar statement for those pairs of individuals that did not interact with each other at all, throughout the period of observation? And there are many such pairs of individuals, as can be seen from Table 6.1, which gives the observed matrix of interactions in a typical colony. Indeed, there are many more pairs that do not interact with each other than there are pairs that do interact with each other. Even more problematic are individuals that simply did not participate in dominance-subordinate interactions with any wasp in their colony; their frequencies of both dominance and subordinate behaviors were zero. Table 6.1 reveals again that there are several such individuals. Are these individuals to be excluded from a description of the dominance relationships of their colonies, or are they, even without themselves displaying dominance or subordinate behavior, somehow part of the system?

Using the data in Table 6.1, we (Chandrashekara and Gadagkar, 1991a) attempted to construct a dominance hierarchy for each colony by simply connecting dominant members of a pair to their subordinate partners by means of arrows. A simple linear dominance hierarchy could not be obtained, which of course is not surprising, because all individuals do not interact with all the other individuals in the colony. Instead, a complex network of dominance-subordinate relationships was obtained for all colonies; it was impossible to assign clear-cut dominance ranks to individual wasps. The network arising out of the information in Table 6.1, which is incidentally of intermediate complexity (in some other colonies, the network is even more complex), is illustrated in Figure 6.1.

Table 6.1. Matrix of dominance-subordinate interactions shown among the members of a colony of *R. marginata.* Note that only a few wasps interact with each other and that most wasps do not interact with most others.

Wasp nos.	1	2	3	4	5	6	7	8	9	10	11	12	13	14Q	15	16	17	18	19	20	21	22	23	24	25	26	27	28	29	30	31	32
1																																
2																																
3																																
4																1								1								
5																																
6																																
7													2																			
8																				1				1								1
9																																
10																																
11						1																										
12																																
13								1																								
14Q																								1				1				
15																												1				
16																		1														

17	1				
18			2		1
19					
20					
21		1			
22	1			1	
23					
24		1		1	1
25					
26					
27					
28			3		
29					
30					
31					
32					

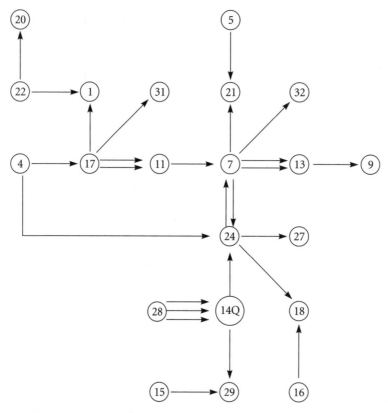

Figure 6.1. The dominance-subordinate network in a typical colony as an example of intermediate complexity. Arrows connect dominant animals to their subordinate partners. Each arrow represents one instance of dominance behavior. Since all animals were observed for the same amount of time, these numbers are directly comparable. Each circle represents a wasp and the number inside is its code. Of 32 individuals, only 22 figure in the dominance-subordinate network shown here; the remaining 10 individuals did not show any dominance or subordinate behavior. (Redrawn from Chandrashekara and Gadagkar, 1991a.)

6.2.   How Does One Construct a Dominance Hierarchy?

How then does one construct a dominance hierarchy and assign unique ranks to all individuals in the colony? With the data contained in a matrix of dominance-subordinate interactions such as those shown in Table 6.1, we should be able to state unambiguously which of the two wasps in a pair of indivduals is higher in rank or that the two have the same rank. In order to infer the relative positions of different individuals in a colony, we need to consider several kinds of information. First of course we must examine the direction of each dominance interaction and ask which wasp is dominant over the other. But that is not enough, because one act of dominance is not the same as 10 acts. We should therefore also consider the frequency of dominance behavior—how many times an individual dominates another. But even that is not enough, because a high frequency of dominance behavior may also be accompanied by an equally high frequency of subordinate behavior. Thus we must also pay attention to an individual's frequency of subordinate behavior. Finally, showing dominance over a high-ranking individual should be weighed differently from showing dominance over a low-ranking individual. Similarly, being subordinate to a low-ranking individual should be weighed differently from being subordinate to a high-ranking individual.

The index of dominance (D) described below, which is a modified form of the index of fighting success developed for red deer by Clutton-Brock et al. (1979), helps effectively deal with all these problems (Premnath et al., 1990).

(6.1)
$$D = \frac{\sum\limits_{i=1}^{n} B_i + \sum\limits_{j=1}^{m} \sum\limits_{i=1}^{n} b_{ji} + 1}{\sum\limits_{i=1}^{n} L_i + \sum\limits_{j=1}^{p} \sum\limits_{i=1}^{n} l_{ji} + 1}$$

where ΣB_i measures the rates at which the subject shows dominance behavior toward colony members and Σb_{ji} measures the sum of the rates at which all individuals dominated by the subject in turn show dominance behavior toward colony members; 1 to m are thus individuals toward whom the subject shows dominance. Similarly, ΣL_i measures the rate at which the subject shows subordinate behavior toward colony members, and Σl_{ji} measures the sum of the rates at which those individuals toward whom the subject shows subordinate behavior in turn show subordinate behavior toward other colony members. Thus 1 to p are the individuals toward whom the subject shows subordinate behavior.

When the index of dominance is calculated for all the 32 individuals in Table 6.1 and the wasps are then arranged in descending order of their dominance indices, the chaos in Figure 6.1 gives way to the order in Fig-

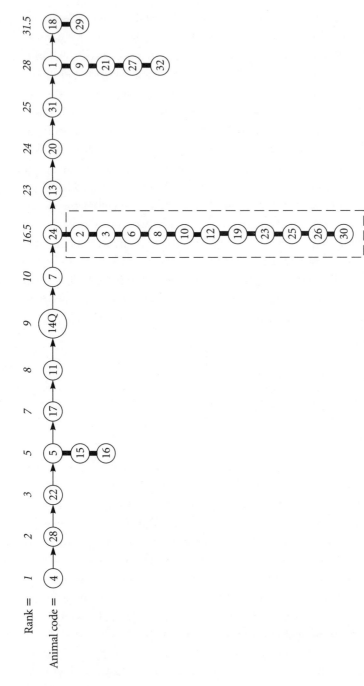

Figure 6.2. The dominance hierarchy in *R. marginata*. The network of dominance-subordinate relationships shown in Figure 6.1 is converted here into a dominance hierarchy by using the index of dominance described in the text and arranging the dominance index scores in descending order. Tied individuals are ranked one below the other. The dotted rectangle encloses those individuals who obtained a dominance index value of 1.0 by virtue of not interacting with any other wasp in the colony. (Data from Chandrashekara and Gadagkar, 1991a.)

ure 6.2. The dominance hierarchy in Figure 6.2 demonstrates an additional advantage of the index of dominance described by equation 6.1: even those individuals that never participated in dominance-subordinate interactions with any wasp in their colonies can be assigned a place in the colony's dominance hierarchy. One can also see from Figure 6.2 that (except for those at rank 16.5, see below), there are relatively few ties, and we can therefore assign a unique rank to most wasps. At rank 16.5, wasp number 24 has a dominance index of 1.0 because both its numerator and denominator in equation 6.1 are 8.0. The remaining 11 wasps tied at rank 16.5 also have a dominance index of 1.0, but that is because they did not participate in any dominance-subordinate interactions at all and merely got the last terms (1) in both the numerator and the denominator of equation 6.1. What this means of course is that individuals not participating in any dominance-subordinate interactions got a higher score on the index of dominance than those that did participate but were more subordinate than they were dominant—had a smaller numerator and a larger denominator in equation 6.1.

6.3.   The Position of the Queen

We have already seen in Chapter 5 that the queens of *R. marginata* are relatively docile sitters in comparison to many of their nestmates, which may be described as aggressive fighters. Using the frequencies of dominance behavior and the dominance index developed above, let us now examine the position of the queen in the system of dominance relationships prevailing in *R. marginata* colonies. And once again, the study of 12 colonies by Chandrashekara and Gadagkar (1991a) comes in handy. In 3 out of the 12 colonies in this study, the queen did not participate in dominance interactions at all, although several such interactions were recorded among her nestmates. This result could not have come about merely because of insufficient sampling effort. In one of these 3 colonies Chandrashekara and I recorded 75 instances of dominance behaviors, 14 of them by a single worker—but none by the queen. In another 3 of the 12 colonies, the queen participated in dominance interactions in her colony but was dominated by one or more of her nestmates. Even in the remaining 6 colonies, where none dominated the queen and where she herself participated in dominance-subordinate interactions, the queen was never the one to show the

Table 6.2.   The position of the queen in the dominance hierarchy of *R. marginata*
colonies. (Data from Chandrashekara and Gadagkar, 1991a.)

Colony code	Number of wasps	Rank of queen
C01	18	1
C02	13	8
C03	10	2
C04	10	4.5
C05	9	3
C06	25	12
C07	32	9
C08	17	2
C09	7	1.5
C10	20	2.5
C11	24	12.5
C12	15	1

highest frequency of dominance behaviors in her colony. Out of the total
number of dominance behaviors shown in a colony, the proportion shown
by the queen was indistinguishable from that shown by one or more of her
nestmates in 3 of these 6 colonies (test of proportions, $p < 0.05$) and significantly less than that shown by at least one of her nestmates in the remaining 3 colonies (test of proportions, $p < 0.05$). In no colony and by no
obvious criterion can we therefore conclude that the queen is at the top of
the behavioral dominance hierarchy of her colony.

Now let us turn to the index of dominance defined by equation 6.1.
When a dominance hierarchy was constructed by computing the index of
dominance for all wasps in each of the 12 colonies, the queen was ranked
number 1 (with the highest value of D) only in 2 colonies. She was tied at
rank 1.5 with one of her workers in one colony, and she was ranked lower
than at least one to eight of her workers in the remaining 9 colonies (Table
6.2). Although the frequencies of dominance behavior did not assign the
top position to the queen in any colony, the fact that we can now assign the
top position (rank = 1) to queens of at least some colonies underscores the
value of using a robust index of dominance such as the one described
above. Nevertheless, these results largely confirm the previously obtained
result that queens of *R. marginata* colonies are not necessarily at the top of
the behavioral dominance hierarchies prevailing in their colonies. Readers

will recall from Chapter 5 that, in contrast to the docile sitter queens of *R. marginata,* queens of *R. cyathiformis* are aggressive fighters. As might be expected, when similar techniques of constructing a dominance hierarchy are applied to *R. cyathiformis,* the queens always come out on top of the dominance hierarchy (Figure 6.3).

6.4.   The Establishment and Maintenance of Reproductive Monopoly

Let us now take up the question of the establishment of reproductive monopoly. Given that queens in established colonies of *R. marginata* are relatively inactive sitters, seldom on the top of the dominance hierarchies of their colonies, do they also begin their careers as inactive sitters or do they need to be overtly aggressive in order to establish their status as queens in the first place? With the help of a student, Sudha Premnath, and a post-doctoral fellow, Anindya Sinha, I addressed this question by comparing the behavior of queens in the following different phases of the colony cycle: the queen of a newly founded colony (founding queen), an individual in the process of challenging the present queen (challenger), a newly installed replacement queen (new queen), and of course a normal queen in an established colony (established queen)(Premnath, Sinha, and Gadagkar, 1994, 1996a).

Studying the behavior of a new queen was the hardest. New colonies of *R. marginata* are initiated at all times of the year and are rather inconspicuous in their early stages—seldom can they be located until after a nest has been built and it contains at least a few eggs, if not larvae. At this stage, perhaps the queen has already established her position as the sole reproductive of the colony and thus become an "established queen"—a docile sitter. In order to observe the behavior of a queen at the earliest stages of the nesting cycle, when she has presumably not yet had the opportunity to establish her status, our only option was to experimentally simulate the founding of new colonies. But how could we do this? We knew that when *R. marginata* colonies are attacked by the predator *Vespa tropica* (see Chapter 4) or are destroyed by other predators or by wind and rain, many or all the wasps abandon their nests and initiate one or more new nests. Each new nest has its own subset of wasps, one of which becomes the

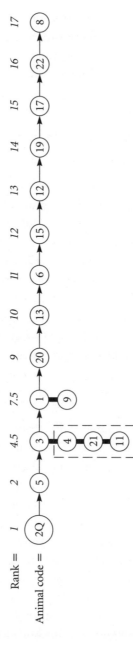

Figure 6.3. A typical dominance hierarchy of *R. cyathiformis* constructed as described for *R. marginata* in section 6.2 and Figure 6.2.

queen. Predation by *V. tropica* on large and conspicuous colonies is fairly frequent, so that many nests of *R. marginata* may actually be initiated in this fashion.

We exploited this phenomenon of renesting after a calamity by simply knocking down large nests and thus giving the wasps an opportunity to renest. Since we were interested in studying the establishment of new queens, we removed the original queen before knocking down the nest. The removal of the original queen also caused the rest of the colony members to fragment into more and smaller subgroups than they usually do in her presence. Twelve such nest knock down experiments were performed. Each nest was subjected to 2 days (10 h each day) of observation prior to nest and queen removal. On the day of nest and queen removal (day 1), all wasps returned (often repeatedly) to the site of their original nest and there performed frequent dominance-subordinate interactions. During these interactions, a single wasp became clearly dominant over all others in every one of the 12 experiments. We called this wasp the alpha individual. In the course of the next 2 or 3 days, a new nest was initiated at or very near the original nest site and the alpha wasp invariably became the queen of the colony in the new nest. During this period, the wasps that did not return to the original nesting site and join the foundress association there, but instead settled at various other places nearby and eventually formed their own colonies with 1 to 10 individuals. Dominance hierarchies were also established in each of these colonies, with the most dominant individual in each colony becoming the queen of that group.

From the 12 nests that were knocked down, we observed that 48 nests were initiated by 1 to 10 individuals. But these nests accounted for only about 51% of the original occupants of the knocked-down nests. The remaining wasps may have initiated nests at other places where we did not manage to find them. Of the wasps that returned and stayed on at the original nesting site or nearby sites that we were able to observe, some 34.5% became egg layers, giving us a large sample of founding queens to study the process of establishment of reproductive monopoly by queens of *R. marginata*. These experiments gave us a clear answer to one of the questions that we began with: How does an *R. marginata* queen, a docile sitter in established colonies, begin her life as a queen and establish her reproductive monopoly over all her nestmates? The answer is simple: by intense overt aggression—qualitatively and quantitatively totally different from

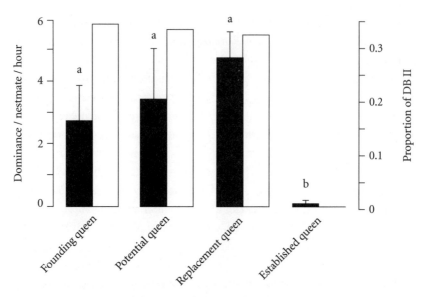

Figure 6.4. Frequency of dominance behavior (black bars) and proportion of DB II (more aggressive and dominant) behaviors (open bars) for queens of *R. marginata* during different stages of the colony cycles. Bars that carry different letters are significantly different from each other. (Data from Premnath, Sinha, and Gadagkar, 1996a.)

the behavior of an established queen. On the first day that groups were formed, the founding queen was intensely aggressive toward all her nestmates, showing a frequency of about 2.7 acts of dominance per hour, as compared to the typical rate of 0.1 acts per hour (Figure 6.4). The queen contributed about 76% of the total dominance behaviors seen on that day. The next most dominant individual showed only about 0.3 acts of dominance per hour and accounted for only about 9% of the total acts of dominance seen on that day. It was also striking that no founding queen was ever dominated by any of her nestmates on the day the subgroups were formed.

The nature of dominance behavior shown by founding queens was also strikingly different from that shown by established queens or other wasps on established colonies. In addition to the repertoire of queens and workers in established colonies that consists of attacking, pecking, chasing, and nibbling, which we labeled category I dominance behavior (DB I), founding queens showed much more intense and longer-lasting aggressive biting and aggressive soliciting behaviors. Frequently the founding queen and her partner in an aggressive act rolled on each other and fell to the ground, occasionally causing injury to the subordinate individual. These more aggressive forms of dominance behavior we labeled category II (DB II). Of the 2.7 dominance acts per hour shown by founding queens, 0.93 (34%) were in DB II category (Figure 6.4). Such intense aggression is seldom seen on mature colonies with established queens except occasionally during queen replacements.

Two of the newly founded nests were monitored until the hatching of the first egg to see if and when the transition between the aggressive behavior of the founding queen and the docile behavior of an established queen would occur. The founding queen's level of dominance behavior fell sharply by about the fourth day, after which she did not ever exhibit the more aggressive, DB II type of behaviors shown by founding queens. The level of dominance behavior shown by the founding queen fell below the mean value for her cofoundresses at about the time of the hatching of the first egg, about 25 days after the group was formed.

It must be mentioned that in one of the nests observed until the hatching of the first egg, three of the four cofoundresses of the queen showed relatively higher levels of dominance behavior than cofoundresses in other nests, 2 to 5 days after group formation. Although the queen laid most of the eggs as expected, each of these three aggressive cofoundresses also managed to lay an egg each. However, the queen ate these eggs and subsequently suppressed all egg laying by her cofoundresses. The three eggs laid by these three cofoundresses are the only examples of egg laying by more than one individual in any nest that we have ever witnessed.

In mature colonies with established queens, the quality and quantity of dominance behavior exhibited by founding queens during the first 4 days of group formation can only be witnessed during queen replacements. Since the timing of queen replacements cannot be predicted, the behavior of the individuals that challenge their queens (potential queens) (six instances) and the behavior of the new queens after replacement (nine instances) were observed opportunistically. Potential queens showed 3.4 dominance acts per hour and replacement queens showed 4.7 acts per hour (Figure 6.4). Potential queens and replacement queens also showed 1.1 acts per hour (33%) and 1.5 acts per hour (32%) respectively of the more aggressive DB II type of dominance behaviors described for found-

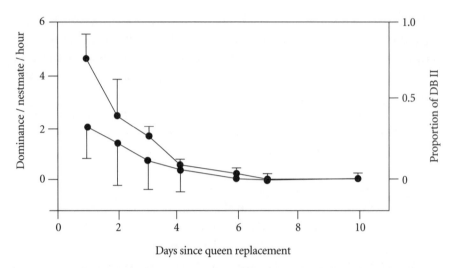

Figure 6.5. Dominance acts per nestmate per hour (upper line) shown by replacement queens and the proportion of DB II behaviors (lower line) from the day of takeover up to 10 days after queen replacement. Means and 1 S.D. are shown for nine nests for days 1–7 and six nests on day 10. (Redrawn from Premnath, Sinha, and Gadagkar, 1996a.)

ing queens (Figure 6.4). The rates of dominance behaviors recorded for founding queens, potential queens, and replacement queens were all significantly different from the corresponding rates for established queens (Mann-Whitney U test, $p < 0.05$). In one nest we were able to observe the change in the dominance behavior of the replacement queen from the day she became the replacement queen up to 10 days later. As in the case of the founding queen, her frequency of dominance behavior fell sharply for the first 4 days after queen replacement and then gradually reached the level characteristic of established queens in mature colonies, by about the tenth day (Figure 6.5). Similarly, the proportion of DB II behavior in the repertoire dropped from above 35% on day 1 to zero by about day 6.

These studies thus solved the paradox of how the relatively inactive, docile sitter queen of *R. marginata* establishes herself as the sole reproductive of her colony: she does so by displaying both dominance behavior that is qualitatively different from that of established queens in mature colonies and significantly higher levels of dominance behavior than do established queens. The strikingly similar behavior of founding queens, potential queens, and replacement queens adds strength to this conclusion. The drop in levels of dominance behavior of founding queens and replacement queens, to reach the low levels indistinguishable from those of established queens, adds further strength to the conclusion that founding queens and established queens differ sharply in the nature of their dominance interactions with their nestmates. But if an established queen of a mature colony is not behaviorally dominant, how does she maintain her reproductive status? At this stage we can only guess that the gradual decrease in the behavioral dominance of *R. marginata* queens may be associated with the onset of some other form of interaction that does not involve overt physical aggression—perhaps the production of a pheromone. Such a transition from a behavioral mechanism of achieving reproductive monopoly to a more subtle, pheromonal mechanism of achieving the same end, postulated by Premnath, Sinha, and Gadagkar (1996a), appears to have a parallel in bumble bees (West-Eberhard, 1977; Hölldobler, 1984; Free, 1987).

The behavior of *R. marginata* founding queens is reminiscent of the behavior of queens of many *Polistes* and *Mischocyttarus* species (Pardi, 1948; West, 1967; West-Eberhard, 1969, 1982a, 1982b, 1986b; Jeanne, 1972; Gamboa and Dropkin, 1979; Strassmann, 1981; Itô, 1985, 1993a; Röseler, 1991). Even in these other species there are some situations in which the dominance-subordinate relationships are rather mild or are nearly absent. But when this happens, the monopoly in egg laying also disappears, and several females in the colony have well-developed ovaries with mature eggs and/or actually lay eggs (Yamane, 1973, 1986; Hoshikawa, 1979; Kasuya, 1981; Itô, 1985, 1986). Consequently, the docile behavior of established queens of *R. marginata*, which enjoy complete reproductive monopoly, seems unique among primitively eusocial species.

6.5. Queen Succession

The rapidity and near certainty with which one of the workers emerges as the undisputed replacement queen within just a day or so of a natural queen loss, or experimental queen removal, cannot help raising a persistent question in the minds of the human observer: How and when is the queen's successor chosen? Is the choice made after loss of the original queen or when her replacement becomes imminent? Alternatively, is the next successor identifiable even in the presence of a healthy queen? What are the characteristics of the most likely successor—is she the oldest, youngest, fattest, most hard working, or most aggressive individual, for example? In addition to the obvious human interest in these issues, their implications for the evolution of division of labor and of sociality itself are profound. Contrast the following two scenarios. In *Polistes exclamans,* old and active foragers have the highest chance of becoming replacement queens, a system dubbed "gerontocracy" by Strassmann and Meyer (1983; see also Hughes and Strassmann, 1988; Hughes, Beck, and Strassmann, 1987). In *Mischocyttarus drewseni* a relatively young nonforager has the highest chance of becoming a replacement queen (Jeanne, 1972). The former scenario must make it relatively easy for selection on workers to encourage working for the welfare of their colonies without necessarily jeopardizing their opportunities to gain direct fitness. The latter scenario, in contrast, must lead to a relatively greater conflict between individual interests and colony interests and hence between avenues for maximizing the direct and indirect fitness components.

Recall that in the study by Chandrashekara and Gadagkar (1992) described in section 5.6, queens of 12 *R. marginata* colonies were removed to ascertain the behavioral caste of potential queens. Each colony was observed over a period of 2 weeks each, before and after the queen was removed, so that during the prequeen removal period we could identify the characteristics of the individual that went on to become the queen in the postqueen removal period. Such an individual was called the potential queen. In that section we were concerned only with the behavioral caste of the potential queen, but here we can use the same data to focus on other characteristics of the successor. For instance, we can analyze the behavioral data obtained during the prequeen removal period and ask how the potential queen was different from all other individuals that did not become queens. During this period, even the observer was quite unaware of which wasp would become the successor queen. Compared to the average value for nonqueens (individuals that did not become queens after the original queen was removed) of their colonies, potential queens spent significantly less time being absent from the nest ($p = 0.023$) and showed significantly higher rates of dominance behavior ($p = 0.034$). It may appear therefore that the time spent in being absent from the nest and the rates of dominance behavior may be good diagnostic features of potential queens. However, comparison of a potential queen with an average nonqueen may be inappropriate because there is considerable interindividual variability among nonqueens. To address this problem, we also compared the values for potential queens with the values for those individuals in their colonies that showed the lowest ("Min." in Figure 6.6) and the highest ("Max." in Figure 6.6) value, for a given variable. Although potential queens spent less time being absent from their nests than the average nonqueen, they spent significantly more time being absent than did nonqueens that had the lowest value in their colony for being absent from the nest. Similarly, although potential queens showed a significantly greater frequency of dominance behavior than the average nonqueen, their rate of dominance behavior was indistinguishable from that of nonqueens showing the highest rates of dominance behavior in their colonies. Indeed, for all variables studied, potential queens either had significantly higher values than, or were statistically indistinguishable from, individuals having the lowest value for that variable in their colony. Similarly, potential queens either had significantly lower values than or were indistinguishable from those individuals that had the highest value for that variable in their colony (Figure 6.6).

Using the method of assigning dominance ranks described in section 6.2, we found that potential queens were the highest-ranking individuals among the nonqueens in only 3 of 12 cases. In the remaining 9 cases, there were 2 to 22 individuals with higher dominance ranks than the potential queen. The ages of the potential queens were statistically indistinguishable from the average age of the nonqueens of their colonies. Potential queens were significantly older than the youngest nonqueens of their colonies and significantly younger than the oldest nonqueens (Figure. 6.6). The potential queen was the oldest nonqueen (1 colony), was one of the 2 similar-aged oldest nonqueens (2 colonies), younger than 1 to 12 nonqueens and equal in age to 1 to 5 nonqueens (8 colonies). In the last 8 cases, the oldest nonqueen was 1 to 112 days (mean ± S.D. = 32.12 ± 37.73) older than the potential queen. Thus it is clear that potential queens do not have any consistent characteristics that may permit us to identify them even in the presence of their predecessors. It may therefore be most appropriate to think of the potential queens in R. *marginata* as "unspecialized intermediates." While interpreting similar results they obtained for the primitively eusocial sweat bee *Lasioglossum zephyrum*, Brothers and Michener (1974) speculated that such unspecialized intermediates might be able to respond quickly to the loss of a queen and succeed her.

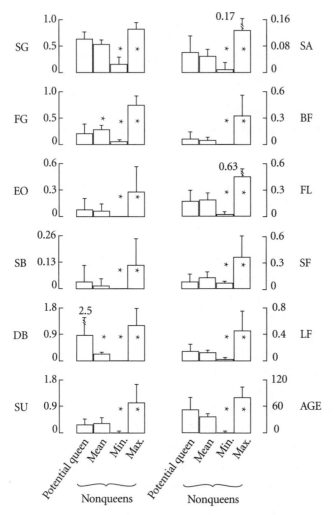

Figure 6.6. Behavioral profiles and ages of potential queens, and mean, min., and max. workers (see the text for definitions). Mean and 1 S.D. from 12 experiments are shown. Statistical comparison is always between potential queens and workers, accomplished with a two-tailed Wilcoxon matched-pairs signed-ranks test. An asterisk in or above a bar denotes that the value is significantly different from that of the potential queen (*n* = 12; 0.002 < *p* < 0.043). SG = sit and groom; FG = forage; EO = extend walls of cells; SB = snatch building material from nestmate; DB = engage in dominance behavior; SU = engage in subordinate behavior; SA = sit with raised antennae; BF = bring food; FL = feed larvae; SF = snatch food from nestmate; LF = lose food to nestmate; AGE = number of days since eclosion. (Redrawn from Chandrashekara and Gadagkar, 1992.)

6.6. The Significance of Worker Aggression

Let us now turn our attention to the third question raised at the beginning of this chapter. In mature colonies with a well-established queen, if she herself does not need overt physical dominance to retain her reproductive monopoly in the colony, why do the workers indulge in such frequent dominance behavior? An obvious possibility is that workers need to indulge in dominance-subordinate interactions among themselves so as to compete effectively with each other in becoming replacement queens. The finding in section 5.4 that queens of *R. marginata* do display extreme, overt physical aggression toward their nestmates at the time of initially establishing their status as the sole reproductives of their colonies adds significant support to this idea. However, as we saw in section 5.6, fighters do not necessarily have a very high chance of becoming replacement queens. Similarly potential queens are not necessarily more aggressive or of higher rank than some of their nestmates that do not become replacement queens. What then is the possible significance of worker dominance behavior? We (Chandrashekara and Gadagkar, 1991) have attempted to answer this question by computing the correlation between the frequency per hour of dominance behavior on the one hand and between other behaviors and anatomical and morphological variables on the other. It turns out that the frequency of dominance behavior is significantly positively correlated with the frequency of subordinate behavior, feeding larvae, extending walls of cells, and building new cells, and the index of ovarian condition, the index of body size, dry weight, and fat content. In addition, as the frequency of dominance behavior increases, the number of individuals receiving that dominance also increases. This suggests that dominance behavior is not necessarily directed toward one or a small number of individuals (perceived as a potential threat), but is distributed over a large number of individuals, with more or less the same rate of dominance shown toward each opponent.

The question of the significance of dominance behavior was also raised in section 5.4 in the context of the role of fighters. I speculated that fighters could be "policemen" whose job is to coerce other individuals to work, especially outside the nest. The data examined in this section also lend some support to such a policing hypothesis. Individuals that show high levels of dominance behavior also show high levels of subordinate behavior. This finding is consistent with the idea that dominance-subordinate behaviors constitute a signaling system, by which the workers inform each other of the colony's needs, rather than a mechanism by which one individual attempts to suppress reproduction by all others. The correlation of dominance behavior with body size is satisfying—policing is a job better done by large-bodied individuals. This idea needs to be investigated further since a previous laboratory study suggested no effects of body size on which wasp would become dominant over which other wasp(s) (Nair, Bose, and Gadagkar, 1990). That individuals showing high levels of dominance behavior distribute their interactions among a large number of recipients also makes sense if the function of dominance is to coerce or signal several nestmates to work for the colony. Wondering how a docile sitter queen maintains her reproductive monopoly, my students and I have speculated that she probably uses a subtle, pheromonal mechanism (Premnath, Sinha, and Gadagkar, 1996a 1996b). The answer to the question of how a docile sitter queen regulates the activities of her workers might well be: she doesn't. But some individuals must and perhaps the workers themselves do so by means of dominance behaviors. And that notion leads to a more explicit consideration of this question.

6.7.   Regulation of Worker Activity

During the period 1983–1990, George Gamboa and his students published a series of papers concerning the mechanism of regulation of worker activity in the primitively eusocial wasp *Polistes fuscatus* (Reeve and Gamboa, 1983, 1987; Gamboa et al., 1990). They made behavioral observations on intact colonies with their queens, colonies deprived of their queens, after reintroduction of a removed queen and even after reintroduction of a chilled and hence behaviorally inactive queen. Their findings were clearcut. In undisturbed colonies, the queen was the most active and most dominant individual, initiating high levels of behavioral interactions with her nestmates and thereby regulating their activity. A particularly elegant result was that the workers synchronized their states of activity and inactivity with those of the queen. In the absence of the queen, the workers went into a lull and the colony virtually came to a standstill. Once the queen was reintroduced, normality returned, but if a chilled, inactive queen was reintroduced, the workers got even more lethargic than they had in her absence. Quite rightly they called the queen the colony's "central pacemaker," and concluded that the mechanisms underlying queen control of worker reproduction may be the same as, or intimately linked to, the mechanism of regulation of worker activity (see also Breed and Gamboa, 1977; Dew, 1983).

I read these papers with fascination but could not believe that their findings would hold for established colonies of *R. marginata*, whose queens are not behaviorally the most dominant or active individuals and do not regulate worker reproduction through overt behavioral means. How then do they regulate worker activity—or do they at all? Of course there was no better way to find out than to repeat with *R. marginata* the experiments that Gamboa and his students had done on *P. fuscatus*. Sudha Premnath and Anindya Sinha and I conducted 13 experiments involving observation (for 10 h each day) of undisturbed nests on day 1, observation after experimental removal of the queen on day 2, and observation after replacement of the queen (not chilled) on day 3 (Premnath, Sinha, and Gadagkar, 1995).

Our results were quite different from Gamboa's but no less clear-cut. But first, some terminology:

Potential queen: The individual that, upon removal of the queen, became extremely dominant and (we know from other experiments) would go on to become the next queen if the original queen was not replaced.

Max. worker: Among individuals other than the queen and the potential queen, that worker having the maximum value for any given variable.

Mean worker: For any given variable, the mean value among all workers other than the queen and potential queen is designated mean worker.

Frequency of initiated interaction: The sum of the frequencies of dominance behaviors, unloading, soliciting, approaching, antennating, and allogrooming another wasp. Values were computed separately for each individual, for each day, after correcting for the proportion of time she spent on the nest.

Frequencies of dominance behavior: As defined earlier, but the values were calculated after correcting for the proportion of time spent by each wasp on the nest, on each day.

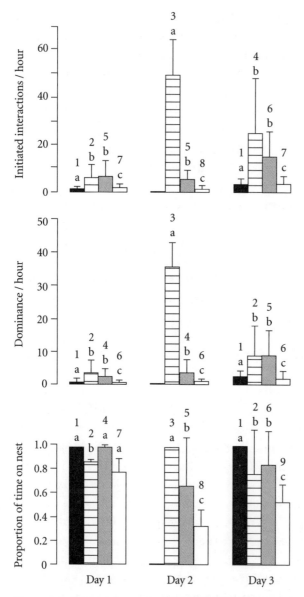

Figure 6.7. Frequencies per hour of initiated interactions (top), frequencies per hour of dominance behavior (middle), and proportion of time present on the nest (bottom) shown by the queen (black bars), potential queen (horizontally marked bars), max. workers (gray bars), and mean worker (open bars) on days 1, 2, and 3 (see the text for a description of the experiment). Bars that carry different letters are significantly different from each other ($p < 0.05$ or less) within each day; bars that carry different numbers are significantly different from each other ($p < 0.05$ or less) among the 3 days. Comparisons are by the two-tailed Wilcoxon matched-pairs signed-rank test. (Redrawn from Premnath, Sinha, and Gadagkar, 1995.)

Activity level: Proportion of time a wasp is active. An individual was considered active unless she was merely sitting, grooming herself, laying eggs, or passively receiving interactions from other wasps.

Measures of foraging: Absence from the nest and frequency of leaving the nest could not legitimately be considered measures of foraging because the proportion of trips when the wasps returned with nothing was very high on day 2 (in the absence of the queen and owing to the extreme aggression by the potential queen) in comparison to days 1 and 3. Hence the sum of the actual frequencies of bringing food and pulp was considered the measure of foraging.

Unloading: The act of receiving food from a forager immediately upon her return.

Association values: We computed the Yule's association coefficient (De Ghett, 1978) as a measure of coordination between the activities of the queen, the potential queen, and other workers. The association coefficient between individuals i and j is given by the formula

(6.2) $Y_{ij} = (ad - bc)/(ad + bc)$

where a is the probability that both individuals i and j are active, b is the probability that individual i is active and j is inactive, c is the probability that individual i is inactive and j is active, and d is the probability that both individuals are inactive. As can be seen from equation 6.2, the values of the association coefficients range conveniently from -1 to $+1$. A value of -1 indicates that individuals i and j are completely out of phase, one being active only when the other is inactive and vice versa. A value of $+1$ indicates that i and j were completely in phase or synchronized in their activities—at any given time either both are active or both are inactive. For each individual in the colony, we computed the mean of its association coefficient with every other individual in the colony.

On day 1, the queen was not the most dominant individual in her colony. Nor did she initiate many behavioral interactions with her nestmates. The potential queen, max. worker, and mean worker all had values significantly higher than the queen for these two variables (Figure 6.7 and 6.8). Instead, the queen was a quiet individual spending significantly more time on the nest than the max. and mean workers. The queen's synchronization of activity with her nestmates, as measured by the Yule's association coefficient,

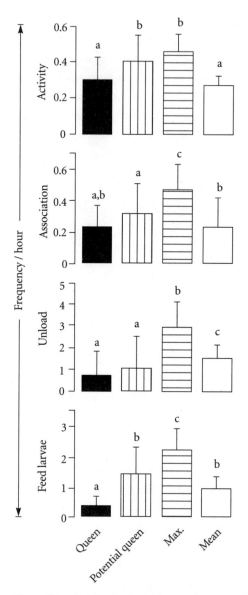

Figure 6.8. Activity level, Yule's association coefficient, frequencies per hour of unloading, and frequencies of feeding larvae by the queen (black bars), potential queens (vertically marked bars), max. worker (horizontally marked bars), and mean worker (open bars) on day 1 (see the text for a description of the experiment). For each variable bars that carry different letters are significantly different from each other ($p < 0.05$ or less). Comparisons are by the two-tailed Wilcoxon matched-pairs signed-rank test. (Redrawn from Premnath, Sinha, and Gadagkar, 1995.)

was indistinguishable from that of an average worker and was significantly lower than that of the potential queen as well as that of the max. worker. The queen was not particularly active in unloading food or in feeding larvae. Her values for these variables were significantly lower than those of both max. workers and mean workers. These findings were consistent with our previous impression of *R. marginata* and with my suspicion that an *R. marginata* queen cannot possibly be described as the central pacemaker of her colony. But the results of the second day of the experiment provided even more striking confirmation of the growing impression that in the regulation of worker activity, *R. marginata* and *P. fuscatus* are in complete contrast.

How did *R. marginata* workers respond to the absence of their queen in terms of regulating their activities? In short, they couldn't care less! As expected from previous experiments, the most striking response to the absence of the queen was that one individual (the potential queen) became extremely aggressive and literally drove all her nestmates away and kept dominating them, whenever they alighted on the nest, during the whole day; as a result the proportion of time spent on the nest went down for everybody except the potential queen. But what about other behaviors? Rather surprisingly, the frequencies with which food was brought to the nest and fed to the larvae did not differ significantly during the three days—in the undisturbed colony with its queen, in the absence of the queen, and after queen replacement (Figure 6.9).

Remarkably enough, approximately the same number of foragers were active in the presence and absence of the queen, and indeed they were active at about the same rates. The proportional contribution to the colony's foraging effort by different individuals was positively correlated between days 1 and 2; this demonstrates even less effect of the absence of the queen. However, the foragers did have a serious problem on day 2. Because the potential queen aggressively drove out almost all the wasps and permitted them to land only briefly, when she dominated them, the foragers had much less help in unloading their loads of food or pulp. The frequency with which incoming foragers were unloaded by wasps sitting on the nests was similar on days 1 and 3 but significantly lower on day 2 (Figure 6.10, upper panel). How did the foragers respond? Contrary to their normal practice, they themselves fed the larvae. The contribution of the foragers to the colony's task of feeding the larvae was indistinguishable on days 1 and 3 but was significantly higher on day 2 (Figure 6.10, lower panel).

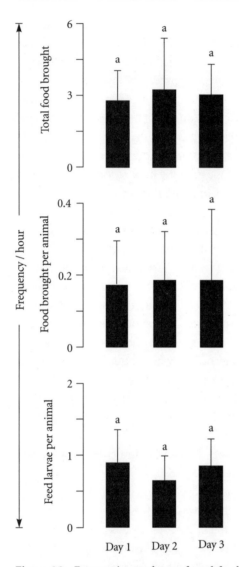

Figure 6.9. Frequencies per hour of total food brought to the nest, of food brought per individual per hour, and of feeding larvae per individual per hour are all not significantly different on days 1, 2 and 3, as can be seen from the identical letters on the bars. Comparisons are by two-tailed Wilcoxon matched-pairs signed-rank test. (Redrawn from Premnath, Sinha, and Gadagkar, 1995.)

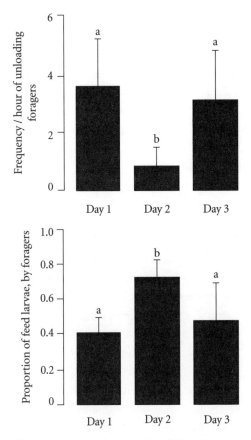

Figure 6.10. Frequencies at which foragers were unloaded (upper panel) and the proportion of the feeding of larvae done by foragers (lower panel) on days 1, 2, and 3. For each behavior, bars with different letters are significantly different from each other. Comparisons are by the two-tailed Wilcoxon matched-pairs signed-ranks test. (Redrawn from Premnath, Sinha, and Gadagkar, 1995.)

When the queen was returned on day 3, she was accepted without any overt aggressive behavior either on her part, on the part of the potential queen, or on any individual's part. Sure enough, the potential queen dramatically reverted to her original self. She lowered her levels of dominance behavior, initiated interactions, and began leaving the nest; her values for these variables on day 3 were statistically indistinguishable from her values on day 1, except for dominance behavior, which was still in the process of coming down to her value on day 1. Equally remarkable is the fact that the queen's behavior on day 3 was indistinguishable from her behavior on day 1.

Dramatic and fascinating as these results are, taken at face value they don't tell us much about how worker activity is regulated, although it is clear that it is not regulated by the queens. In sections 5.4 and 6.6, I suggested a relationship between dominance behavior shown by workers and the regulation of worker activity. There are several hints in this study that point to this possibility that dominance-subordinate interactions among workers constitute the mechanism of regulation of worker activity. On day 1, the frequency of dominance received by the foragers (2.06 ± 3.99) was significantly greater than that received by nonforagers (0.85 ± 0.49; $p < 0.05$). Besides, the frequency of dominance received by a forager was significantly correlated with her foraging rate ($\tau = 0.20$, $p = 0.02$, $n = 67$). Most telling, however, is the result that the fraction of the total dominance received by a forager is positively correlated with her fractional contribution to the colony's foraging efforts. When these results are viewed in conjunction with the finding that foragers are not unloaded efficiently on day 2 and that they respond to this situation by feeding larvae themselves, we have a plausible theory for the regulation of worker activity (Premnath, Sinha, and Gadagkar, 1995).

The dominance-subordinate interactions exhibited by the workers should perhaps be viewed as a system of signals informing the extranidal workers of the hunger levels of the colony's adults and larvae. Fighters who perform most of the dominance behaviors also specialize in intranidal tasks, including feeding the larvae, and hence they should possess maximum information about larval hunger levels. When foragers are not unloaded efficiently, they themselves feed the larvae and thereby obtain first-hand information about larval hunger levels and the colony's need for food. That dominance behaviors in established colonies are rather mild and ritualized perhaps makes them suitable for serving as signals of hunger rather than for serving the function of suppressing reproduction by the recipient. It has been well established in honey bee colonies that the efficiency with which a nectar forager is unloaded gives her a signal of whether or not the colony is in need of more nectar. If foragers returning with nectar are ignored by unloader bees and they have to wait around with nectar in their crops, they are less likely to bring nectar again for some time. But if they are immediately attended to and their nectar load is received with eagerness, they will continue to forage for nectar (Seeley, 1995). Is it so unreasonable then to suppose then that if a returning *R. marginata* forager is attacked, pecked, chased, or nibbled upon her return from a foraging trip, or even when she is idling on the nest, she should consider this a signal that the colony members (larvae and adults) are hungry?

Conspecific aggression is so widespread in all solitary species that it may be the perfect pre-adaptation needed to signal hunger levels to foragers in incipient societies. An early step in social evolution might thus be the ability to use mild forms of aggression to signal hunger levels (DB I) and severe forms (DB II) to suppress conspecifics during the establishment of a new colony or the establishment of a new queen in an old colony. If by regulation of worker activity we mean simply regulation of idling versus foraging, then we seem to have a reasonable understanding of how it is achieved in *R. marginata*. Regulation of foraging for food versus pulp or water may occur at some crude level by the efficiency of unloading or may not occur at all to any appreciable degree.

I began this chapter with several questions. Let me end with a brief recapitulation of the questions and possible answers. How does an *R. marginata* queen establish herself as a queen? By overt and extreme physical aggression. How does the queen maintain her reproductive monopoly? Not by overt aggression, but perhaps by the production of a pheromone. What individual succeeds the queen? An unspecialized intermediate among the workers. Why do the workers indulge in dominance-subordinate interactions? At least partly to convey larval and adult hunger signals to foragers. Which wasps regulate worker activity? The worker themselves, not the queen. For information on some aspects of dominance behaviors not investigated here, I refer readers to Roseler et al. (1980, 1984, 1985), Downing and Jeanne (1985), Pfennig and Klahn (1985), Downing (1991b), Theraulaz et al. (1992), Bull et al. (1998), O'Donnell (1998a), and Bloch and Hefetz (1999).

7   AGE AND DIVISION OF LABOR

 o social insect researcher can resist the temptation to compare his species with that enduring model organism, the honey bee. One aspect of the honey bees that lends itself to useful comparison with any social insect species is age polyethism. In her lifespan, some 6 weeks or so, a worker honey bee sequentially performs four principal tasks—cleaning, feeding the larvae, storing food (including building the comb, if necessary), and foraging (Seeley, 1982, 1995; Winston, 1987). This system of age polyethism provides honey bees with a flexible mechanism for division of labor, compared with the polyethism based on morphologically differentiated worker castes seen in many ants and termites (Wilson, 1968; Oster and Wilson, 1978; Seeley, 1985, 1995; Bourke and Franks, 1995). Because primitively eusocial species lack morphological caste differentiation, even between queens and workers, let alone between workers specializing in different tasks, age polyethism would appear to be a convenient way for them to organize work in their colonies. However, the empirical evidence for age polyethism in primitively eusocial species has been weak (Naug and Gadagkar, 1998b). On theoretical grounds too, one would not expect polyethism facilitating work organization in the colony to evolve very easily in primitively eusocial species. In highly eusocial species such as ants or honey bees, workers have few or no reproductive options of their own, so that, especially in the presence of a healthy queen, their best option is always to enhance colony efficiency and productivity. Patterns of physical or age polyethism that may require workers to sacrifice their personal welfare or their almost negligible reproductive abilities can thus evolve relatively easily.

In primitively eusocial species, by contrast, workers have substantial reproductive options—they can leave to found their own nests or can replace the existing queens in their present colonies. In such situations, workers must experience a significant conflict between catering to individual interests and to colony interests. To the extent that polyethism requires sacrifice of the workers' personal interests, this must make it difficult for it to evolve. I suspect that such a theoretical expectation has contributed its share to slowing down the pace of empirical research on age polyethism in primitively eusocial insects (but see O'Donnell and Jeanne, 1995a, 1995b for arguments about how age polyethism can evolve without seriously compromising workers' interests). I remember having age polyethism in my long list of possible research topics for new Ph.D. students, but it was always somewhere near the bottom of the list, where it lay safely unclaimed until Dhruba Naug chose it. And thus began our investigation of age polyethism in *R. marginata* (Naug and Gadagkar, 1998a, 1998b, 1999).

7.1. Age Polyethism

Our data come from 6 hours of observations per day on four colonies for periods ranging from 2 to 3 months each. The tasks performed by the wasps can be unambiguously classified into two major categories: tasks performed on the nest (intranidal tasks) and tasks performed outside the nest (extranidal tasks). Feeding larvae and building (the nest) constitute two major intranidal tasks, and bringing pulp (for building) and bringing food constitute two major extranidal tasks. We found clear evidence of a significant effect of age on task performance. Of 39 wasps that were seen to perform both intranidal and extranidal tasks, none performed any extranidal tasks (bringing pulp or bringing food) before performing at least some intranidal tasks (feeding larvae and building). A clear effect of age on the choice of tasks by wasps is evident even when we used a finer classification of tasks. Feeding larvae, building, bringing pulp, and bringing food, in that order, was the most preferred sequence in which the four tasks were performed as the wasps advanced in age (Figure 7.1). Wasps performing their first act of feeding larvae were significantly younger than those performing their first act of building. Similarly, the age of first performance of building was significantly lower than the age of first performance of bringing pulp, and the age of first performance of bringing pulp was significantly less than the age of first performance of bringing food (Figure 7.2).

Young workers

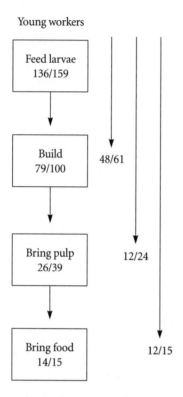

Old workers

Figure 7.1. Temporal polyethism in *R. marginata:* the most common sequence in which the four tasks feeding larvae, building, bringing pulp, and bringing food are performed. The numbers within each box represent the observed proportions of wasps that performed the task in that position, independent of what they did at any other position. Out of 159 wasps that did at least one task, 136 did feed larvae as their first task. Of 100 wasps that did at least two tasks, 79 did build as their second task, and so on. Each of these proportions is significantly different from the proportion expected (1/4) if wasps were taking up tasks at random ($G = 258.86, 128.32, 29.91,$ and 32.04 respectively, $p < 0.001$). The numbers at the heads of the arrows represent the observed proportions of wasps that followed the particular sequence corresponding to the arrow. Out of 61 wasps that performed only two tasks, 48 followed the sequence feed larvae and build; of 24 wasps that performed only three tasks in their lives, 12 followed the sequence feed larva, build, and bring pulp, and so on. Since there are four tasks to choose from, the expected probability of a sequence with two tasks is 1/12 and that of a sequence with three or four tasks is 1/24. The observed proportions of sequences with two, three, and four tasks were significantly different from those expected at $p < 0.001$ ($G = 177.61$, $G = 44.02$, and $G = 61.51$, respectively). (Redrawn from Naug and Gadagkar, 1998b.)

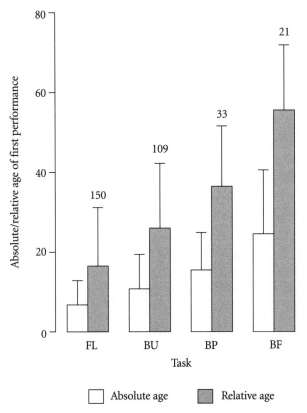

Figure 7.2: Age polyethism in *R. marginata:* mean (± S.D.) age of first performance for each task in terms of absolute age (open bars) and relative age (gray bars). FL = feed larvae; BU = build; BP = bring pulp; BF = bring food. The sample size for each task is given above each bar. Multiple comparisons of mean ages of first performance using the Tukey-Kramer method indicated significant differences across tasks ($p <$ 0.05). The first performance of a task was significantly influenced by absolute age (one-way ANOVA; $F = 33.47$, $p < 0.0001$) and relative age ($F = 49.12$, $p < 0.0001$). Mean ages for successive tasks were also significantly different ($p < 0.05$) when subjected to a Mann-Whitney U test. (Redrawn from Naug and Gadagkar, 1998b.)

For a more detailed analysis of the effect of age on task performance we used two measures of task performance and two measures of age:

Probability of task performance (PTP) is the probability that a worker of a given age will perform a given task relative to other tasks it performs (Seeley, 1982) and is calculated as:

(7.1) $$P_{ij} = \frac{n_{ij}}{\sum_{i=1}^{k} n_{ij}}$$

where P_{ij} is the probability of performance of task i by a member of age j; n_{ij} is the total number of times task i was performed by a member of age j; and k is the number of tasks being considered.

Frequency of task performance (FTP) is the rate (number of times per h) at which a worker of a given age performs a task.

Absolute age is simply the number of days since the eclosion of a given wasp.

Relative age is the ranked age of a wasp in her colony and is a measure of her position in the age distribution of the colony. Age ranks were computed on each day of observation for each wasp, based on the absolute ages of all wasps alive on that day. The youngest wasp was assigned a rank of 1 and wasps of successive ages were assigned successively higher ranks with the oldest wasp obtaining the highest rank, equal to the number of individuals in the colony on that day. Wasps of the same age were assigned tied ranks. The range of ranks was then scaled between 0 and 100 according to the equation

(7.2) $$SR_i = \frac{R_i - 1}{N - 1} \times 100$$

where SR_i = the scaled rank of wasp i; R_i = the rank of wasp i; and N = the number of individuals in the colony. This scaling corrected for differences in the number of individuals present in the colony on different days. We used polynomial equations to regress each measure of task performance on each measure of age, and where necessary, we used the Wilcoxon matched-pairs signed-rank test to compare the variances explained (r^2) by different regression equations.

Thus we regressed PTP and FTP for intranidal and extranidal tasks and also separately for feeding larvae, building, bringing pulp, and bringing food against the absolute and relative ages of the wasps (Naug and Gadagkar, 1998b). For brevity here I will illustrate our results, complete with all data points and the regression statistics, only for extranidal tasks, chosen as an example (Figure 7.3). In addition, Figure 7.4 provides a quick,

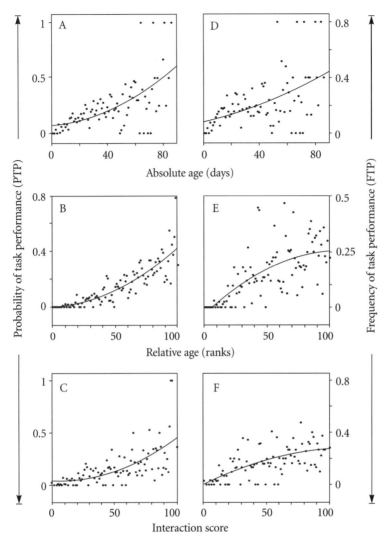

Figure 7.3. Probability of task performance (PTP) and frequency of task performance (FTP) as a function of absolute age, relative age, and interaction scores fitted with second-order polynomial regression lines. Data points represents the mean value for all individuals in that age class in five colonies. A: Extranidal tasks with respect to absolute age and PTP; $y = 0.07 + 0.001x + (5.47 \times 10^{-5})x^2$, $r^2 = 0.38$, $p < 0.0001$. B: Extranidal tasks with respect to relative age and PTP; $y = (3.84 \times 10^{-4}) + (5.21 \times 10^{-4})x + (3.73 \times 10^{-5})x^2$, $r^2 = 0.75$, $p < 0.0001$. C: Extranidal tasks with respect to interaction score and PTP; $y = 0.04 - (5.76 \times 10^{-4})x + (4.70 \times 10^{-5})x^2$, $r^2 = 0.43$, $p < 0.0001$. D: Extranidal tasks with respect to FTP and absolute age; $y = 0.08 + 0.002x + (1.92 \times 10^{-5})x^2$, $r^2 = 0.24$, $p < 0.0001$). E: Extranidal tasks with respect to FTP and relative age; $y = -0.03 + 0.005x - (2.44 \times 10^{-5})x^2$, $r^2 = 0.48$, $p < 0.0001$. F: Extranidal tasks with respect to FTP and interaction score; $y = 0.003 + 0.004x - (2.008 \times 10^{-5})x^2$, $r^2 = 0.34$, $p < 0.0001$. (Redrawn from Naug and Gadagkar, 1998b.)

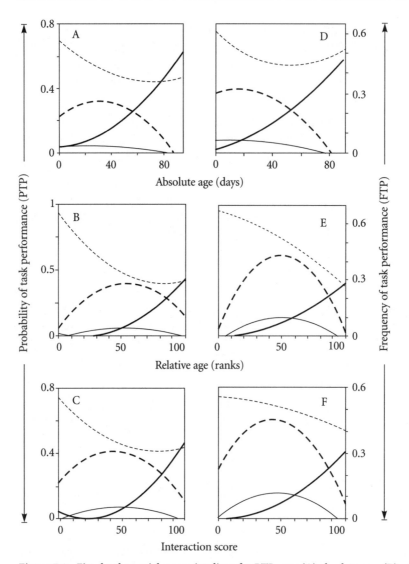

Figure 7.4. Fitted polynomial regression lines for PTP over (A) absolute age, (B) relative age, and (C) interaction score and FTP over (D) absolute age, (E) relative age, and (F) interaction score, for all four tasks. The fitted lines for all tasks are plotted together on the same scale to discern the overlap of performance of different tasks across age. Thin broken lines refer to feeding larvae, thick broken lines to building, thin solid lines to bringing pulp, and thick solid lines to bringing food. Note that there is more overlap in the age of individuals engaged in building and bringing pulp, although this overlap lessens when the tasks feeding larvae and bringing food are considered. (Redrawn from Naug and Gadagkar, 1998b.)

bird's-eye view of all the 32 fitted polynomial regression lines without the data points. The reader should refer to Naug and Gadagkar (1998b) for details. In every case we found a significant influence of age (both absolute and relative) on both PTP and FTP. In general, the probability and frequency of performing intranidal tasks declined with age and those of extranidal tasks increased with age. Among the two intranidal tasks, the pattern for feeding larvae was qualitatively similar to the pattern seen for intranidal tasks—it declined with age. The other intranidal task, building, peaked in middle age. Extranidal tasks increased with age. Again the pattern of variation of one of its components—bringing food—mirrors the pattern seen for extranidal tasks as a whole. The other component, bringing pulp, like building, peaks at about middle age. The qualitative similarity of the patterns observed, irrespective of whether we used PTP or FTP, absolute age or relative age, was striking.

Relative to what we know of other primitively eusocial species, *R. marginata* thus seems to exhibit rather strong age polyethism. Indeed, the pattern of age polyethism we found in *R. marginata* is strikingly similar to the pattern of age polyethism seen in honey bees. Perhaps the most significant conclusion arising from our demonstration of age polyethism in *R. marginata* is that, contrary to the theoretical reservations expressed at the beginning of this chapter, age polyethism can indeed evolve even before workers have lost their reproductive options. The observed sequence of task performance bears a logical biological interpretation (Naug and Gadagkar, 1998b). The postponement of the inevitably more risky extranidal tasks for later ages and the devotion of the early ages to the less risky intranidal tasks is not surprising—that colony-level selection will favor such a pattern has often been recognized (see, for example, Jeanne, 1986a). In *R. marginata* both individual-level and colony-level selection are expected to operate, but this pattern will be equally favored by both. That feeding larvae precedes building among intranidal tasks and bringing pulp precedes bringing food among extranidal tasks may have to do with the relatively lower levels of skill and experience required for feeding larvae as compared with those required for building the nest and for scraping twigs for cellulose fibers as opposed to hunting for live prey. So far I have emphasized the qualitative similarity of the patterns of change of task performance with age, irrespective of whether we use PTP or FTP on the one hand and absolute age or relative age on the other. Now let me focus on dissimilarities. There are subtle and interesting quantitative differences in the patterns. Relative age is a consistently better predictor of task performance than absolute age. What is the significance of this result?

7.2.   The Controversy

In recent times the phenomenon of age polyethism has come under severe criticism on many grounds, at least some of them quite valid. These arguments are well summarized by Bourke and Franks (1995). Some of the major arguments are: (1) Claims of empirical demonstration of age polyethism are often based on studies made over short periods of time on age cohorts or on laboratory colonies with disturbed size and age structures. (2) Correlation of task performance with age is not evidence of a causal link between the two. (3) Age-dependent task allocation constitutes an inflexible mechanism, making it difficult for social insect colonies to adaptively respond to internal contingencies (skewed age distribution, altered demands, and so on) and external contingencies (bad weather, unexpected abundance of food, and so on). As an alternative to the idea of age as a causative factor, a "foraging-for-work" hypothesis has been put forward (Tofts and Franks, 1992) that argues that task performance simply depends on which tasks individuals encounter. The assumption is that tasks are spatially ordered—as they indeed appear to be in the ant *Leptothorax unifasciatus* (Franks and Sendova-Franks, 1992) and in honey bees (Camazine, 1991)—and that workers have age-dependent spatial fidelity zones—as they indeed appear to do in *Leptothorax unifasciatus* (Sendova-Franks and Franks, 1995a, 1995b; see also Gordon, 1996, 1999).

Under these assumptions, even if workers perform whatever task they encounter, without explicit reference to their age, a correlation of task performance with age can result "spuriously." Objection (1) does not hold in the case of our study because, relative to the lifespans of workers (27 + 21 days; Naug and Gadagkar, 1998b), we performed long-term (2–3 months) observations of natural, free-foraging colonies. In *R. marginata*, the queen does not lay eggs at a single place, the workers do not sort brood, and both the queens and workers move about the entire nest surface during every day of their life. The foraging-for-work hypothesis can thus hardly be applicable. But it is true that correlation with age does not prove that age causes the observed pattern of task performance and that age per se cannot provide sufficient flexibility to deal with contingencies. In this context, clearly relative age is superior to absolute age in predicting the pattern of task allocation. If wasps perform tasks depending on their relative age in the colony rather than their absolute age, there will be greater flexibility—a colony may not have 10-day-old workers or 40-day-old workers for example, but all colonies will always have a youngest worker and an oldest worker. If the youngest workers are programmed to feed larvae, those in the middle of the colony's age distribution are programmed to build and bring pulp, and the oldest workers are programmed to bring food, much of the strength in the argument against the inflexibility of age disappears. Besides, in a work organization based on relative age, what any individual does depends greatly on what every other individual is doing—this provides an in-built mechanism for self-regulation.

7.3.　The Flexibility of Age Polyethism

If the workers indeed used their relative age in their colonies, rather than their absolute ages, to choose their tasks, *R. marginata* colonies should be able to respond to situations of skewed age distributions by reallocating tasks. It is well known that honey bees can do so. In a colony consisting of only young individuals, honey bee workers begin to forage well below the typical age for foraging in normal colonies—these are called precocious foragers. Conversely, in a colony consisting of only old workers when no bee that normally does nursing is available, some foragers revert to nursing duties—these are called overaged nurses. In the language of Huang and Robinson, (1992) workers go through a certain rate of behavioral development in normal colonies, progressing from one task to another. In young-worker colonies behavioral development is accelerated to yield precocious foragers, and in old-worker colonies behavioral development is retarded to yield overaged nurses.

Naug and I have conducted experiments to see if *R. marginata* can behave in the same way. We created five young-worker colonies by removing all workers older than 7 days before the start of behavioral observation and also all those that eclosed during the period of observation, that lasted from 2 to 3 weeks. Thus we observed colonies consisting of a 7-day cohort of individuals whose absolute ages ranged from 1 to 24 days during the period of observation. We compared the young-worker colonies with the en-

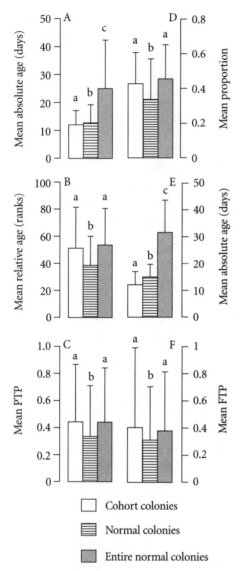

Cohort colonies

Normal colonies

Entire normal colonies

Figure 7.5. Mean (± S.D.) (A) absolute ages of all individuals, (B) relative ages of all individuals, (C) PTP of all individuals, (D) proportion of foragers, (E) absolute age of foragers, and (F) FTP of all individuals, in the young-cohort colonies (open bars), the corresponding age group in normal colonies (horizontally hatched bars), and the entire normal colonies (gray bars). One-way ANOVA followed by multiple comparisons of means by the Tukey-Kramer method was used to test the effect of age distribution on the parameters given above. Within each graph, bars with different letters are significantly different from each other ($p < 0.05$), while those with the same letters are not significantly different ($p > 0.05$). (Redrawn from Naug and Gadagkar, 1998a.)

tire normal colonies (using the data set described at the beginning of this chapter) as well as with that subset of workers from natural colonies whose ages corresponded to those of the wasps in the young-worker colonies (ages 1–24 days). The results were as expected for honey bees: we found precocious foragers in the young-worker colonies (Naug and Gadagkar, 1998a). Both the age and the proportion of foragers was significantly influenced by age distribution. The mean age of foragers in the young-cohort colonies (12.04 ± 4.88 days) not only was significantly lower than the mean age of foragers in the normal colonies (31.54 ± 12.01 days) but also was significantly lower than the mean age of foragers from among the individuals 1–24 days old in normal colonies (14.94 ± 0.18 days) (Figure 7.5). The proportion of foragers in the young-worker colonies was significantly higher than the proportion of foragers among the individuals 1–24 days old in the normal colonies and statistically indistinguishable from the proportion of foragers in the entire normal colonies (Figure 7.5). The same results hold for PTP and FTP also. The PTP and FTP for foraging was significantly higher in the young-worker colonies than in the group aged 1–24 days in the normal colonies and not different from the PTP and FTP in entire normal colonies (Figure 7.5). Thus young-worker colonies of *R. marginata*, like honey bees, can readjust their work allocation and respond adaptively to skewed age distributions. Experiments are currently in progress to create old-worker colonies and see if they will produce over-aged nurses. Our finding of precocious foragers in particular and the general pattern of the readjustment of task allocation in young colonies strengthen the conclusion that work organization in *R. marginata* is based on relative age.

7.4.   The Assessment of Relative Age

But how does a wasp know its relative position in the age distribution in the colony? Whatever the mechanism, it must involve some form of interaction with an individual's nestmates during which information about the relative age of the interactants may potentially be gathered. There are three major forms of adult-adult interaction seen in *R. marginata*: dominance-subordinate interactions, food exchange, and a behavior we call soliciting. The former two have well-defined, specific functions not explicitly connected with age. We therefore consider them unlikely to be involved in the assessment of age. Soliciting involves mouth-to-mouth contact between two individuals, without any obvious dominance or exchange of food. Soliciting is the most frequent form of interaction between adult wasps and occurs more or less randomly across different age classes and behavioral castes. It appears almost like a form of greeting and may be comparable to the trophallaxis seen among honey bee workers. Presumably some exchange of saliva takes place that may result in a transfer of information, as is well known in honey bee trophallaxis (Winston, 1987).

Since we were interested in considering the possibility that wasps assess their relative ages by such interactions, we computed an interaction score for each individual. This is given by the ratio between the number of interactions that an individual has with wasps younger than herself and the total number of interaction with all wasps on that day. It turns out that such an interaction score is strongly correlated with relative age, and thus it is possible for the wasps to assess their relative ages by means of such interindividual interactions. We then regressed PTP and FTP directly on interaction score, rather than on absolute or relative age. Remarkably enough, the interaction score is about as good a predictor of task performance as absolute or relative age. Thus we postulate that the wasps in a colony participate in interindividual interactions, assess their relative position in the age distribution of the colony, and appropriately adjust their choice of task.

7.5.   The Activator-Inhibitor Model

It is thus easy to see that a pattern of work organization based on relative age provides a system of age polyethism flexible enough that all required tasks are performed even if a colony has a skewed age distribution. What is harder to see is the possible proximate mechanism that makes all this possible. Faced with a more or less similar situation, Huang and Robinson (1992) proposed the so-called activator-inhibitor model for age polyethism in honey bees. This model also assumes that worker-worker interactions modulate age-correlated behavioral changes and hence it is attractive to consider for *R. marginata*. Huang and Robinson (1992) postulated an interplay between an intrinsic activator (A) that promotes behavioral development and an inhibitor (I) that is transferred among workers during behavioral interactions and retards behavioral development. The production of both activator and inhibitor are assumed to increase with the age

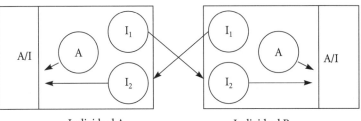

Figure 7.6. The activator-inhibitor model for age polyethism (Huang and Robinson, 1992). Each individual has three pools, A, I_1, and I_2. The pools A and I_1 contain an activator and an inhibitor respectively, the synthesis of which is coupled with and increases with the age of the individual. Social interactions result in the transfer of the entire quantity of accumulated inhibitor from pool I_1 of one individual to the pool I_2 of the other and vice versa. The inhibitor so lost from pool I is replenished instantaneously. The resultant A/I ratio determines the task profile of the individual. The inhibitor in pool I_1 does not interact with the activator in pool A and hence an individual cannot inhibit itself. (Redrawn from Naug and Gadagkar, 1999.)

of the worker. We translated this verbal model of Huang and Robinson (1992) into a numerical computer-simulation model to see if it could suggest a plausible proximate mechanism for age polyethism in *R. marginata* (Naug and Gadagkar, 1999).

The assumptions employed in the simulation model are described in Figure 7.6. The following empirical information was derived from the colonies used for demonstrating the influence of age on task performance described at the beginning of this chapter. The number of individuals per colony was 22.12 ± 10.82. The wasps ranged from 1 to 80 days in age. The age distribution was best described by the negative exponential function $0.55e^{-0.48age}$ ($R^2 = 0.86$). The wasps participated in 9.25 ± 4.17 solicits per day. For the simulations, we therefore generated 1000 colonies with 25 wasps per colony, by randomly picking individuals between 1 to 80 days of age using the above-mentioned negative exponential function. We assumed that 250 interindividual interactions (25 individuals × 10 interactions/individual) took place randomly between the members of a colony, as per the model described in Figure 7.6. After all these interactions, the total accumulated quantity of inhibitor in the I_2 pool of each wasp was computed. In the empirical studies, individuals younger than 6.23 ± 5.30 days were found not to perform any tasks (defined as idlers). Those older than this but younger than 17.84 ± 12.89 days were found to perform only intranidal tasks, which consisted largely of feeding larvae; these were defined as nurses. Still older individuals performed extranidal tasks, principally foraging, although these individuals continued to perform intranidal tasks also. These older wasps were defined as foragers here (not to be confused with the behavioral castes of foragers in the principal component analysis in Chapter 5). Thus the A/I ratio obtained from the simulation for 6-day-old individuals (0.018) and for 18-day-old individuals (0.056) were set as the threshold for nursing and foraging respectively. In other words, individuals with A/I ratios less than 0.018 were classified as idlers, those with values between 0.018 and 0.056 were classified as nurses, and those with values above 0.056 were classified as foragers. The simulation thus permitted us to determine the ages as well as the proportions of idlers, nurses, and foragers.

Our first result from the simulation was the demonstration of age polyethism. The mean age of idlers, nurses, and foragers turned out to be 4, 14, and 37 days respectively. The proportions of these categories were 0.28, 0.36, and 0.36. A crucial aspect of this model is that it is expected to provide a mechanism by which age polyethism becomes sufficiently flexible so that the colony can respond to contingencies such as a skewed age distribution. To see if the model would indeed do so, we simulated not only colonies with individuals ranging in age from 1 to 80 (typical colonies) days but also colonies with individuals ranging in age from 1 to 10 days (young-worker colonies) as well as colonies with individuals ranging in age from 70 to 80 days (old-worker colonies) (for more details see Naug and Gadagkar, 1999). The different age distributions in the three kinds of colonies (typical, young-worker, and old-worker) significantly influenced both the mean ages and the proportion of idlers, nurses, and foragers (Figure 7.7). Nevertheless, the flexibility of age polyethism was very clear. Young-worker colonies had precocious foragers with a mean age of 8 days, and old-worker colonies had overaged nurses with a mean age of 75 days. These finds must be contrasted with the mean age of 14 days for nurses and 37 days for foragers in colonies with typical age distribution. Young-worker colonies had fewer idlers than typical colonies, and old-worker colonies had none—not surprising, since colonies with a skewed age distribution had a limited work force. Besides, young-worker colonies had a higher proportion of foragers and old-worker colonies had a higher proportion of nurses than normal colonies—this too makes biological sense, since precocious foragers and overaged nurses cannot be expected to be as efficient as their counterparts in typical colonies.

Emboldened by this success we went on to see if, without any additional assumptions, this model can also explain how colonies respond to other contingencies, such as an altered task demand. A higher than normal larva/adult ratio can produce a high demand on workers. But a high larva/adult ratio also means more interaction between larvae and adults, leaving that much less time for the adults to interact with each other. Thus we simulated changes in demand by altering the rate of adult-adult interactions. We simulated colonies with 5 (corresponding to high-demand situations), 10 (corresponding to intermediate-demand situations), and 15 (corresponding to low-demand situations) interactions per individual. These different rates of interindividual interaction significantly influenced the mean age and proportion of idlers, nurses, and foragers (Figure 7.8). Remarkably enough, the mean ages of idlers, nurses, and foragers decreased with increasing demand, meaning that workers worked harder to meet the increased demand. The proportion of idlers and nurses decreased with increasing demand but that of foragers increased. The decrease in the proportion of idlers and increase in the proportion of foragers is clearly useful in dealing with a high-demand situation. But the decline in the proportion of nurses may seem perplexing because an increased larva/adult ratio must increase the demand not only for foraging but also for nursing. We think that the clue to this riddle is that while nurses seldom participate in foraging tasks, foragers routinely combine foraging and nursing duties. Indeed, there appears to be some evidence that *R. marginata* foragers increasingly combine foraging with nursing duties under conditions of high demand (see also section 6.7). In summary this simple activator-inhibitor model seems to account for all observed empirical results concerning age polyethism in normal colonies as well as in colonies with altered age distributions or altered task-demand levels.

What is the evidence for the presence of the proposed activator and inhibitor molecules? In *R. marginata* none as yet, but there is reasonable evidence for them in honey bees. The obvious candidate for the activator molecule is the juvenile hormone (JH) (Fahrbach and Robinson, 1996; Fahrbach, 1997; Robinson and Vargo, 1996). There is plenty of evidence that JH is involved in *Apis mellifera*. Young bees performing intranidal

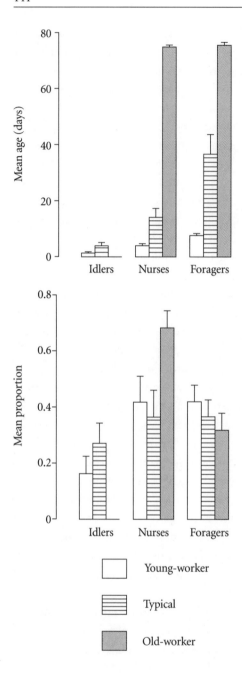

Figure 7.7 (facing page). Above: Mean (± S.D.) ages of the three task groups, idlers, nurses, and foragers, in relation to colony age distributions. Age distribution of the colonies significantly influenced the mean age of idlers (one-way ANOVA; $F = 4821.53$, $p < 0.0001$), nurses ($F = 382412.27$, $p < 0.0001$), and foragers ($F = 67446.61$, p < 0.0001). Multiple comparisons of means within each task group with the Tukey-Kramer method indicated significant differences across age distributions ($p < 0.01$). Below: Mean (± S.D.) proportions of the three task groups, idlers, nurses, and foragers, in relation to colony age distributions. Age distribution of the colonies significantly influenced the mean proportion of idlers (one-way ANOVA; $F = 1289.54$, $p < 0.0001$), nurses ($F = 4050.52$, $p < 0.0001$), and foragers ($F = 718.55$, $p < 0.0001$). Multiple comparisons of means within each task group with the Tukey-Kramer method indicated significant differences across age distributions ($p < 0.01$). (Redrawn from Naug and Gadagkar, 1999.)

tasks have low titers of JH, while older bees performing extranidal tasks have high titers (Robinson, 1987, 1992; Robinson et al., 1989). JH appears to play a causal role in modulating task performance in honey bees and also in the swarm-founding wasp *Polybia occidentalis.* Injection or external application of JH can produce precocious foragers, for example. Precocious foragers and overaged nurses produced by the alteration of colony demography have JH titers appropriate for the task they are performing, though inappropriate for their age (Jaycox, 1976; Rutz et al., 1976; Robinson, 1987; Robinson et al., 1992; Huang and Robinson, 1992, 1996; O'Donnell and Jeanne, 1993). Although the inhibitor is not yet identified even in honey bees, there appears to be evidence for social inhibition of behavioral development. Worker bees kept in isolation precociously synthesize JH at high rates (Huang and Robinson, 1992, 1996). Huang, Plettner, and Robinson (1998) have shown that physical contact with other bees is necessary for social inhibition of behavioral development. They have also shown that removing the mandibular gland of bees renders them less inhibitory or completely uninhibitory, which suggests mandibular glands as a possible source of the potential inhibitor.

Two major conclusions emerge from the work described in this chapter. One, a fairly sophisticated, almost honey bee like, age polyethism can evolve in a primitively eusocial species such as *R. marginata,* where workers still retain reproductive options. Several other species of primitively eusocial wasps also show some form of age polyethism (Jeanne, 1991a). Two, the seemingly unbridgeable gap between the hypothesis of age as

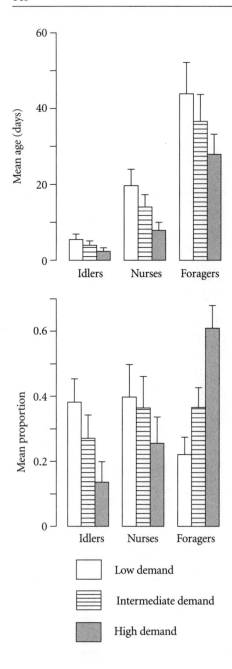

Figure 7.8 (facing page). Above: Mean (± S.D.) ages of the three task groups, idlers, nurses, and foragers, in relation to colonies with different demand levels (brood/adult ratios). The demand levels of the colonies significantly influenced the mean age of idlers (one-way ANOVA; $F = 1677.86$, $p < 0.0001$), nurses ($F = 3101.14$, $p < 0.0001$), and foragers ($F = 1297.31$, $p < 0.0001$). Multiple comparisons of means within each task group with the Tukey-Kramer method indicated significant differences across demand levels ($p < 0.01$). Below: Mean (± S.D.) proportions of the three task groups, idlers, nurses, and foragers, in relation to colonies with different demand levels (brood/adult ratios). The demand levels of the colonies significantly influenced the mean proportions of idlers (one-way ANOVA; $F = 3199.39$, $p < 0.0001$), nurses ($F = 643.20$, $p < 0.0001$), and foragers ($F = 10173.13$, $p < 0.0001$). Multiple comparisons of means within each task group with the Tukey-Kramer method indicated significant differences across demand levels ($p < 0.01$). (Redrawn from Naug and Gadagkar, 1999.)

a causative agent in behavioral development and the criticism that age polyethism is potentially inflexible (Tofts and Franks, 1992; Franks and Tofts, 1994; Robinson, Page, and Huang, 1994; Franks, Tofts, and Sendova-Franks, 1997; Robson and Beshers, 1997; Traniello and Rosengaus, 1997; Calderone, 1998) can indeed be bridged by exploring relative age rather than absolute age as the agent of work organization. I refer the reader to O'Donnell (1996, 1998a, 1998c, 1998b), Downing and Jeanne (1987, 1988, 1990) for aspects of polyethism and division of labor in primitively eusocial wasps not investigated here. I also refer the reader to Bonabeau, Theraulaz, et al. (1997), Page (1997), and Detrain, Deneubourg, et al. (1999) for a discussion of the role of self-organization in division of labor—something that is yet to be studied in R. marginata. I must also mention that in swarm-founding wasps, polyethism, task partitioning, and work organization reach levels of sophistication approaching those of honey bees and other highly eusocial species (Jeanne, 1986a; 1986b; Jeanne, Downing, and Post, 1988; O'Donnell and Jeanne, 1990, 1996; O'Donnell, 1996, 1998c).

PART III

The Evolution of Eusociality

8   THE THEORETICAL FRAMEWORK

$\mathcal{L}$ET US move on to a more explicit consideration of the problem of the evolution of eusociality and see what we can learn by using *R. marginata* as a model system. In Chapters 4–7, I described those aspects of the social biology of *R. marginata* that we must know about before beginning such an exercise. Here I will outline the theoretical framework that will guide us from now on. In Chapters 9–14, I will provide additional empirical findings, especially those used explicitly to test models for the evolution of eusociality.

8.1.   What Do We Mean by Understanding the Evolution of Eusociality?

Grafen (1991) called the philosophy of behavioral ecology a phenotypic gambit, where we assume quite brazenly (for we know it cannot be literally true) that phenotypic characters of interest are determined by the simplest genetic system—perhaps even a haploid locus at which each allele produces a distinct phenotypic character—and that enough mutations occur to produce the required variations. Natural selection is then expected to act on this variation and favor or disfavor specific variants as appropriate to the existing environmental conditions. The phenotypes of interest to us are (1) a selfish wasp that prefers to rear her own offspring and (2) an altruistic worker that cares for another wasp's offspring rather than rear her own. The question is: Under what conditions will the selfish and the altruistic wasps respectively be favored by natural selection?

It is useful to make a distinction between the origin of eusociality and its maintenance. By the maintenance of eusociality I mean the conditions under which a selfish mutation cannot spread in a population of altruists so that eusociality is stably maintained. It seems reasonable to turn to the highly eusocial species when we are dealing with stably maintained populations of altruists. Thus the question is: Why don't honey bee, ant, and termite workers become selfish, stop working for their colonies, and revert back to solitary life? An answer to this question has seldom been attempted. The problem is not that we have reduced the underlying genetics to an unrealistic level but that another assumption of the phenotypic gambit may be false. This is the assumption that enough mutations occur to produce the required variation in characters. Because the reproductive and worker castes are morphologically differentiated and because caste determination is essentially irreversible in highly eusocial species, a worker cannot easily revolt against the queen and either drive her away and replace her or leave to found a new nest of her own. The transition from the solitary to the highly eusocial must have occurred millions of years ago (the current estimate for ant sociality is about 132 million years; Agosti, Grimaldi, and Carpenter, 1997), and the highly eusocial species are now probably caught in an evolutionary cul-de-sac where it may be nearly impossible for mutations to take them back to solitary existence. Social organization and caste determination have been perfected by such a series of elaborate physiological mechanisms that it is hard to imagine that one or more mutations could retrace these steps (Wilson, 1971; but see Chapter 18).

The fact remains, therefore, that we cannot rule out the possibility that workers in most highly eusocial species remain altruistic simply because they have irretrievably lost their reproductive options and have no choice but to work for their colonies, even if the worker strategy is no longer advantageous—even if the worker strategy is potentially invadable by a selfish alternative. In other words, there is not sufficient variability in the populations of highly eusocial species for us to study competition between selfish and altruistic alternatives. This does not of course mean that there are no conflicts in highly eusocial species. Bourke and Franks (1995) in their excellent monograph *Social Evolution in Ants* list at least 12 kinds of conflicts in ant societies—conflict over sex allocation, male production, queen number, kin class of adopted queens, allocation of reproduction between multiple queens, queen quality, queen replacement, investment in maintenance versus reproduction, investment per worker, caste determination, worker activity levels, and male-female conflict. Indeed, the study of conflict in highly eusocial species, especially ants and bees, has contributed a great deal to the testing of several subtle predictions of inclusive fitness theory and has significantly enhanced our understanding of insect sociobiology. But the question of why eusociality is stably maintained in highly eusocial species is not so easily amenable to empirical investigation because virtually all known highly eusocial species are stably eusocial (but for some possible exceptions, see Chapter 18).

By contrast, the origin of eusociality concerns the conditions under which altruists can invade populations of selfish, solitary individuals. To discover these conditions, we would ideally like to be able to trace the evolutionary past of the eusocial species. But that is not possible because there are no fossils, and even if there were they would tell us precious little about behavior, altruism, and social organization. The next best thing is to turn to the primitively eusocial species to see if they can shed light on the origins of eusociality (or perhaps some noneusocial, group-living species; Cowan, 1991). Fortunately, the primitively eusocial species are certainly in no evolutionary cul-de-sac. As we have seen, caste differentiation is not morphological and workers have clear reproductive options; they can replace their queens or leave to start their own solitary-foundress nests. Why then do they not always do so? This is a promising question because we can rule out the trivial alternative hypothesis that they have no option but to continue to work for their colonies. Here we have abundant opportunities to study competition between selfish and altruistic alternatives.

Before we proceed further, however, two clarifications are in order. No claim is being made that primitively eusocial species have achieved eusociality more recently in evolutionary time than highly eusocial ones. Nor is it being claimed that primitively eusocial species are on their way to becoming highly eusocial (see Packer and Knerer, 1985, who issue a similar caveat while classifying levels of sociality in sweat bees). But it is perhaps reasonable to assume that highly eusocial species have passed through stages that might have resembled today's primitively eusocial species. If that is reasonable, we may explore the factors that tilt the balance in favor of altruistic workers, rather than selfish solitary workers, in the primitively eusocial species today and assume that similar factors may have been responsible for the origin of eusociality.

8.2.   The Paradox of Altruism

As we have seen, eusocial species are characterized by the presence of altruism. Reproductive caste differentiation often results in one or a small number of individuals functioning as fertile reproductives and the rest of the individuals serving as sterile workers. Workers who remain sterile and devote their lives to the welfare of their colonies and to assisting their queens to reproduce can be said to behave altruistically because they lower their own direct fitness and enhance the fitness of their queens. Thus altruism may appear paradoxical from the point of view of Darwinian natural selection because that theory usually prepares us to expect competitive selfishness, not altruism.

Hamilton (1964a, 1964b) argued that the paradox of altruism was no paradox at all if we realize that natural selection is dependent on changes in relative frequencies of alleles without regard to the pathway through which those changes are brought about. It follows then that producing offspring is only one way to increase the representation of an individual's genes in the population. Aiding genetic relatives, which also carry copies of one's genes, identical by descent, is an alternate, equally legitimate way of doing so. And since the probability of occurrence of copies of an individual's genes, identical by descent, in any class of genetic relatives, including offspring, can be relatively easily calculated, we can compute a composite quantity called inclusive fitness. Inclusive fitness, then, is the sum of an individual's own fitness plus all its influence on fitness in its relatives other than direct descendants.

Given this defintion, Hamilton formulated an elegant rule, now known as Hamilton's rule, that specifies the condition under which an "altruist" allele will spread in a population relative to an alternate, selfish allele. The condition is usually represented algebraically as:

$$(8.1) \qquad B/C > 1/r$$

where B is the benefit to the recipient of the altruism, C is the cost to the altruist, and r is the coefficient of genetic relatedness between altruist and recipient—the probability that genes present in the altruist are also present in the recipient. This expression can be conveniently rewritten as

$$(8.2) \qquad B \cdot r > 1 \cdot C$$

meaning that the benefit to the recipient, devalued by the genetic related-ness between altruist and recipient, should be greater than the cost to the altruist. Even more conveniently for our purpose, it may be expressed as

(8.3) $B/C > r_o/r_i$ or $B.r_i > C.r_o$

where r_i is the genetic relatedness of the altruist to the recipient's offspring and r_o is the genetic relatedness of the altruist to his or her own offspring. Here B can be thought of as the additional numbers of offspring produced by the recipient owing to the help given by the altruist, and C can be thought of as the number of his own offspring given up by the altruist in order to help the recipient.

Hamilton's rule—or inclusive fitness theory or the theory of kin selec-tion, as it is variously called—has ushered in a truly new era in the study of animal behavior and behavioral ecology. Not surprisingly, social insects have been at the forefront of this new field of sociobiology (Wilson, 1971, 1975; Hölldobler and Wilson, 1990; Trivers, 1985; Bourke and Franks, 1995; Crozier and Pamilo, 1996).

8.3. Inclusive Fitness Theory as a Unifying Theme

Inclusive fitness theory offers a unifying theme to evaluate the multitude of factors that might be responsible for favoring altruistic workers over selfish solitary nesters. Inclusive fitness theory predicts that workers do not become selfish and start their own solitary-foundress nests because

(8.4) $\Omega > W$

where Ω and W are the inclusive fitnesses of a worker and a solitary found-ress respectively. The question then is why should, or under what cir-cumstances would, the inclusive fitness of workers be greater than that of a solitary foundress? To understand this let us break up inclusive fitness into its constituent components and rewrite inequality 8.4. The inclusive fitness of workers and solitary foundresses can be broken up into at least three components (Gadagkar, 1990a, 1991b; Queller, 1989), so that we have the prediction that workers should be favored over solitary foun-dress if

(8.5) $\rho \beta \sigma > r b s$

where ρ is the coefficient of genetic relatedness of a worker to the brood she rears; ß is the intrinsic productivity of a worker, defined as the number of individuals she can rear to adulthood provided she survives for their entire developmental period; and σ is the demographic correction factor for a worker, defined as that factor by which a worker's intrinsic productivity should be devalued because she might die before the brood under her care complete development. The corresponding parameters for a solitary foundress are r, b, and s. Clearly at least three classes of factors can contribute to inequality 8.2: those that help make $\rho > r$, $ß > b$, and $\sigma > s$ respectively.

8.4.   Predisposition to Eusociality

One reason why the inclusive fitness of workers may be greater than that of solitary foundresses is that workers may have access to brood that are more closely related to themselves than a solitary foundress would be to her own offspring. The haplodiploid genetic system in the Hymenoptera, coupled with an ability to bias investment in favor of female brood, can make this possible (Hamilton, 1964a, 1964b; Trivers and Hare, 1976; see also Gadagkar, 1990c, 1991c; Bourke and Franks, 1995; Crozier and Pamilo, 1996). This is sometimes called the haplodiploidy hypothesis, but I will call it genetic relatedness predisposition or simply genetic predisposition and discuss it in Chapters 9 and 10.

Another reason why the inclusive fitness of workers may be greater than that of solitary foundresses is that workers may be able to rear more brood per worker than a solitary foundress, probably because there is better protection from parasites, predators, and conspecific usurpers in a group's nest than in a solitary nest (see, for example, Litte, 1977, 1979, 1981; Lin and Michener, 1972; Gamboa, 1978; West-Eberhard, 1978a; Suzuki and Murai, 1980; Gadagkar, 1985a; Itô, 1986; Strassmann, Queller, and Hughes, 1988). I will call this ecological predisposition and discuss it in Chapter 11.

Workers may also rear more brood per worker because individuals who opt for worker roles may be subfertile; that is, they may be individuals whose b would be relatively small if they became solitary foundresses but whose ß as workers would be relatively high if they became workers (see, for example, West-Eberhard, 1975; Craig, 1983; Gadagkar et al. 1988,

1990). I will call this physiological predisposition and discuss it in Chapter 12.

Yet another reason why the inclusive fitness of workers may be greater than that of a solitary foundress is that workers may have lower mortality rates than solitary foundresses, and, more important, the consequence of similar mortality rates may be quite different for workers that function in groups than for solitary foundresses (Queller, 1989; Strassmann and Queller, 1989; Gadagkar, 1990a, 1991b; Nonacs, 1991; Field et al., 2000). I will call this demographic predisposition and discuss it in Chapter 13.

The use of the word "predisposition" needs some justification. To predispose means to render susceptible or liable, beforehand. This connotation is perfectly valid in the case of genetic predisposition caused by the relatedness asymmetries created by haplodiploidy. Since nearly all known hymenopterans are haplodiploid and eusociality is restricted to some hymenopteran groups, haplodiploidy can safely be inferred to be a more primitive character than eusociality. Hymenopterans may thus be legitimately said to be potentially genetically predisposed to the evolution of eusociality.

The situation with ecological, physiological, or demographic predisposition is somewhat different. We do not know if variations in fertility, lifespans, time taken to attain reproductive maturity, and so on, can really be said to be more primitive or more advanced characters of the Hymenoptera than eusociality. As a matter of fact, we know almost nothing about the distribution of these factors among different groups in the Hymenoptera. Nevertheless I think we can justifiably use the term predisposition even for ecological, physiological, and demographic factors as long as we are dealing with primitively eusocial species. After all, we will be looking at primitively eusocial species, where workers behave altruistically in spite of having the option of direct reproduction, and we will be equating the factors that favor altruistic workers over solitary nesters in these species today with factors that might have been responsible for the origin of eusociality. So any ecological, physiological, or demographic asymmetries that we may discover in today's primitively eusocial species can therefore be considered predispositions to eusociality.

Another clarification, this time regarding the phrase "genetic predisposition," is in order. That phrase is sometimes used to emphasize that some character has a genetic basis rather than being solely influenced by the environment. For example, Page and Robinson (1992) have demonstrated a genetic predisposition for undertaking behavior in honey bees. Here I am using "genetic predisposition to eusociality"instead of the more cumbersome "genetic relatedness predisposition" to refer to the possibility that asymmetries in genetic relatedness in primitively eusocial species, such as R. marginata, favor workers over solitary foundresses and thus predispose them to the evolution of eusociality.

Here is yet another clarification, regarding the contrast between solitary foundresses and workers. I sometimes use the words "worker" and "joiner" interchangeably, depending on the context. In multiple-foundress nests, all individuals except one foundress become workers. We do not really know whether the queen is the real "foundress" and the others are actual "joiners." I rather doubt it. A group of female wasps get together and one of them, not necessarily the first one, becomes the queen. When a wasp decides to be in a group, she faces a high probability of becoming a worker. Hence I contrast the "solitary-founding" strategy and "worker" strategy and ignore the small probability that a joiner might actually become the queen of her group.

In summary, we will attempt to understand the evolution of eusociality by exploring the potential genetic, ecological, physiological, and demographic asymmetries between workers and solitary foundresses in R. marginata.

9 GENETIC PREDISPOSITION I: INTRACOLONY GENETIC RELATEDNESS

$\mathcal{I}$N ADDITION to proposing the theory of inclusive fitness, Hamilton (1964a, 1964b) realized its immediate and powerful application to understanding the evolution of eusociality in the Hymenoptera. The haplo-diploid genetic system of hymenopterans creates asymmetries in genetic relatedness such that in an outbred population a female is related to her full sisters by 0.75 and is, like any diploid female, related to her offspring only by 0.5. Thus full sisters are genetically more valuable than offspring, so that even if the benefit and cost terms in Hamilton's rule (inequality 8.3) are equal, the inclusive fitness of workers that rear full sisters is greater than that of solitary foundresses that rear their own offspring. In other words, more copies of an altruist allele will be represented in future generations if a female hymenopteran bearing such an allele rears full sisters rather than an equal number of offspring. Thus the relatedness asymmetry created by haplodiploidy can by itself potentially promote the evolution of eusociality. The efficiency of workers in passing on copies of the altruist allele decreases if workers rear mixtures of brothers and sisters because they are less related to their brothers than they are to their sisters and offspring. Here I am using the so-called life-for-life relatedness values rather than the regression relatedness values (see Bourke and Franks, 1995, for clear definitions of these terms). Trivers and Hare (1976) first showed that this difficulty can be offset if workers skew their investment in favor of sisters.

We do not know if *R. marginata* workers are capable of effecting a female-biased investment ratio—perhaps we will never know. Sex ratio in social insect colonies is correctly defined as the ratio of reproductive females (as opposed to workers) to males. In *R. marginata*, there is no morphological distinction between queens and workers, nor are potential

queens distinguishable from future workers by the timing of their production, as is the case in many temperate-zone *Polistes* species (Reeve, 1991).
As we saw in Chapter 4, female *R. marginata* eclosing at any time of the
year have a fair chance of becoming queens. Whether a given individual
becomes a queen or a worker depends a great deal on the social circumstances she finds herself in. Since there is no simple way to distinguish between queens and workers, there is no simple way to estimate sex ratio
in *R. marginata*. However, sex-biased investment can only help enhance
workers' inclusive fitness if they have opportunities to rear female brood to
which they are more related than they are to their offspring. Hence we can
at least see who rears brood with a greater degree of relatedness to themselves—workers or solitary foundresses. I must emphasize that this is the
limited question I am asking in this chapter. I am not asking, for example,
whether haplodiploidy promotes the evolution of worker behavior in *R.
marginata*. I shall return to that question in Chapter 14.

But given that *R. marginata* is haplodiploid, why should we doubt that
workers have opportunities to rear more closely related brood than solitary
foundresses? There are two phenomena that can lower worker-brood genetic relatedness: polyandry (multiple mating by queens) and polygyny
(multiple queens per colony).

9.1.   Polyandry

In collaboration with a colleague, M. S. Shaila, and a post-doctoral fellow,
K. Muralidharan, I obtained evidence for multiple mating in *R. marginata*
by comparing the genotypes of mothers and daughters and inferring the
genotypes of fathers using electrophoretically variable enzyme loci (Muralidharan, Shaila, and Gadagkar, 1986). Small pre-emergence colonies
were collected in and around Bangalore. The nest and its brood were discarded and the adults were individually identified with unique spots of
paint and released into laboratory cages for nest initiation, as described in
Chapter 3. At that stage of our study, we were not sure that there would be
a single egg layer in a colony. We therefore took special precautions to be
certain of the maternity of the wasps eclosing in the laboratory cages. After
the wasps brought from nature initiated a new nest and began laying eggs
in the laboratory cages, we made *ad libitum* observations to identify the

egg layer. Once the egg layer was identified, all other wasps were removed from the cage. Any eggs laid until then, while we were not watching, were also removed. Now the egg layer was alone in her cage and we could be certain that all offspring eclosing from the nest would be her offspring. When the egg layers produced adult offspring, daughters were removed from time to time and subjected to electrophoresis (no males were produced under these conditions). We usually left behind one or two daughters to assist the mother in nest building and brood care because even if the daughters laid any eggs they would be detected because they would develop into males. When it appeared that the egg layer in any cage might die, the experiment was terminated and the egg layer and any remaining daughters were electrophoresized. In addition, male and female wasps were opportunistically obtained from "leftovers" in other experiments and used to standardize the electrophoretic techniques and to assess the number of alleles segregating at each locus in our experimental system. Very briefly, single wasps were homogenized in 100 μl of 0.1M phosphate buffer (pH 7.1) and centrifuged in a Beckman microfuge. We electrophoresized the supernatant on vertical polyacrylamide (7.5%) slab gels using standard methodology (Shaw and Prasad, 1970), except that we used Tris-Hel buffer (pH 7.1) as a gel buffer. Gels were stained for nonspecific esterases in the manner described by Shaw and Prasad (1970).

The initial standardization experiments revealed the presence of three nonspecific esterase loci, which we designated a, b, and c. Males always showed a single band at each locus, while females always showed a single band at locus a and one or two bands at loci b and c, suggesting the presence of at least two alleles at b and c. The two alleles at the dimorphic loci were designated f and s for the fast- and slow-moving bands respectively. We did not have enough data on allele frequencies from a randomly drawn sample to determine the actual number of matings. But a comparison of the genotypes of mothers and daughters at any variable locus can be used to infer the genotypes of the father/s and thus detect polyandry, if it occurs. By this procedure, we did detect polyandry in R. marginata. Our estimate that R. marginata queens mate with one to three males is a minimum estimate and the consequent calculation of genetic relatedness between the daughters of a female of 0.53 is a preliminary estimate, and probably an overestimate (Muralidharan, Shaila, and Gadagkar, 1986; Gadagkar,

Table: 9.1.   Evidence for multiple mating in *R. marginata*. (Modified from Muralidharan, Shaila, and Gadagkar, 1986; Gadagkar, 1990b.)

Experiment code number	Genotype of mother	Genotype of daughters (no. of individuals)	Minimum number of matings	Inferred genotype of father(s)	Mean genetic relatedness between daughters[a]
M11	$b^i b^i c^i c^i$	$b^i b^i c^i c^i$ (6)	3	$b^i c^i$	0.48
		$b^i b^i c^i c^i$ (3)		$b^i c^i$	
		$b^i b^i d^i c^i$ (1)		$b^i d^i$	
M13	$b^i b^i c^i c^i$	$b^i b^i c^i c^i$ (5)	2	$b^i c^i$	0.61
		$b^i b^i d^i c^i$ (1)		$b^i d^i$	
M15	$b^i b^i c^i c^i$	$b^i b^i c^i c^i$ (7)	3	$b^i c^i$	0.46
		$b^i b^i c^i c^i$ (2)		$b^i c^i$	
		$b^i b^i d^i c^i$ (3)		$b^i d^i$	
M23	$b^i b^i c^i c^i$	$b^i b^i c^i c^i$ (1)	1	$b^i c^i$	0.75
		$b^i b^i c^i c^i$ (2)			
Weighted mean					0.53

a. calculated using the expression $1/2 \ (1/2 + \sum_{i=1}^{m} f_i^2)$ where f_i is the proportion of daughters fathered by the ith male and m is the number of males.

1990b; see Table 9.1). Nevertheless, the fact that evidence of multiple mating and mixing of sperm can be obtained by assaying the first 6–12 daughters of an egg layer is very suggestive.

9.2.   Serial Polygyny

Just as polandry can lead to lowering of intracolony genetic relatedness because of the production of different patrilines, polygyny can lower intracolony genetic relatedness by producing different matrilines within the same colony. Polygyny can be of two kinds. The first involves the simultaneous presence of multiple egg layers in a colony and may be called simultaneous polygyny. The implications of simultaneous polygyny for social evolution in the Hymenoptera have received considerable attention (Hölldobler and Wilson, 1977; West-Eberhard, 1978a, 1990; Bourke, 1988; Ross and Carpenter, 1991; Strassmann et al., 1991). The second form of polygyny involves frequent queen replacements and is often referred to as serial polygyny. *Mischocyttarus drewseni* (Jeanne, 1972; see also West-Eberhard, 1978a) and *Ropalidia variegata jacobsoni* (Yamane, 1986) are known to exhibit serial polygyny. It is very clear that simultaneous polygyny does not occur in *R. marginata*, a strictly monogynous species—there is never more than one egg layer in any colony at any given time. As described in section 4.2, one of the options available to a female *R. marginata* is to stay back in her natal nest for a while as a worker and then, at an opportune moment, to drive the original queen away and take over the colony as its next queen. Obviously, then, serial polygyny does occur in *R. marginata*.

My students and I, however, had not paid any attention to serial polygyny and its implications for the evolution of eusociality until we were jolted into doing so by a curious set of circumstances. The organizers of the twelfth congress of the International Union for the Study of Social Insects (IUSSI) canceled the symposium I was to speak in and transferred all the papers to another symposium on polygyny. I had to speak on polygyny in *R. marginata* if I were to speak at all. I quickly met with several of my students, including K. Chandrashekara, Swarnalatha Chandran, and Seetha Bhagavan, to see what I could say about serial polygyny in *R. marginata* based on data already on hand.

It turns out that *R. marginata* is an excellent model to study serial polygyny and its implications for the evolution of eusociality. For all our experimental nests, we routinely maintain nestmap and census records. Because eggs laid in a cell hatch into larvae and pupate in the same cell, the cell number is the address of an individual wasp from the egg to the adult stage. As soon as an adult wasp ecloses, it is given a unique paint mark so that, in effect, we have all individuals uniquely identified from egg to death of the adult. Because there is only one egg layer at any given time and because we discover a queen turnover usually within a day or so of the event, we know the genealogical relationships between every pair of individuals, be they eggs, larvae, pupae, or adults. The perennial, indeterminate nesting cycle of *R. marginata* allows us to witness a succession of queens on the same nest.

For the IUSSI talk, we were able to pull together nestmap and census data on four colonies for periods ranging from 261 to 606 days, study the system of serial polygyny, and tease out its implications for the evolution of eusociality (Gadagkar et al., 1990, 1991a, 1993a, 1993b). During the observation period, the number of queens ranged from 2 to 10 per colony. The mean tenure of queens was about 80 days, with a large variation ranging from a mere 7 days to as much as 236 days. The queens were 4 to 78 days old at the time they took over and were 37 to 262 days old at the time

Table 9.2.   Serial polygyny in *R. marginata*. Data are from four colonies. Note that the first two
rows refer to an average of four colonies while the remaining rows refer to an average
of 21 queens. (From Gadagkar et al., 1993b.)

	Minimum	Maximum	Mean	S.D.
Duration of study per colony (days)	261	606	418.50	132.10
Number of queens per colony	2	10	5.25	2.95
Tenure of a queen (days)	7	236	79.71	72.36
Age of queen at takeover (days)	4	78	35.18	19.28
Age of queen at the end of tenure (days)	37	262	103.52	65.31
Number of eggs laid by a queen	9	2207	308.38	487.93
Number of larvae produced by a queen	2	1625	210.67	362.54
Number of pupae produced by a queen	0	679	97.76	162.00
Number of females produced by a queen	0	360	60.10	89.64
Number of males produced by a queen	0	136	16.00	33.19
Number of adults produced by a queen	0	394	76.10	111.81
Number of eggs laid per day by a queen	1.00	10.22	3.08	1.96
Prop. of eggs laid by a queen that became adults	0.00	0.68	0.25	0.16

they were replaced. The productivities of queens similarly varied over a
wide range of values (Table 9.2).

9.2.1.   The Pedigrees

Because of strict monogyny at any given time, we were able to assume that
the eggs laid during the tenure of a queen, including replacement queens,
were her offspring, and for the reasons given above we could assign all
eggs, larvae, pupae, and adult wasps, at any time, to a particular matriline.
Thus we could construct pedigrees for queens in all four colonies (Figure
9.1). The pedigrees vary greatly in complexity. In colony T01, each queen is
the daughter of her immediate predecessor queen. At the other extreme is
the pedigree in colony T08, where new queens were daughters, sisters,
nieces, or cousins of their immediate predecessor queens (Table 9.3). These
are merely the relationships between new queens and their immediate predecessors. If we considered each queen and all her predecessors in that colony, the inter-queen relationships would of course be even more distant.

9.2.2.   Worker-Brood Genetic Relatedness

As far as the consequences of serial polygyny for levels of intracolony genetic relatedness is concerned, we should really be looking at worker-

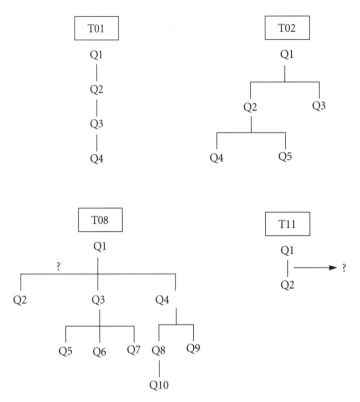

Figure 9.1. Pedigrees of queens in the four colonies studied. The relationship between queen 1 (Q1) and queen 2 (Q2) is unknown for colonies T08 and T11 because both Q1 and Q2 were among the wasps present on the nest at the time of its collection and transplantation. (Redrawn from Gadagkar et al., 1993b.)

brood relatedness, which depends on the extent of overlap between workers that are offspring of a given queen, brood that are offspring of another queen, and the relatedness between the two queens. From the nestmap and census data we could readily determine the extent of such overlap.

Considering data from all four colonies, we found that the extent of overlap between workers of each matriline and brood of any other matriline was such that brood were classified as brothers, sisters, nieces and nephews, cousins, cousins' offspring, mother's cousins, mother's cousins' offspring, and even mother's cousins grand-offspring. I must confess that we were totally unprepared to find that *R. marginata* colonies would have such a complicated genetic structure. The observed relationships, along

Table 9.3.   Genetic relationships between successive queens and between workers
            and brood observed in the four colonies. (From Gadagkar et al., 1993b.)

Relationship between queens and their immediate predecessors	Relationship between workers and brood
a. Daughters	1. Sisters (0.75 or 0.53)
b. Sisters	2. Brothers (0.25)
c. Nieces	3. Nieces and nephews (0.375 or 0.265)
d. Cousins	4. Cousins (0.1875 or 0.1325)
	5. Cousins' offspring (0.0938 or 0.0663)
	6. Mother's cousins (0.0938 or 0.0663)
	7. Mother's cousins' offspring (0.0469 or 0.0331)
	8. Mother's cousins' grand-offspring (0.0234 or 0.0165)

Note: Values of relatedness given in parentheses are for single mating (r for sisters $= 0.75$) or multiple mating (average r for sisters $= 0.53$), based on electrophoretic data from Muralidharan, Shaila, and Gadagkar, 1986, and Gadagkar, 1990b.

with the coefficient of genetic relatedness under single and multiple mating by queens, are given in Table 9.3. At different times in the life of these four colonies, the worker-brood genetic relatedness could have attained a maximum value of 0.75 and a minimum value of 0.0165.

To assess the impact of these different values of worker-brood relatedness, we need to know the frequencies with which each kind of worker-brood genetic relationships listed in Table 9.3 actually occur in the colony. To estimate this we computed, at weekly intervals, the proportions of workers and brood belonging to each matriline and the relationship between each brood matriline and each worker matriline. On any given day, the mean worker-brood genetic relatedness was calculated as

$$(9.1) \qquad \bar{r} = \sum_{i=1}^{n} \sum_{j=1}^{n} w_i b_j r_{ij}$$

where w_i is the proportion of workers that belong to the i_{th} matriline, b_j is the proportion of brood that belong to the jth matriline, r_{ij} is the coefficient of genetic relatedness between workers belonging to the ith matriline and brood belonging to the jth matriline, and n is the number of matrilines. Since we had no way to sex the brood, we simply computed two sets of worker-brood relatedness, assuming first that all brood were female and next that all brood were male.

Worker-brood genetic relatedness varied considerably with time (Figure 9.2). As expected, worker-female brood relatedness dropped after a queen turnover because workers who previously were rearing sisters had to begin to rear nieces, cousins, or other more distantly related brood, depending on the genetic relatedness between the old and the new queen. Rather interestingly, a queen turnover sometimes led to an increase in worker–male brood relatedness, because workers who were rearing brothers ($r = 0.25$) now began to rear nephews ($r = 0.375$, under single mating) if the new queen was the daughter of the old queen. Under the assumption of multiple mating, the pattern of variation in worker-brood relatedness was qualitatively similar, although the values, as expected, were uniformly lower (Gadagkar et al., 1993b). For a more sophisticated approach to estimating relatedness values from these data, see Haridas and Rajarshi (1997, 2000).

9.2.3. Evaluating the Observed Levels of Relatedness

It is clear from section 9.2.2 that serial polygyny leads to a reduction in worker-brood relatedness in *R. marginata*. What are the consequences of such reduction for the evolution of worker behavior? If reduced worker-brood relatedness lasts for very short periods or occurs only when the colonies are very small, its impact may not be very significant. We therefore computed for each colony a grand mean value of worker-brood relatedness by using the equation

$$(9.2) \qquad \bar{\bar{r}} = \frac{\sum_{i=1}^{s} (W_k + B_k)\bar{r}_k}{\sum_{k=1}^{s} (W_k + B_k)}$$

where W_k is the total number of workers present on the k_{th} day, B_k is the total number of brood present on the k_{th} day, and $\bar{r}_k$ is the mean worker-brood relatedness on the k_{th} day, computed with equation 9.1. The values obtained for each colony, under single and multiple mating, are given in Table 9.4. The worker–female brood values range from 0.22 to 0.65 and those for worker–male brood range from 0.18 to 0.29.

How low are these values? Are they low enough to permit the conclusion that the genetic asymmetries created by haplodiploidy are lost through serial polygyny and polyandry? Taking into consideration the observed levels of polyandry and serial polygyny, can we say whether workers rear more

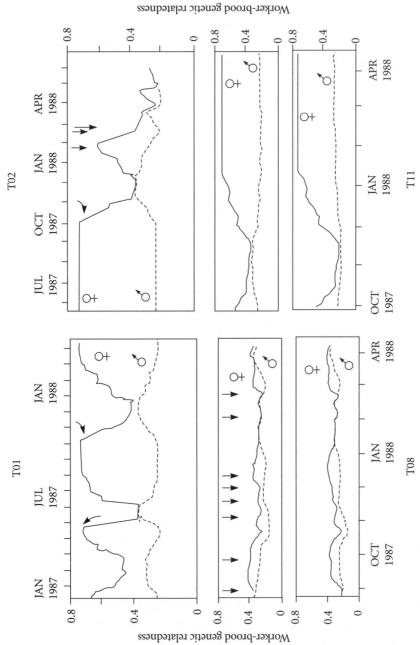

Figure 9.2. Variation in mean worker-brood genetic relatedness computed at weekly intervals in four colonies. Arrows indicate queen turnovers. Queens are assumed to be mated singly. For colonies T08 and T11, the upper panel assumes the unknown relationship between queens 1 and 2 to be that of daughters, while the lower panel assumes the relationship to be that of sisters. (Redrawn from Gadagkar et al., 1993b.)

closely related brood ($\rho > r$) or solitary foundresses rear more closely related brood ($\rho < r$)? Before we can answer this question we need to make
some assumptions about the sex investment ratios achieved by the workers. One possibility is that workers skew their investment in male and female brood in the ratio of their relatedness to the two classes of brood
(Trivers and Hare, 1976). If they do this we can compute ρ as the resultant
weighted mean relatedness (the relatedness of each sex weighted by its
abundance) to the brood. Thus if workers could skew their investment in
the ratio of their relatedness to the brood then the weighted mean relatedness to the brood would be given by

$$(9.3) \qquad \bar{r}_w = \frac{\left(f\dfrac{f}{m}\right) + (m)}{\left(\dfrac{f}{m}\right) + 1}$$

where $\bar{r}_w$ is the weighted mean relatedness to the brood, f is the relatedness
to female brood, and m is the relatedness to male brood. Thus ρ is calculated as r_w and a value greater than 0.5 is expected to tilt inequality 8.5 in
favor of eusociality. Alternatively, one may argue that, during the origins of
sociality, workers may even be selected to rear all-female broods. Thus any
value of worker-female brood relatedness greater than 0.5 can be thought
to tilt inequality 8.5 in favor of the evolution of eusociality.

Note from Table 9.4 that neither the weighted mean relatedness to
brood (ranging from 0.2 to 0.38) nor the relatedness to female brood
(ranging from 0.2 to 0.46) is ever greater than 0.5 under multiple mating.
For comparison, the values under a hypothetical single mating system are
also shown. Note that even under the unlikely assumption of single mating, four out of six values of weighted mean relatedness and two out of six
values of relatedness to female brood are lower than 0.5. We can thus conclude that in R. marginata polyandry and serial polygyny break down the
genetic asymmetries created by haplodiploidy enough that inequality 8.5
cannot be satisfied unless $\beta > b$ or $\sigma > s$. By our definition then (see Chapter 8) R. marginata is not genetically predisposed to the evolution of
eusociality, because $\rho < r$. I must point out that in a variety of social insect
species, intracolony genetic relatedness values have turned out to be much
lower than previously suspected (for reviews see Gadagkar, 1990f, 1990c,
1991c).

Table 9.4. Effects of serial polygyny in *R. marginata* on worker-brood genetic relatedness. (From Gadagkar et al., 1993b.)

Colony	Number of queens	Relationship between successive queens	Single mating			Multiple mating (relatedness between sisters = 0.53)[a]		
			Grand mean genetic relatedness of workers to		Weighted mean relatedness of workers to brood[b]	Grand mean genetic relatedness of workers to		Weighted mean relatedness of workers to brood[b]
			Female brood	Male brood		Female brood	Male brood	
T01	4	Known	0.65	0.28	0.54	0.46	0.25	0.38
T02	5	Known	0.53	0.28	0.44	0.38	0.24	0.32
T08	10	All but one known; one unknown relationship assumed: daughters	0.35	0.28	0.32	0.25	0.20	0.23
		All but one known; one unknown relationship assumed: sisters	0.32	0.24	0.29	0.22	0.18	0.20
T11	2	Assumed: daughters	0.63	0.29	0.52	0.45	0.26	0.38
		Assumed: sisters	0.57	0.23	0.47	0.40	0.21	0.34

a. Data from Muralidharan, Shaila, and Gadagkar, 1986; Gadagkar, 1990b.

b. Weighted mean relatedness to brood obtained if workers skew investment in sisters and brothers in the ratio of their relatedness to them (Gadagkar, 1990c, 1991c).

9.2.4. Simulation Models

In an attempt to generalize these conclusions beyond the four colonies studied to *R. marginata* in general, and also to other species of social insects exhibiting serial polygyny, we constructed simple simulation models of serial polygyny. Queens were assumed to lay one egg every day. The three parameters used in the model and their numerical values derived empirically from studies of *R. marginata* are shown in Table 9.5. In the first model, we merely considered the mean values of these parameters, a queen tenure of 80 days, a brood developmental period of 62 days and a worker lifespan of 31 days. New queens were assumed to be either daughters or sisters of their predecessor queens and queens were assumed to mate singly or multiply. As before, a mean relatedness between sisters of 0.53 derived from electrophoretic data (section 9.1) was used in the case of multiple mating. The model clearly mimics the empirical data. The grand mean worker-brood relatedness and the weighted mean relatedness of workers to brood (Table 9.6) as well as variation with time in the mean worker-brood relatedness (Figure 9.3) qualitatively resemble the empirical data (Table 9.4 and Figure 9.2). Our conclusion from the simulation model is therefore identical to that from the empirical analysis. Under either method of computing ρ, and whether new queens are sisters or daughters of their predecessors, the evolution of worker behavior cannot be attributed to genetic predisposition on account of haplodiploidy. Serial polygyny in *R. marginata* is thus a powerful reducer of worker-brood genetic relatedness.

The model described above ignored the variation seen in the empirically derived values of the parameters used in the model. In the next step, we attempted to incorporate this variation with the use of Monte Carlo techniques. To do this, we reconstructed a normal distribution for each of the

Table 9.5. Parameters used in simulation models of serial polygyny in *R. marginata*. (From Gadagkar et al., 1993b.)

Parameter	Mean ± S.E. (values derived from natural colonies)
Queen tenure	79.7 ± 15.8 days
Brood development period	61.5 ± 1.4 days
Worker lifespan	31.2 ± 3.4 days

Table 9.6. Results of a simulation model for serial polygyny in *R. marginata*. (From Gadagkar et al. 1993b.)

Relationship between successive queens	Single mating			Multiple mating (relatedness between sisters = 0.53)[a]		
	Grand mean genetic relatedness of workers to		Weighted mean relatedness of workers to brood[b]	Grand mean genetic relatedness of workers to		Weighted mean relatedness of workers to brood[b]
	Female brood	Male brood		Female brood	Male brood	
Daughters	0.53	0.32	0.45	0.38	0.26	0.33
Sisters	0.42	0.21	0.35	0.30	0.18	0.26

a. Data from Muralidharan et al., 1986; Gadagkar, 1990b.
b. Weighted mean relatedness to brood obtained if workers skew investment in sisters and brothers in the ratio of their relatedness to them (Gadagkar, 1990c, 1991c).

parameters of the model using the mean and standard error values in Table 9.5. We then repeated each simulation 1000 times by using random values from the normal distribution for each of the parameters and obtained a distribution of worker-brood relatedness values. The results of these simulations indicate that even though we took into consideration the natural variation in queen tenure, brood developmental period, and worker lifespans, all or most of the thousand worker-brood relatedness values fell below the threshold required for haplodiploidy-induced genetic predisposition for the evolution of worker behavior. These results reinforce the conclusion that eusociality in R. *marginata* cannot simply be attributed to the genetic asymmetries created by haplodiploidy, unless we make the most unlikely assumption that queens are always singly mated and new queens are always daughters of their predecessors.

9.2.5. *Dissecting the Model*

In addition to generalizing the empirical results, the simulation models can help us gain some insight into the mechanism by which serial polygyny brings about the reduced values of worker-brood genetic relatedness. With this in mind we dissected the model by varying one of the three parameters at a time. This allowed us to determine the threshold value of each parameter above which haplodiploidy can cause genetic predisposition for the evolution of worker behavior. With an increase in queen tenure, the relatedness to female brood as well as the weighted mean relatedness to brood increased. In the case of multiple mating, given a brood developmental period of 62 days and a worker lifespan of 31 days, there was no value of queen tenure that yielded a weighted mean relatedness greater than 0.5. There were some values of queen tenure that yielded a relatedness to female brood of 0.5 or more, but these were unrealistically high—465 days if new queens were daughters of their predecessors and 715 days if new queens were sisters of their predecessors.

Keeping the queen tenure fixed at 80 days and the worker lifespan at 31 days, we then explored the consequences of varying the brood developmental period. As expected, worker-brood relatedness decreased when the brood developmental period increased. Under multiple mating, there was no value of brood developmental period when the worker–brood developmental period was higher than either of the required thresholds. Similarly, when queen tenure and brood developmental period were held constant at

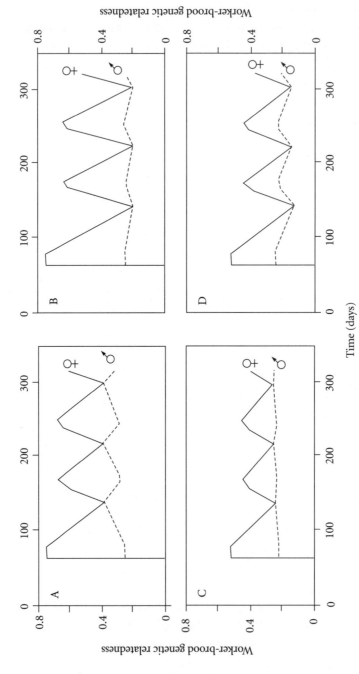

Time (days)

Figure 9.3. Variation in worker-brood genetic relatedness derived from a simulation model using empirically derived mean values of 80 days as the queen tenure, 62 days as the brood developmental period, and 31 days as the worker lifespan. Note that worker-brood relatedness is zero for the first 62 days, when no workers have yet eclosed. Queens are assumed to mate singly in A and B. In C and D, queens are assumed to mate multiply. New queens are assumed to be daughters of previous queens in A and C while they are assumed to be sisters of previous queens in B and D. (Redrawn from Gadagkar et al., 1993b.)

80 and 62 days respectively, worker-brood relatedness decreased with increased worker lifespan. Once again under multiple mating, there was no value of worker lifespan where worker-brood relatedness was higher than either of the required thresholds. These results added to the robustness of the conclusion that haplodiploidy does not cause the expected genetic predisposition for the evolution of worker behavior in R. marginata. The values of the relevant parameters empirically derived for R. marginata are not just marginally different from those required for genetic predisposition—they are very different indeed. What did dissection of the model tell us about the mechanism by which serial polygyny brings about such a reduction in worker-brood genetic relatedness? It became clear from these simulations that short queen tenure and long brood developmental periods and worker lifespans permit serial polygyny to reduce worker-brood genetic relatedness below the thresholds required for the evolution of eusociality through genetic predisposition.

The simulation model can also be used to derive conclusions that may have wider applicability to other species exhibiting serial polygyny. With this in mind we delineated regions in the parameter space where haplodiploidy leads to genetic predisposition for the evolution of worker behavior from those regions where it does not. First, we conducted an extensive search in the entire parameter space of queen tenure, brood developmental period, and worker lifespan and found that the simulation model can be adequately described by two parameters: (1) queen tenure and (2) the sum of brood developmental period and worker lifespan. This makes it convenient to map the two-dimensional parameter space into regions where haplodiploidy does (white regions in Figure 9.4) and does not (gray regions in Figure 9.4) lead to a genetic predisposition for the evolution of worker behavior. This exercise shows that haplodiploidy does not lead to a genetic predisposition for the evolution of worker behavior in R. marginata unless queens mate singly, new queens are always daughters of their predecessors, and a relatedness to female brood greater than 0.5 is sufficient to drive social evolution.

The equations (given in the caption to Figure 9.4) describing the lines that delineate the regions where haplodiploidy does and does not lead to genetic predisposition for the evolution of worker behavior were obtained by repeating the simulations for all pairs of value of the two parameters. These equations can be used to test the role of haplodiploidy in other spe-

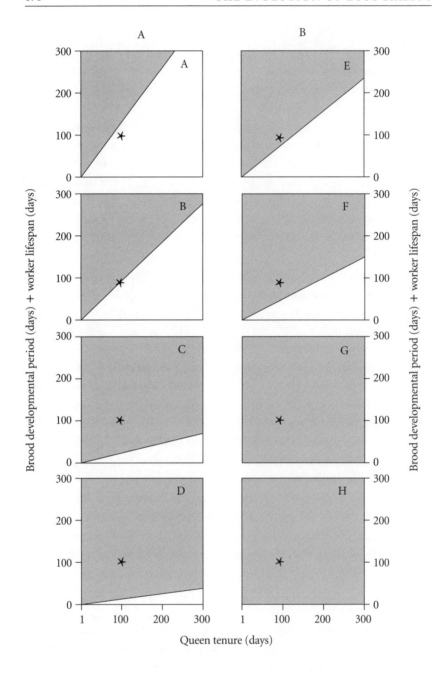

Figure 9.4 (facing page). Regions in the queen tenure and brood developmental period plus worker lifespan parameter space where haplodiploidy cannot (gray region) and can (white region) lead to genetic predisposition for the evolution of worker behavior. In panel A, the condition for genetic predisposition for the evolution of worker behavior is that the relatedness of workers to female brood be greater than 0.5. In panel B, the corresponding condition is that workers should obtain a weighted mean relatedness to brood of two sexes of at least 0.5 after they bias investment in female and male brood in the ratio of their relatedness to them. A and E: Queens mate singly and new queens are daughters of their predecessors. B and F: Queens mate singly and new queens are sisters of their predecessors. C and G: Queens mate multiply and new queens are daughters of their predecessors. D and H: Queens mate multiply and new queens are sisters of their predecessors. The lines separating the gray and white regions are given by the equations (A), $Q \times 1.33 = B + W$; (B), $Q \times 0.9 = B + W$; (C), $Q \times 0.02 = B + W$; (D), $Q \times 0.13 = B + W$; (E), $Q \times 0.8 = B + W$; and (F), $Q \times 0.5 = B + W$; where Q is the queen tenure, B is the brood development period, and W is the worker lifespan. The position of *R. marginata* is shown in each part by an asterisk to illustrate that haplodiploidy does not lead to genetic predisposition for the evolution of worker behavior in this species except in case A. (Redrawn from Gadagkar et al., 1993b.)

cies where data on queen tenure, brood developmental period, and worker lifespan and on the level of multiple mating are available. For example, Figure 9.4A shows that for haplodiploidy to promote worker behavior, the sum of brood developmental period and worker lifespan must be no more than 1.33 times the queen tenure.

We therefore conclude from the results presented in sections 9.1 and 9.2 that genetic asymmetries created by haplodiploidy are sufficiently broken down to render them ineffective in promoting worker behavior—there is not sufficient genetic predisposition to tilt the inclusive fitness in inequality 8.5 in favor of workers. Indeed, it is solitary founding that should be favored.

9.3. Movement of Foundresses between Nests

In addition to polyandry and polygyny, there is yet another phenomenon that must serve to lower intracolony genetic relatedness: the movement of wasps from one nest to another. Although we are not yet in a position to quantify its effects on genetic relatedness, we should take note of it and be aware that the values of worker-brood genetic relatedness computed in the previous two sections may be overestimates; recall that the relatedness between sisters (as a result of multiple mating) computed in section 9.1 was already thought to be an overestimate.

Post-emergence colonies do not generally receive nonnestmate joiners. On rare occasions, we have seen wasps that did not eclose on a nest land on it and be very aggressively repelled. We have assumed that the alien wasps must have landed on the wrong nest by mistake. But, also on rare occasions, we have seen a nonnestmate land on a nest without causing any aggression on the part of the residents. The significance of these events is unclear. But in any case they are undoubtedly very rare and unlikely to be of much consequence for intracolony genetic relatedness.

Pre-emergence colonies are another matter altogether; they quite frequently receive and accept joiners. To some extent we were able to quantify the magnitude of this phenomenon in the study of 145 pre-emergence nests by Shakarad and Gadagkar (1995) described in Chapter 4. All adult wasps on all the nests were marked with unique paint spots. When an unmarked individual or a wasp known to belong to one of the remaining 144 nests was seen on a nest, that wasp was termed a joiner and that nest was said to have received a joiner. Of the 676 wasps recorded during that study, 217 (32.1%) were seen to join previously established nests. Of the 145 nests studied, 69 (47.6%) received at least one joiner each. Nests received joiners in their egg, larval, and pupal stages. Of the 217 joiners, 191 were unmarked individuals so we do not know where they came from. They could have come from the same original nest from which the wasps they are joining came or they may have come from different nests. However, 26 joiners were individuals we had marked on some nest other than the nest they joined, and they thus represent unambiguous evidence of movement of foundresses from one nest to another. We found that at least 16 nests consisted of wasps we had originally marked on 2 different nests, at least 3 nests consisted of wasps that we had originally marked on 3 different nests, and at least 1 nest consisted of individuals we had originally marked on 4 different nests. We have speculated that the observed movement of wasps from one nest to another suggests that they are not sorting according to genetic lines (Shakarad and Gadagkar, 1995). The logic was that one might expect identification of genetic relatives to be a more definite, one-time process not requiring movement from nest to nest. However, if wasps were assessing their potential of becoming future egg layers in different groups, then they might have been engaging in a trial-

and-error process requiring movement between nests. This is just one example of several situations we encountered throughout our study where we were reminded that factors other than genetic relatedness (such as the probability of becoming future egg layers) might be important in promoting worker behavior in *R. marginata*. Estimates of intracolony genetic relatedness (made by using both allozyme markers and microsatellite markers) have often led to a similar conclusion that factors other than relatedness are more important in molding the evolution of altruism in primitively eusocial wasps (see, for example, Hughes et al., 1993; Strassmann et al., 1989).

10 GENETIC PREDISPOSITION II: KIN AND NESTMATE RECOGNITION

$\mathcal{I}$n the previous chapter, I showed that worker-brood genetic relatedness in *R. marginata* is considerably lowered owing to polyandry and serial polygyny. I concluded therefore that if the cost and benefit terms in Hamilton's rule are equal, or if $\sigma \beta = sb$ in inequality 8.5, the payoff to workers will be less than the payoff to solitary foundresses. Hence haplodiploidy, through its effects on genetic relatedness, by itself cannot promote the evolution of worker behavior. In drawing such a conclusion I did, however, make the implicit assumption that when workers are faced with brood of diverse levels of relatedness, they dispense care indiscriminately. But if workers can discriminate between different levels of relatedness to different members of their colony (brood and/or adults), they could increase their payoff by being selectively altruistic toward more closely related brood and nestmates. In other words, intracolony kin discrimination can render the observed low levels of worker-brood genetic relatedness consistent with the haplodiploidy hypothesis.

The logical course of action was therefore to test the assumption that workers indiscriminately care for different classes of genetic relatives in their colonies. This is not an easy hypothesis to test, especially in the context of worker-brood interaction. The main observable worker-brood interaction that has a clear functional significance is the feeding of larvae by the workers. To establish differential feeding of larvae belonging to different patrilines or matrilines is fraught with serious technical difficulties. Patrilines cannot be identified in the absence of visible larval phenotypes. Adults move from cell to cell, and they often do so with a visible lump of solid food in their mouths and one can see that the food disappears by the time the workers have visited a few larval cells. Thus one can make the reasonable inference that adults are feeding the larvae. But it is almost impossible, through behavioral observations, to estimate the quantity of food delivered to different larvae. Even if it was possible, the interpretation of differential feeding is not easy. The optimum quantity of food given to different larvae depends on their age, sex, hunger level, and perhaps even their future caste (see Chapter 12). None of these factors is easily measured in larvae, and even if they can be measured, as in the case of age, it is difficult to partition variation in larval feeding between the various factors and thus determine if larvae belonging to different matrilines or patrilines are fed differently. Even in the honey bees, where sample sizes of larvae, of workers, and of feeding bouts can be large and the desired number of patrilines can be created by artificial insemination of virgin queens of known genotypes with sperm from males of known genotypes, such an approach has proved very difficult (Noonan, 1986; Visscher, 1986). In *R. marginata* this approach seemed impossible.

However, George Gamboa and his students had developed a simple triplet assay to ask a related question: Can adult workers in *Polistes* wasps discriminate between their nestmates and nonnestmates (Gamboa, Reeve, and Pfennig, 1986; Gamboa et al., 1986, 1987; Shellman and Gamboa, 1982)? Their answer was in the affirmative, but for such discrimination to occur it was necessary for the wasps to be exposed to their nest or their nestmates after eclosion. It seemed that such nestmate discrimination was itself worth investigating in *R. marginata*. If studies on nestmate discrimination can also help throw light on the question of intracolony kin recognition, so much the better. But to make this possible it is necessary to explore the mechanism of nestmate discrimination.

10.1. The Mechanism of Nestmate Discrimination

For nestmate or kin recognition to be possible, every wasp must carry a label on her body and a template in her brain. She should compare the label on any encountered individual with the template in her brain and thus be able to discriminate between individuals whose labels are similar to her template and those whose labels are relatively dissimilar. It is reasonable to assume that labels consist of a specific blend of cuticular hydrocarbons or other chemicals on the surface of wasps and that templates must consist of a knowledge/memory of the smell of that particular blend of chemicals. A wasp may acquire labels in two ways. Labels can be produced by the metabolic machinery of the individual wasp and thus could carry some information about its genotype. I call such labels "self-produced." Alternatively, labels can be acquired by an individual from sources outside its body, such as other individuals, the nest, the food, or other materials encountered, and thus carry no information about its genotype. I call such labels "not self-produced." Similarly, templates may be acquired in two ways. The template can be innately specified—an individual either may be born with the right template or may acquire it by smelling itself—and thus be "self-based." Alternatively, the template can be acquired when an individual smells something external to itself—other individuals, the nest, and so on—and thus be "nonself-based." Understanding the ontogeny of labels and templates can help understand whether the mechanism of nestmate discrimination can also permit intracolony kin recognition (Gadagkar, 1985b). If labels are self-produced and templates are self-based, a wasp would be able to discriminate between a closely related individual and a distantly related individual even though both of them belong to the same nest. But if labels are not self-produced and/or templates are nonself-based, all members of a colony can have common or overlapping labels and templates, making it difficult for them to discriminate between nestmates with different levels of genetic relatedness.

In a paper in 1985, I proposed an experimental procedure to determine the ontogeny of labels and templates to make feasible the use of nestmate discrimination to determine the possibility of intracolony kin recognition (Gadagkar, 1985b). What we need is a behavioral assay that will permit us to say whether or not two nestmates recognise each other outside the context of the nest. The assay should permit us to recognize the hypothetical possibility that individual A recognizes individual B although B does not recognize A. Because we need to allow for the possibility that wasps may acquire their labels and templates from sources outside their bodies, we need to have wasps that have had the opportunity to acquire labels and templates if that is how they get them. I will call such wasps "experienced." We also need wasps that have been denied the opportunity to acquire labels and templates from external sources and that can therefore have normal labels and templates only if the labels are self-produced and the templates are self-based. I call such wasps "naïve." Experienced wasps should possess normal labels and templates irrespective of whether they are self-based or not. Hence experienced wasps should recognize each other as nestmates (Figure 10.1, panel I). In an experiment with two naive nestmates, if neither can recognize the other, we can conclude that the naive individuals lack the labels, templates, or both—that is, the labels are not self-produced, the templates are nonself-based, or both (Figure 10.1, panel II).

It is important to emphasize that either missing labels or missing templates is sufficient for neither naive individual to recognize the other. Both naive individuals may possess perfectly normal templates but may lack labels. Despite the presence of templates, the absence of labels will prevent them from recognizing each other just as the absence of templates will even if they both possess appropriate labels. Thus it could be that the labels are not self-produced but the templates are self-based or that the labels are self-produced and the templates are nonself-based. In panel III of Figure 10.1, the experienced individual recognizes the naive one but the naive individual cannot recognize the experienced individual. The experienced individual is expected to possess normal labels and templates and the naive individual may possess neither. Because the experienced individual can recognize the naive individual, the latter must possess an appropriate label, even though it is naive—thus labels must be self-produced. Conversely, the naive individual fails to recognize the experienced individual, and therefore the former must lack a template—thus the templates must be nonself-based. In panel IV, because the experienced individual does not recognize a naive individual, the latter must not possess an appropriate label—thus the labels must not be self-produced. Similarly, because the naive individual recognizes the experienced one, the former must possess a normal template—thus the templates must be self-based. By a similar logic, we can conclude from panel V that labels are self-produced and templates self-based and from panel VI that labels are not self-produced and templates are not self-based. So here was an experimental approach to study nestmate discrimination and test the possibility of intracolony kin recognition.

Thus began a long series of fairly tedious experiments to assess the nestmate discrimination and intracolony kin-recognition abilities of *R. marginata* workers. The experiments my students and I planned required many hours of observations of marked wasps. For instance, the first set of experiments (Venkataraman et al., 1988) involved 1104 hours of observation of 94 triplets of wasps during which 15,706 interindividual interac-

WASP A	RESULT OF BEHAVIORAL ASSAY	WASP B	CONCLUSION
Experienced		Experienced	
I (A)	Recognition ⇄ Recognition	(B)	Behavioral assay is working
Naive		Naive	
II (A)	No recognition ⇠⇢ No recognition	(B)	Either label not self-produced or template nonself-based or both
Experienced		Naive	
III (A)	Recognition ⇢ No recognition	(B)	Label self-produced but template nonself-based
Experienced		Naive	
IV (A)	No recognition ⇠ Recognition	(B)	Label not self-produced but template self-based
Experienced		Naive	
V (A)	No recognition ⇠⇢ No recognition	(B)	Label not self-produced and template nonself-based
Naive		Naive	
VI (A)	Recognition ⇄ Recognition	(B)	Label self-produced and template self-based

Figure 10.1. An experimental approach to distinguish between the roles of labels and templates in kin recognition. A and B are two wasps that may or may not recognize each other as close genetic relatives depending on their rearing conditions. Knowing the effect of rearing conditions on the ability or inability of wasps to recognize relatives, one can infer the ontogeny of the labels and the templates (see the text for details). (Redrawn from Gadagkar, 1985b.)

tions were recorded. This work required trained observers and a very willing trio volunteered, consisting of a graduate student, Arun Venkataraman, and two assistants, Swarnalatha and Padmini Nair.

10.2. The Triplet Assay

To assay nestmate discrimination we used a large number of *R.marginata* nests collected from four different localities within a radius of 150 km from Bangalore. Following Shellman and Gamboa (1982), we made our triplet assay consist of two wasps that belonged to the same nest (nestmates) and one that belonged to a different nest (nonnestmate), marked with unique spots of paint of a single color. We ensured that the two nests from which individuals were drawn for an experiment were collected from sites at least 8 km away from each other. All wasps were held individually in ventilated plastic boxes 22 × 11 × 11 cm since the time of collection or since their eclosion (see Chapter 3).

As soon as the nests were brought to the laboratory, all adults were separated from the nest, the males (if any) were discarded, and the females were isolated in individual plastic boxes. All eggs and larvae were removed from the nests and discarded. The nests containing only the pupae were then cut into approximately two equal pieces and the two nest fragments were maintained in two separate plastic boxes. Wasps used for the experiments were subjected to one of the following four treatments.

Treatment I: Wasps were removed from the nest at the time of collection and isolated in individual plastic boxes for 8–45 days.

Treatment II: Eclosing wasps were allowed to remain on their respective nest fragments for 4–28 days, after which they were kept in individual boxes for 3–62 days.

Treatment III: Eclosion was continuously monitored, day and night, and eclosing wasps were removed immediately (within 1–2 min of their eclosion) and isolated in individual plastic boxes for 6–48 days.

Treatment IV: Wasps were artificially removed from their pupal cases (about 24 h before their expected time of natural eclosion), covered with tissue paper, placed in a petri dish, and allowed to complete their development in an incubator maintained at 26° ± 2°C. Such wasps spent 11–49 h in the incubator before they were isolated in individual boxes for 6–62 days.

One hour before the commencement of the experiment, the three required wasps were introduced into a single new ventilated plastic box. During the experiment, behavioral interactions between all pairs of individuals were recorded continuously for 3 hour at a time. Four such three-hour observation periods were completed during a 2-day period. In all observations, the observer was unaware which individuals were nestmates and which non-nestmates. We recorded 15 kinds of behavioral interactions and ranked them qualitatively in increasing order of tolerance (or decreasing order of aggression). Our ranking was based on more than 200 hours of observations of R. marginata colonies in the laboratory and in nature (Table 10.1).

Behavioral interactions recorded during the observations were classified into six categories labeled a-f, as shown in Figure 10.2. Nestmate discrimi-

Table 10.1.   The 15 behaviors observed, ranked by increasing order of tolerance. (From Venkataraman et al., 1988.)

Rank	Behavior
1	Aggressive bite: the most extreme form of aggression seen in this species; sometimes leads to injuries
2	Attack: a ritual act of aggression where the dominant animal climbs onto the subordinate and attempts to bite its mouthparts
3	Peck
4	Chase
5	*Aggressive mutual antennation: a kind of sparring contest
6	Nibble: relatively mild with little chance of injury
7	Crash: the crashing of one wasp into a sitting wasp that results in one or both falling to the ground; very brief and appears much milder than behaviors 1–6
8	*Falling fight: two animals grappling with each other and falling to the ground; very brief and appears much milder than behaviors 1–6
9	Avoiding
10	Soliciting
11	*Mutual approach with withdrawal
12	Approach I: the other withdraws
13	Approach II: the other does not withdraw
14	Antennation
15	*Mutual Antennation

Note: * indicates bidirectional behaviours

nation was assessed by examining differences in tolerance between nest-mates and nonnestmates. To facilitate this, tolerance indices T_a-T_f were computed as in this example:

$$(10.1) \qquad T_a = \sum_{i=1}^{15} p_i r_i$$

where p_i is the proportion of the ith behavior in a and r_i is the tolerance rank (Table 10.1) of the ith behavior. From the six dyadic tolerance indices T_a-T_f thus obtained, two additional tolerance indices were derived as follows: tolerance among nestmates, $T_x = T_a + T_b$; and tolerance among nonnestmates, $T_\alpha = (T_y + T_z)/2$, where $T_y = T_c + T_d$ is the tolerance between one pair of nonnestmates and $T_z = T_e + T_f$ is the tolerance between the second pair of nonnestmates.

In the first four sets of experiments, both nestmates and nonnestmates were from *treatment I, II, III, and IV*, respectively. In the fifth set of experiments, the two nestmates were from treatment II and the nonnestmate was from treatment IV. Conversely, in the sixth set of experiments, the two nestmates were from treatment IV and the nonnestmate was from treatment II.

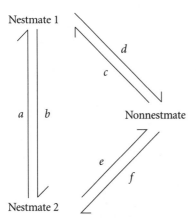

Figure 10.2. The triplet assay. The behavioral interactions seen in all experiments classified into six categories, designated *a, b, c, d, e, f,* such that all interactions initiated by nestmate 2 toward nestmate 1 are assigned to a, all interactions initiated by nestmate 1 toward nestmate 2 are assigned to b, and so on. (Redrawn from Venkataraman et al., 1988.)

Table 10.2. Comparison of tolerance indices by the Wilcoxon matched-pairs signed-ranks and Mann-Whitney U tests. (Based on Venkataraman et al., 1988.)

Treatment	No. of expts.	Tolerance among nestmates	Tolerance among nonnestmates	T^a	p^b	U^c	p^d
Wasps present on the nest at the time of collection and subsequently isolated for several days (treatment I)	14	17.23 ± 3.97	12.80 ± 2.97	12	0.006	32	0.001
Wasps eclosed in the laboratory but exposed to half of their nest and a subset of nestmates (treatment II)	16	17.44 ± 2.67	15.28 ± 3.82	34	0.039	79	< 0.05
Wasps isolated from their nest and nestmates immediately after eclosion (treatment III)	16	13.86 ± 4.41	11.96 ± 3.44	44	0.107	82	< 0.05
Wasps isolated from their nest and nestmates before their natural eclosion (treatment IV)	17	17.87 ± 2.60	17.06 ± 2.72	63	0.261	118	> 0.05
Tolerance of isolated wasps (treatment IV) by exposed wasps (treatment II)	16	7.84 ± 3.56	8.32 ± 2.17	64	0.418	122.5	> 0.05
Tolerance of exposed wasps (treatment II) by isolated wasps (treatment IV)	15 (13[e])	8.94 ± 2.31	8.19 ± 2.76	30	0.139	90	> 0.05

a and b. Statistic and p value respectively for the Wilcoxon matched-pairs signed-ranks test.
c and d. Statistic and p value respectively for Mann-Whitney U test.
e. Sample size is 13 only for the Wilcoxon matched-pairs signed-ranks test.

10.3. Evidence for Nestmate Recognition and against Intracolony Kin Recognition

The triplet assay showed clearly that wasps present on the nest at the time of collection (treatment I) strongly discriminated nestmates from non-nestmates: nestmates were significantly more tolerant of each other than they were of nonnestmates (Table 10.2, row 1). In this experiment, nest-mates may have had the benefit of familiarity (by virtue of having been present on the same nest at the time of collection), genetic similarity, and past exposure to common environmental cues to aid them in recognition. Similarly, nonnestmates could have had unfamiliarity, genetic dissimilar-ity, and past exposure to different environmental cues to aid them in dis-crimination. From this experiment we cannot tell which of these factors played any role in recognition of nestmates versus nonnestmates but from subsequent experiments we can do so. Among wasps subjected to treat-ment II also, nestmates were significantly more tolerant of each other than nonnestmates were of each other (Table 10.2, row 2). Here nestmate wasps had the benefit of genetic similarity (by virtue of belonging to the same nest) and exposure to common cues (by virtue of eclosing and remaining on fragments of the same nest) but were not familiar with each other (be-cause they eclosed on different, separated fragments of the nest). Familiar-ity is therefore unnecessary for nestmate discrimination. The assay thus satisfied panel I of Figure 10.1.

At this stage my students and I took special precautions to rule out the possibility that factors other than nestmateship could account for the ob-served differences in tolerance between nestmates and nonnestmates. Al-though we matched the three wasps used in each experiment as closely as possible, for body size (= fresh weight) and age (only for treatments II, III, and IV), we tested the hypothesis that tolerance was influenced by any re-maining differences in age and body size. There was clear evidence (1) that the older or smaller of the two wasps available was not treated significantly differently from the younger or smaller one; (2) that wasps did not treat individuals older or larger than themselves significantly differently from those younger or smaller than themselves; and (3) that wasps did not treat individuals closer to themselves in age or body size significantly differently than individuals more distant from themselves in age or body size. With these results, we had greater confidence in concluding that wasps in our experiments were indeed responding to and discriminating between nest-mateship and nonnestmateship and not to residual differences in age or body size.

Treatments I and II showed that nestmate discrimination is based on genetic similarity and/or exposure to common cues. But which of these is more important? We had hoped that treatment III would answer this question, but that was not to be. Treatment III, where the wasps were isolated from their nest fragments and nestmates within 1–2 min of their eclosion, yielded inconclusive results: tolerance between nestmates was significantly different from that between nonnestmates by one statistical test but not significantly different by another statistical test (Table 10.2, row 3). If nestmate discrimination is based on assessment of genetic similarity and cues acquired from the nest were unnecessary, there is no reason why wasps subjected to treatment III should not discriminate nestmates from nonnestmates as strongly as those subjected to treatment II. If discrimination is based not on genetic similarity but on cues gained from the nest, the wasps of treatment III should not discriminate between nestmates and nonnestmates. We might conclude from this intermediate result that both genetic similarity and cues derived from the nest are involved in the nestmate recognition system of *R. marginata* (see, for example, Gamboa et al., 1986 for a similar conclusion). But that is not a very satisfying conclusion. Even if both genetic and environmental cues are involved in nestmate discrimination, we must at least attempt to say which is more important. Hence we considered the possibility that wasps subjected to treatment III are not completely deprived of nest-derived cues. Newly eclosed wasps may derive some cues by exposure to their nests within 1–2 min of their eclosion, or perhaps wasps derive some of the required cues after completing their development but just before eclosion. To rule out these possibilities, we repeated the experiment with wasps subjected to treatment IV, where the wasps were removed from their pupal cases even before eclosion and allowed to complete their development in an incubator. These wasps clearly did not have the ability to discriminate between nestmates and nonnestmates, by either statistical test (Table 10.2, row 4). It is thus clear that nest-derived cues (present in treatment II but not in treatment IV) are necessary and sufficient for nestmate discrimination and that genetically derived cues (present in treatments II and IV) are unnecessary and insufficient.

In section 10.1, I discussed why such a result as the one obtained so far can be obtained even if only the labels are not self-produced or only the templates are nonself-based—it is not necessary for both labels and templates to be acquired from outside the individual. To see if exposure to the nest is indeed necessary for acquiring both labels and templates (and not just for acquiring one of them), we need to explicitly test the ability of experienced wasps to recognize naive wasps and the ability of naive wasps to recognize experienced wasps (panels 3 and 4 in Figure 10.1). From experiments involving treatments II and IV, it is clear that wasps subjected to treatment II are experienced and wasps subjected to treatment IV are naive, by the definitions in section 10.1. Thus we conducted experiments where an exposed wasp was required to discriminate between an isolated nestmate and an isolated nonnestmate and an isolated wasp was required to discriminate between an exposed nestmate and an exposed nonnestmate. The predictions of the various outcomes of such experiments are shown in Figure 10.3.

The experimental design is sufficiently powerful to discriminate between all four possibilities, labels self-produced and template self-based, labels self-produced and templates nonself-based, labels not self-produced and templates self-based, and labels nonself-produced and templates nonself-based. The results that we did obtain (rows 5 and 6 in Table 10.2) show clearly that exposed wasps cannot discriminate between isolated nestmates and isolated nonnestmates, and isolated wasps cannot discriminate between exposed nestmates and exposed nonnestmates. For successful discrimination between nestmates and nonnestmates to occur, both the discriminating wasps and the discriminated wasps need to be exposed. From this we concluded that labels are not self-produced and templates are nonself-based. As I argued in section 10.1, this means that the system of nestmate discrimination in R. marginata cannot permit intracolony kin recognition. We must conclude therefore that intracolony kin recognition cannot help restore the lost primacy of asymmetries in genetic relatedness potentially created by haplodiploidy in promoting the evolution of workers in R. marginata. However, a system of nestmate discrimination based on acquiring exogenous odors (nonself-produced labels) and post-eclosion learning of templates (nonself-based templates) is eminently suited to serve as an efficient mechanism of nestmate discrimination to keep away alien conspecifics that may be potential usurpers or social parasites in-the-

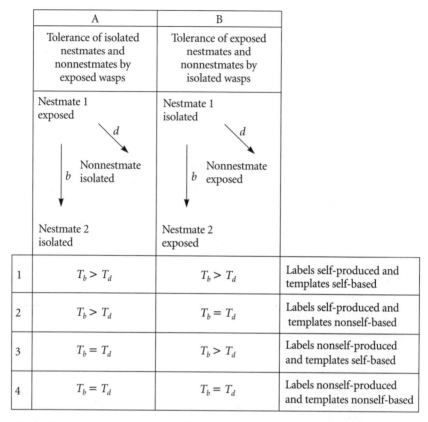

	A	B	
	Tolerance of isolated nestmates and nonnestmates by exposed wasps	Tolerance of exposed nestmates and nonnestmates by isolated wasps	
	Nestmate 1 exposed d b Nonnestmate isolated Nestmate 2 isolated	Nestmate 1 isolated d b Nonnestmate exposed Nestmate 2 exposed	
1	$T_b > T_d$	$T_b > T_d$	Labels self-produced and templates self-based
2	$T_b > T_d$	$T_b = T_d$	Labels self-produced and templates nonself-based
3	$T_b = T_d$	$T_b > T_d$	Labels nonself-produced and templates self-based
4	$T_b = T_d$	$T_b = T_d$	Labels nonself-produced and templates nonself-based

Figure 10.3. Experimental design of mixed triplets aimed at distinguishing between self-produced and nonself-produced labels and self-based and nonself-based templates. In panel A the ability of an "exposed" wasp to discriminate between a nestmate and a nonnestmate both of which are "isolated" is assessed. Conversely, in panel B the ability of an "isolated" wasp to discriminate between a nestmate and a nonnestmate both of which are "exposed" is assessed. The six possible types of interactions between the three wasps in these triplets are similar to those in Figure 10.2, but only the relevant interactions, b and d, are labeled. Discrimination is again assessed by comparison of tolerance indices calculated for b and d, as in the previous set of experiments. (Redrawn from Venkataraman et al., 1988.)

making (Venkataraman et al., 1988, 1990). I should perhaps mention here that, contrary to the case with our studies with *R. marginata*, considerable progress has been made in understanding the chemistry of the labels used in nestmate discrimination in other primitively eusocial wasps (Dani, Fratini, and Turillazzi, 1996; Singer and Espelie, 1996; 1997; Singer, 1998).

10.4.   Nestmate Discrimination in a Seminatural Context

All the experiments described in section 10.3 were done under artificial conditions with three wasps placed in a plastic jar and allowed to interact with each other, in the absence of a nest and even in the absence of food. Subsequently, we repeated these experiments using a slightly more natural setting for the wasps (Venkataraman and Gadagkar, 1992). As before, we collected pairs of nests from sites more than 10 km away from each other, to derive nestmates and nonnestmates. One nest from each pair was cut into three parts. The part containing relatively more brood was fixed in a wood and wire-mesh cage 45 × 45 × 45 cm, and the adults present on the nest at the time of collection were released into this cage. Most of these re-leased adults returned to the nest fragment, repaired damaged cells, and continued tending brood. In about 50 days, an apparently normal nest was thus established in the cage. The two other fragments were maintained separately to obtain "exposed" and "isolated" wasps. The wasps obtained from these two fragments were termed "relatives" of the wasps on the re-generated nest in the cage. A second nest in each pair of nests was cut into two fragments, which were maintained separately to obtain "exposed" and "isolated" nonrelatives of the wasps in the nest established in the cage. Note that "exposed" wasps were always exposed to a fragment of the nest and nestmates belonging to the same nest from which they had eclosed.

On the day of the experiment four individuals (which were by then 26–52 days old) of each category (exposed relatives, isolated relatives, exposed nonrelatives, and isolated nonrelatives) were simultaneously released into the cage containing the regenerated nest. All introduced individuals were marked with unique spots of paint, although the observer was unaware of the identity of the introduced wasps. The observations commenced an hour after the introduction of wasps into the cage and were carried out from 0830 to 1200 and again from 1430 to 1800 for 5 consecutive days in the first week after introduction and again for 5 consecutive days in the second week after introduction. In the third week observations were made for 6 consecutive days either from 0830 to 1200 or from 1430 to 1800. All occurrences of interactions (see Chapter 3) between the wasps resident on the nest before introduction and the introduced wasps were recorded, dur-ing 15 5-min observation blocks per session (0800–1200 or 1430–1800).

These observations were focused on the nest (the nest itself and any area within a radius of 9 cm from the nest periphery) because we were interested only in recording interactions between residents and introduced wasps. But some introduced wasps seldom approached the nest. Hence separate, 5-min focal animal sampling sessions (see Chapter 3) were conducted once for each introduced wasp per observation session (0800–1200 or 1430–1800). Here even those interactions between resident wasps and introduced wasps that occurred away from the nest (anywhere in the cage) were recorded. The entire experiment was repeated three times on three different pairs of nests. All this amounted to 1144 5-min all-occurrence sessions and 1002 5-min focal animal sampling sessions and a total of 178 h and 50 min of observations.

Tolerance/aggression shown by the resident wasps toward the different categories of introduced wasps was compared. In addition to the tolerance index discussed in section 10.2 (here, tolerance index I), a second, relatively more objective tolerance index (here, tolerance index II) and an aggression index were computed from these data (see Venkataraman and Gadagkar, 1992 for definitions). This experiment yielded several interesting results.

1. No introduced wasps of any category were ever accepted on any of the nests. This result is in agreement with observations in nature, where post-emergence colonies do not normally accept foreign wasps (section 9.3). It should be noted that the introduced wasps were not attacked (let alone killed) as long as they stayed away from the nest.

2. In the vicinity of the nest, the resident wasps were more tolerant of exposed relatives than they were of exposed nonrelatives or of isolated nonrelatives, among the introduced wasps. Tolerance toward isolated relatives was intermediate and not significantly different from that shown toward all other categories of wasps.

3. While away from the nest and its vicinity, the resident wasps made no distinction between the four different categories of introduced wasps.

4. In one of the experiments, four introduced wasps (indeed, exactly one from each category) got together and constructed their own sep-

arate nest on which an exposed relative became the queen. Note that
the four wasps participating in this "satellite" nest belonged to two
different parent nests. Even more interesting was the observation
that three resident wasps from the main nest later joined the satellite
nest as workers.

Thus the results of this set of experiments were by and large consistent
with the conclusion drawn in section 10.3. But because these experiments
were conducted in a more natural context, especially because they were in
the context of a nest, they are more satisfying. In addition, the facts that
no introduced wasp was accepted onto the nest, that none was killed, and
that introduced wasps belonging to two different nests (including exposed
wasps from two different nests), as well as the resident wasps, got together
to establish their own separate nest once again remind us that genetic relatedness may be relatively unimportant in the lives of primitively eusocial
wasps such as *R. marginata.*

10.5. Differential Aggression toward Conspecifics

The experiments described in the previous section were conducted in a
more natural setting than the triplet assay, but the fact remains that the introduced wasps were removed from their nest of eclosion, either before
their natural eclosion (isolated wasps) or 10–20 days after eclosion, and
maintained in isolation for almost a month before the experiment. This
lack of social experience may have made the introduced wasps somewhat
unnatural and the resident wasps may not have recognized them adequately, either as potential cooperators or as potential competitors. Hence
we next performed an even more natural experiment, extending the natu-
ralness this time not only to the resident wasps but also to the introduced
wasps.

To do this we created two different laboratory colonies by collecting a
pair of nests from sites separated by 10 km or more. On the day of the experiment, adult wasps present on one of the colonies were introduced into
the cage containing the other, intact colony. Thus the introduced wasps in
this experiment, in contrast to those in the previous experiment (section
10.4), had normal social experience and thus may have better mimicked

the occasional landing of alien wasps on *R. marginata* nests. Our observation that aliens occasionally landing on post-emergence nests are sometimes aggressively repelled and at other times readily accepted (see section 9.3) had remained perplexing, and a chance to understand this phenomenon constituted an additional motivation for this experiment.

The most striking result of this experiment was that the queen from among the introduced wasps was selectively attacked, dismembered, and killed within a few hours of her introduction. Unfortunately, this special attention that the alien queen received in the first such experiment could not be confirmed in the next two experiments: in one case the queen died a day before she and her nestmates were to be introduced, and in the other case the colony from which the introduced wasps were derived was probably queenless. Of the 34, 19, and 28 wasps introduced in the three experiments, 2, 6, and 0 respectively were accepted, 2, 4, and 7 respectively were killed, and the remaining wasps were allowed to live in the cage without either being killed or being accepted onto the nest.

This result provided an excellent opportunity to understand why the resident wasps accepted some aliens, killed others, and remained indifferent to the rest. All the accepted individuals were less than 8 days old. To see if the probability of being killed was influenced by what the introduced wasps had been doing on their original nests, before introduction, we used the behavioral profiles of the introduced wasps as determined before introduction. Their behavioral profiles were constructed in the same way we established behavioral caste differentiation (see Chapter 5). Using the method of logistic regression analysis, we tested the proportion of time spent by the introduced wasps in different behaviors (while in their original nest) as predictors of the probability of their being killed after being introduced into a cage containing an unrelated nest. Only one regression coefficient was significant: a negative coefficient associated with being absent from the nest (Table 10.3). We obtained a similar result when we divided the introduced wasps into three classes: those that had spent less than 25% of their time away from their nest, those that had spent 25–50% of their time away from the nest, and those that had spent more than 50% of their time away from the nest. Nearly 70% of those that had spent less than 25% of their time away from the nest were killed, while wasps that had spent more than 25% of their time away from their nest had less than a 20% chance of being killed. Thus wasps that spent more time absent

Table 10.3.   Logistic regression analysis linking the probability that an introduced wasp is killed with the proportion of time it had spent being absent from its natal nest prior to introduction. (Data from Venkataraman and Gadagkar, 1993.)

Independent variables	Logistic regression coefficients	S.E.	Test Statistic Z
Being absent from the nest	−7.04	3.07	−2.29[a]
Sitting and grooming	2.62	2.24	1.17
Walking	15.46	8.66	1.92
Sitting with raised antennae	1.40	5.11	0.28
In cells	3.91	11.40	0.34

a. $p < 0.05$.

from their original nests had a low probability of being killed. Conversely, wasps that had spent relatively more time on their original nests had a high probability of being killed when introduced into a cage containing an unrelated nest.

Our twin results that young wasps less than 8 days old are accepted into unrelated colonies and that wasps who spend less time away from their original nests (more time on their original nests) are killed have an attractive evolutionary explanation. Young wasps may be more welcome as joiners either because they are more easily suppressed and molded into the required roles or because they have poorly developed ovaries, or both. Wasps spending more time on their original nests may be killed either because they have better-developed ovaries (see Chapter 5) or because they are not good foragers, or both. Recall that foragers have poorly developed ovaries. The proximate cues that the resident wasps use in identifying potential foragers and nonforagers or discriminating between potential egg layers and non–egg layers may be directly linked to their ovarian status—they may simply smell differently. Alternatively, individuals that do not forage much and that spend most of their time in their nests and that incidentally have well-developed ovaries would also be expected to have acquired much stronger recognition labels from their nests and nestmates. Note that both proximate cues (smell related to ovarian development or smell related to their nests) are consistent with the ultimate evolutionary explanation of why some individuals are accepted and others are killed.

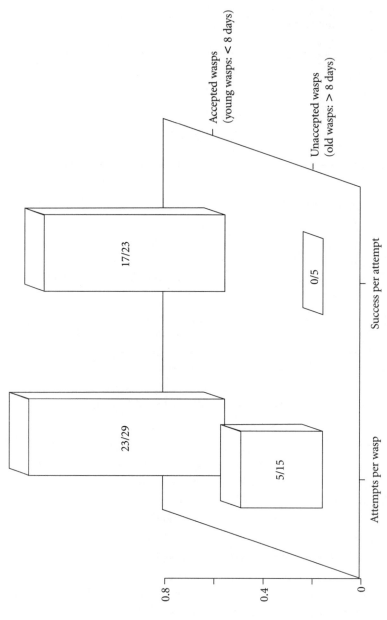

Figure 10.4. Young introduced wasps attempt to join alien nests and get accepted. The bars on the left represent the number of attempts per wasp, for wasps 8 days old or younger and for wasps older than 8 days. The bars on the right represent the number of successful attempts/total attempts for wasps 8 days old or younger and for wasps older than 8 days. Wasps 8 days old or younger make significantly more attempts per individual than wasps older than 8 days (Monte Carlo test, $p <$ 0.02); wasps 8 days old or younger also have a higher success rate per attempt than wasps older than 8 days (Monte-Carlo test, $p <$ 0.001). (Redrawn from Venkataraman and Gadagkar, 1995.)

10.6. Factors Affecting the Acceptance of Alien Conspecifics

My students and I undertook two additional studies to investigate in greater detail the factors affecting the acceptance of alien conspecifics by *R. marginata* colonies. In one study (Venkataraman and Gadagkar, 1995), we focused specifically on the role of age. Using four laboratory cages as before, we introduced 44 foreign wasps of known ages (18 different ages ranging from 0 to 29 days)—all isolated nonrelatives, so that there was no confounding effect of nonnestmateship cues. This experiment confirmed that alien conspecifics below the age of 8 days had a finite chance of being accepted but that individuals older than that had zero chance of being accepted. Monte-Carlo simulations demonstrated that the observed probability of acceptance of individuals younger than 8 days was significantly greater than the observed probability of acceptance (zero) for older individuals. The probability of acceptance fell sharply with age. Introduced wasps were not immediately accepted; it took 0–14 days for them to be accepted. Even within the age group of 0–8 days, relatively older individuals had to wait longer than relatively younger individuals. It appeared that age at introduction was more important than age on the day of acceptance, suggesting that the resident wasps assessed the ages of introduced wasps flying about in their cage even before they accepted them onto their nests.

This study provided an opportunity to investigate the role of behavior of the introduced wasps in influencing the probability of their acceptance. Introduced wasps sometimes briefly landed on the nest and were immediately repelled. We defined these happenings as attempts on their part to join the nest. We obtained clear evidence that young wasps (< 8 days old) made more attempts to join the nest and were more successful per attempt. Conversely, old wasps (> 8 days old) made fewer attempts and were less successful per such attempt (Figure 10.4). But why should old wasps make fewer attempts to join? Either they themselves perceive joining as undesirable or the resident wasps are so aggressive toward them outside the nest that they are deterred from making any attempts to join. The latter possibility is supported by our observation that the rates of aggressive behavior shown toward the introduced wasps outside the nest is significantly correlated with the age of the introduced wasps.

Two graduate students, H. S. Arathi and Mallikarjun Shakarad, and I decided to repeat this experiment involving introduction of young unre-

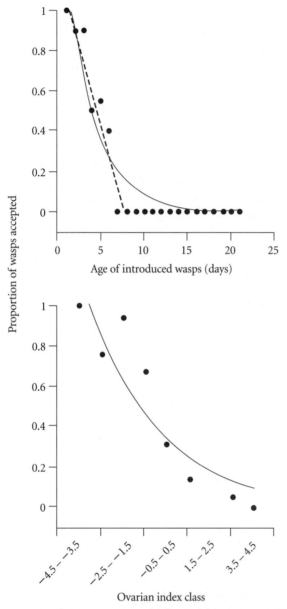

Figure 10.5. Above: The proportion of alien wasps of different ages accepted on the recipient nests decreased with increasing age. Beyond the age of 6 days none of the introduced wasps was accepted ($Y = 1.52 * e^{(-0.28x)}$; $r^2 = 0.91$; solid line). Linear fit for wasps 1 to 8 days only: $Y = 1.22 - 0.15X$; $r^2 = 0.93$; dashed line. Below: The proportion of introduced wasps belonging to different ovarian index classes accepted onto the recipient nests decreased with increasing ovarian index values ($Y = 0.33 * e^{(-0.29x)}$; $r^2 = 0.84$). (Redrawn from Arathi, Shakarad, and Gadagkar, 1997a.)

lated wasps into laboratory colonies with three objectives in mind (Arathi, Shakarad, and Gadagkar, 1997a): (1) We wished to contrast the relative importance of age and ovarian development in the probability of acceptance. (2) We wished to address a legitimate criticism of a reviewer of a previous paper (Venkataraman and Gadagkar, 1995) that the older the introduced wasp, the longer she would have been sitting in our plastic boxes and therefore the greater her chances of acquiring some odor from the plastic box and consequently the greater chance of her being rejected as smelling strange. (3) We wished to examine the possibility of using this procedure of getting unrelated wasps accepted into alien colonies to create genetically mixed laboratory colonies with which to investigate cooperation and conflict in genetically mixed groups of wasps. To increase the probability of acceptance of introduced wasps and to facilitate better observation of the interactions between resident and introduced individuals, the wasps were individually released on the nest. But the cage in which the nest was located was open so that both resident wasps and introduced wasps were free to leave the nest or even the cage itself if they chose.

The second objective was easy to achieve. Instead of holding the wasps in convenient plastic boxes we held them, just for this experiment, in inconvenient, but clean, 500-ml glass beakers with holes for ventilation and showed that there was no difference in our results: the probability of acceptance fell sharply with age (Figure 10.5A). There was a similarly negative correlation between ovarian development and probability of acceptance (Figure 10.5B). In this experiment we needed to assess the ovarian condition of accepted and rejected wasps. Introduced wasps that remained on the nest for the entire duration (10 h) of our observations were termed "accepted." Those that left following high levels of aggression from the resident wasps—this happened 20 min to 2 h after introduction—were termed "rejected." Both accepted and rejected wasps were removed at the end of observations and dissected, to ascertain the state of development of their ovaries.

To aid in our delineation of the relative roles of age and ovarian condition, we used, in addition to the probability of acceptance, other behavioral variables associated with acceptance and rejection. The behaviors shown by the resident wasps toward the introduced wasps can be classified into tolerant and intolerant behaviors. Intolerant behaviors included aggressive biting, attacking, engaging in a falling fight (defined in Table

10.1), sitting on, holding in mouth, nibbling, pecking, chasing, and avoiding. The tolerant behaviors included approaching, antennating, and allogrooming. Because all individuals, both accepted and rejected ones, stayed
on the nest for at least 20 min, we used data on behavioral interactions between the resident wasps and the introduced wasps during the first 20 min
after introduction to compute values for the behavioral variables. The variables we computed are: number of intolerant acts received, number of resident wasps showing intolerant acts, number of tolerant acts received, and
number of resident wasps showing tolerance. The age of introduced wasps
was measured as days since their eclosion from their natal nests, and the
state of development of their ovaries was measured by the composite index
of ovarian development described in Chapter 4. We used logistic regression analysis to assess the influence of the independent variables (age and
ovarian index) on the probability of acceptance and linear regression models to assess their influence on the other behavioral variables described
above. In regression models considering one independent variable at a
time (age or ovarian index), all the dependent variables were significantly
influenced both by age and by ovarian condition. Thus younger individuals and those with poorly developed ovaries had higher probabilities of acceptance, received smaller number of intolerant acts from a smaller number of residents, and received a higher number of tolerant interactions by a
larger number of resident wasps. Conversely, older introduced wasps and
those who had better-developed ovaries not only had a lower probability
of acceptance but also received a higher number of intolerant behaviors
from a larger number of resident wasps and a smaller number of tolerant
behaviors from a smaller number of residents (Table 10.4).

That age and ovarian index have very similar effects both on the behavioral interactions received by the introduced wasps and on the probability
with which they were accepted is not surprising. There is a strong and significant positive correlation between age and ovarian index (Figure 10.6).
But we used the method of partial correlation analysis and the method of
multiple regression analysis to tease out the relative roles of age and ovarian condition. For all variables studied above, the partial correlation coefficients were significant when the ovarian index was held constant but was
not significant when age was held constant (Table 10.5). This indicates
that age has the primary, direct effect and that ovarian index acts merely
through its correlation with age. Similarly, in multiple regression models,

Table 10.4. Regression models (logistic and linear) to examine the possible effects of age and ovarian index (considered separately) of introduced wasps on their probability of acceptance and on the behaviors of resident wasps. (Modified from Arathi, Shakarad, and Gadagkar, 1997a.)

Dependent variable	Independent variable	Estimate	S.E.	t_s	p
Probability of	Age	−0.95	0.17	−5.53	< 0.01
acceptance[a,b]	Ovarian index	−1.68	0.20	−5.83	< 0.01
Probability of	Age	0.73	0.23	−3.12	< 0.01
acceptance[a,c]	Ovarian index	−0.71	0.24	−2.88	< 0.01
No. of intolerant acts	Age	7.91	1.16	6.83	< 0.01
received	Ovarian index	17.19	2.79	6.16	< 0.01
No. of residents	Age	0.63	0.13	5.00	< 0.01
showing intolerant acts	Ovarian index	1.81	0.28	6.41	< 0.01
No. of tolerant acts	Age	−0.06	0.01	−5.45	< 0.01
received	Ovarian index	−0.15	0.02	−6.01	< 0.01
No. of residents showing	Age	−0.07	0.01	−6.43	< 0.01
tolerance	Ovarian index	−0.17	0.03	−6.33	< 0.01

a. Logistic regression.
b. All introduced wasps included in the analysis.
c. Only wasps ≤ 6 days old included in the analysis.

when age and ovarian index were considered simultaneously, the coefficient associated with age was always significant while the coefficient associated with ovarian index was always insignificant (Table 10.6).

This finding was surprising. We had expected ovarian condition of the introduced wasps to be the primary predictor of both the behavioral response and the acceptance or rejection by the resident wasps, and expected that age was a spurious correlate of this response. But the results we obtained were contrary to this expectation. How then do these results alter our perception of the proximate and ultimate factors that govern acceptance and rejection? If age per se (rather than ovarian condition) influences acceptance and rejection, perhaps we should go back to the possibility discussed earlier that when a wasp gets older she may acquire larger quantities of recognition labels. It must be emphasized that the introduced wasps in our experiment were isolated wasps that spent varying numbers of days in isolation; so they could not have acquired different extents of recognition labels from their nests or nestmates. There is, however, some evidence that even an isolated individual can exhibit age-dependent varia-

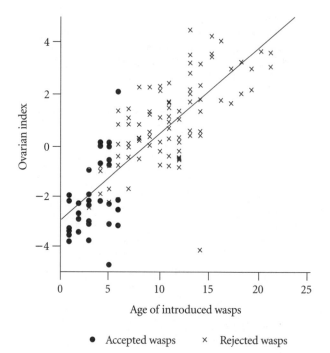

Figure 10.6. Regression of the age of introduced wasps on their respective ovarian index values, indicating that age and ovarian index are significantly positively correlated ($Y = -2.86 + 0.34X$; $r^2 = 0.64$; $n = 132$; slope significantly different from zero, $p < 0.01$). (Redrawn from Arathi, Shakarad, and Gadagkar, 1997a.)

tion in its cuticular chemistry (Jackson and Bartlet, 1986; Morel, Vander Meer, and Lavine, 1988; Waldman, 1988; Singer and Espelie, 1992).

But what disadvantage could there be of accepting older individuals if it is not their better ovarian development? In other words, what advantage could there be of accepting younger individuals if it is not that they have underdeveloped ovaries? Again we should perhaps go back to the possibility discussed earlier that younger wasps, as opposed to older ones, may be more easily molded into desired behavioral roles by the resident wasps.

These experiments also fulfilled our third objective. The confirmation of results obtained earlier using a large sample size—we introduced 132 wasps into 12 different recipient colonies and obtained the acceptance of 57 wasps onto unrelated colonies—certainly boosted our confidence that this technique could be developed as a paradigm for creating genetically

Table 10.5.   Partial correlation analysis to examine the relation between age and
              ovarian index of introduced wasps and behaviors of resident wasps.
              (Modified from Arathi, Shakarad, and Gadagkar, 1997a.)

Variables	Partial correlation coefficient	t	p
$r_{c,a.b}$	0.25	3.02	< 0.01
$r_{c,b.a}$	0.12	1.33	> 0.05
$r_{d,a.b}$	0.31	3.69	< 0.01
$r_{d,b.a}$	0.03	0.31	> 0.05
$r_{e,a.b}$	0.23	2.65	< 0.01
$r_{e,b.a}$	0.12	1.34	> 0.05
$r_{f,a.b}$	0.20	2.34	< 0.01
$r_{f,b.a}$	0.14	1.65	> 0.05

Note: $r_{c,a.b}$ is to be read as partial correlation between age of the introduced wasps and rates
of aggressive acts received, with ovarian index being kept constant. Note that all partial
correlations with *a.b* are significant, while all those with *b.a* are not significant, indicating that
any influence of ovarian index on the dependent variables is only mediated through age.
Subscripts: *a*, age of the introduced wasps; *b*, ovarian index of the introduced wasps; *c*, rates of
aggressive acts received by the introduced wasps; *d*, number of residents showing aggression
towards the introduced wasps; *e*, rates of tolerance acts received by the introduced wasps; *f*,
number of residents showing tolerance toward the introduced wasps.

mixed colonies and investigating behavior and social organization under
conditions of low intracolony genetic relatedness.

10.7.   Cooperation among Unrelated Individuals

Everything we had learned so far about *R. marginata* had consistently
given the impression that asymmetries in genetic relatedness play little or
no role in promoting the evolution of worker behavior. Polyandry creates
multiple patrilines within a colony and serial polygyny creates multiple
matrilines. Movement of wasps between nests, at least in the pre-emergence phase, further increases intracolony genetic variability. The wasps do
not seem to have the ability to discriminate between different levels of relatedness among colony members. At the same time, adult wasps have
multiple reproductive options: they can leave to found new nests, join
nests already founded by other wasps, or overthrow the queen. All of this
clearly suggests that although most wasps die as sterile workers and gain
nothing more than the indirect component of inclusive fitness, it must
be true that adult wasps are strongly selected to maximize their chances

Table 10.6. Multiple regression models (logistic and linear) used to examine the
possible effects of age and ovarian index (considered simultaneously) of
introduced wasps on their probability of acceptance and on the
behaviors of resident wasps. (Modified from Arathi, Shakarad, and
Gadagkar, 1997a.)

Dependent variable	Independent variable	Estimate	S.E.	t_s	p
Probability of	Age	−0.79	0.19	−4.01	< 0.01
acceptance[a,b]	Ovarian index	−0.40	0.26	−1.51	> 0.05
Probability of	Age	−0.57	0.26	−2.17	< 0.01
acceptance[a,c]	Ovarian index	−0.37	0.27	−1.36	> 0.05
Rates of aggressive acts	Age	5.69	1.91	2.98	< 0.01
received	Ovarian index	6.54	4.48	1.46	> 0.05
Number of residents	Age	1.72	0.47	3.67	< 0.01
showing aggression	Ovarian index	0.05	0.24	0.24	> 0.05
Rates of tolerant acts	Age	−0.11	0.04	−2.61	< 0.01
received	Ovarian index	−0.02	0.02	−1.29	> 0.05
Number of residents	Age	−0.04	0.02	−2.29	< 0.01
showing tolerance	Ovarian index	−0.09	0.05	−1.68	> 0.05

a. Logistic regression.
b. All introduced wasps included in the analysis.
c. Only wasps < 6 days old included in the analysis.

(however small) of direct reproduction. Consider the possibility that a
wasp pursues the strategy of working for her colony, helping to bring it to a
healthy state with a good worker force, and then driving the queen away
and inheriting and utilizing the entire resources of the colony to aid her
own reproduction. It matters little whether the colony, at the time she in-
herits it, consists of her close relatives or of distant relatives. If this is possible, then we should not expect any simple rule linking cooperation with
high levels of relatedness and conflict with low levels of relatedness. Indeed, intracolony relatedness levels should play little or no role in modulating the efficiency of social organization, division of labor, and cooperation. Creating genetically mixed colonies, by introducing young unrelated
conspecifics, and then comparing such colonies with undisturbed normal
colonies, which can be expected to be relatively genetically homogenous,
would be a powerful way of testing the effect of genetic relatedness, if any,
on social organization, division of labor, and cooperation.

This is just what we (Arathi, Shakarad, and Gadagkar, 1997b) next set
out to do. We created 12 genetically mixed colonies by introducing young

(< 24 h old) wasps eclosing on unrelated nests into observation colonies set up in the laboratory. As before, the donor and recipient nests were collected from well separated localities. We took two additional precautions. First, we made observations on the recipient colonies before introduction so that social organization before introduction could be compared with that afterward. Second, we matched introduction of unrelated wasps carefully with eclosion of new individuals in the donor colonies so that the introduced wasps had matched (in age) partners among the natal wasps, which could be treated as controls. Eighty-five wasps were introduced and all were accepted into their foster colonies, without any aggression. As expected, there was no evidence of intracolony kin recognition. We compared the behavioral profiles of individual wasps before and after introduction (for those wasps present during both phases) and failed to find any significant differences. This result suggests that introduction of the unrelated wasps and the consequent reduction in intracolony genetic relatedness did not perceptibly alter the behavior of the wasps present before and after the introduction.

When we first performed this experiment, we had no idea what the introduced wasps would do in their foster colonies. It was fascinating to find that the introduced wasps did everything that their natal counterparts did and in similar proportions too. We compared the time budgets of introduced wasps with those of their natal counterparts and failed to detect any significant difference. We compared rates of interactions among kin and nonkin and again failed to find any significant differences. To make a more robust comparison by considering all behaviors simultaneously, we subjected the time budgets of all wasps to principal components analysis and plotted the position of the wasps in the coordinate space of the first two principal components (as described in Chapter 5). Here we could compute the distance (in the parameter space of the principal components) between all possible pairs of natal wasps, between pairs of introduced wasps, and between natal wasps and introduced wasps. The former two can be considered as kin-kin distances and the latter as a kin-nonkin distance, in the behavioral parameter space. We compared kin-kin distances with kin-nonkin distances and failed to find any significant differences. Thus we found no evidence of task specialization between natal wasps and introduced wasps. As far as we could make out, social organization, division of labor, and cooperation were unaffected in the genetically mixed colonies.

Then something happened that greatly enhanced our appreciation of the extent to which the introduced wasps were intergrated into their foster colonies. Of the 12 nests in the study, 2 experienced spontaneous queen replacements, and we were most intrigued to find that in both cases it was one of the introduced wasps that took over as the replacement queen. We then decided to induce queen replacements (by removing the existing queen) in the remaining colonies that were still under observation. There were 6 such colonies, and thus we witnessed queen replacements (spontaneous or induced) in 8 genetically mixed colonies. Of the eight replacement queens (one per colony of course), three were introduced wasps, three were natal wasps, and two were from among the original wasps present on the nest before introduction of genetically unrelated wasps (which we call "resident" wasps). There was no significant difference between the probabilities with which introduced (3/85) and natal (3/69) wasps became replacement queens ($G = 0.049$; $p > 0.05$).

But a proper comparison should take into account the relative numbers of introduced and natal wasps that were available on each nest on the day of queen replacement. This we did, using a Monte Carlo simulation. In the simulation, each event of queen turnover was simulated 10,000 times by assigning the new queen to the natal or introduced category, depending on the proportion of the two categories of wasps available at the time of that queen replacement event. The simulations showed that we cannot rule out the possibility that new queens are chosen from different categories of wasps merely on the basis of their availability. We concluded therefore not only that introduced wasps were accepted into their foster colonies, but that they became well integrated and behaviorally indistinguishable from the natal wasps, eventually becoming foragers and even having a fair chance of becoming replacement queens. The results of this experiment clearly show that the advantage of working for the welfare of a colony and eventually taking over a reproductive role may well be one of the factors that drives the evolution of worker behavior in R. marginata.

Nevertheless, we could not yet rule out the possibility that there is a subtle form of intracolony kin discrimination leading, perhaps, to quantitative differences in cooperation and productivity. In the next study (Arathi and Gadagkar, 1998), we therefore set out to explicitly test the possibility that poor cooperation and lower productivity constitute a cost of living in colonies with high genetic variability. We did this by setting up pairs of wasps in plastic boxes and closely examining cooperation among them and measuring their joint productivity. We set up 91 nestmate pairs and 85 nonnestmate pairs of wasps and followed them until the eclosion of the first adult offspring or until the death of one of the wasps. As before, the nestmates had eclosed from the same nest and the nonnestmates had eclosed from nests collected at least 8 km away from each other. We compared nestmate pairs and nonnestmate pairs for the proportion of successfully initiated nests, for their productivity, and for the rates at which they attained their productivity, only to find no demonstrable differences between them (Figure 10.7). It appears therefore that there is no subtle form of kin discrimination leading to lowered efficiency of cooperation among nonrelatives.

For this conclusion to be valid, however, we needed to demonstrate that both wasps indeed cooperated in the nestmate boxes as well as in the nonnestmate boxes. If only one wasp in each pair was involved in nest building and brood care, these would be equivalent to single-foundress nests and there would be no reason to expect systematic differences between the nestmate boxes and the nonnestmate boxes. To confirm that nest building and brood care were indeed a result of cooperation by both members of each pair, we conducted behavioral observations on 28 nestmate pairs and 27 nonnestmate pairs. We found no significant differences between nestmate queens and nonnestmate queens in any of the 10 behaviors we compared, the same behaviors we used to establish and compare behavioral castes in Chapter 5. Each pair had a queen and a worker and we found no significant differences between queens and workers (except, as expected, in the time spent in sitting) either in the nestmate pairs or in the nonnestmate pairs. As in the previous experiment involving genetically mixed colonies, we used principal components analysis to compare and contrast the behavioral profiles of nestmates and nonnestmates and queens and workers. In the coordinate space of the first two principal components, queen-worker distances among nestmates was not significantly different from queen-worker distance among nonnestmates. We also compared queen-queen distances and worker-worker distances (nestmate versus nonnestmates) and found no effect of nestmateship on these distances.

These results suggest that the behavioral relationship between queens and workers did not depend on whether they were nestmates or nonnestmates. Thus both members of each pair participated in nest building and

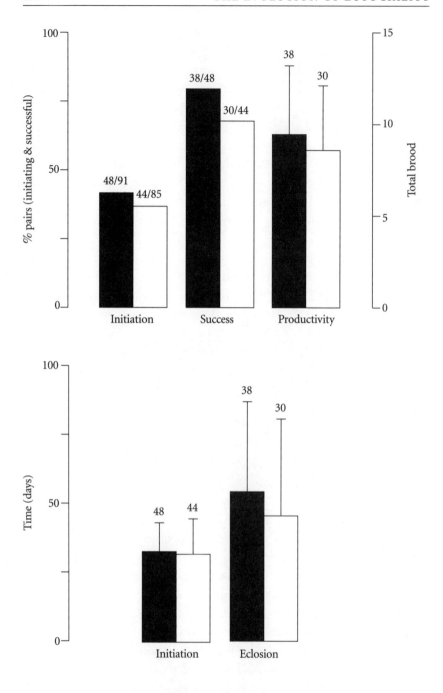

Figure 10.7 (facing page). Comparisons of nestmate and nonnestmate pairs for nest initiation, productivity, and developmental time. Black bars = nestmate pairs; open bars = nonnestmate pairs. The numbers above the bars are the sample sizes. Above: The percentage (primary Y axis) of nestmate pairs that initiated nests is not significantly different from the percentage of nonnestmate pairs that did so ($G = 0.017$; $p > 0.05$). The percentage of these that successfully produced an adult offspring was not significantly different between nestmate and nonnestmate pairs ($G = 1.43$; $p > 0.05$). The total brood (secondary Y axis) at the time of eclosion of the first adult in nests of nestmate pairs was not significantly different from that of nonnestmate pairs (Mann-Whitney U test, $U = 485$; $p > 0.05$). Below: The time taken to initiate nests was not significantly different between the nestmate and nonnestmate pairs (Mann-Whitney U test, $U = 1023$; $p > 0.05$). Similarly, the developmental period from egg to adult was also not significantly different between the nestmate and nonnestmate pairs (Mann-Whitney U test, $U = 589$; $p > 0.05$). (Redrawn from Arathi and Gadagkar, 1998.)

brood care. The lack of any significant differences between the success and productivities of the nestmate pairs and the nonnestmate pairs should be seen in the light of this evidence of cooperation between the members of each pair. Even though the nestmates in these experiments may not have been full sisters because of polyandry and serial polygyny in the colonies from which they eclosed, it is reasonable to expect that two wasps eclosing from the same nest would be more closely related to each other than two wasps eclosing from nests collected from well-separated sites. The absence of any significant differences between nestmate pairs and nonnestmate pairs in their rates of nest initiation, success, productivity, and developmental periods can therefore be interpreted to mean that there is no apparent cost to living with distantly related individuals. Thus the message of the experiments discussed in this and the previous chapter was clear: look for factors other than genetic relatedness that might promote the evolution of worker behavior in *R. marginata*. We did so (see Chapters 11–13) but did not entirely abandon the idea that genetic relatedness might have some influence on the evolution of sociality in *R. marginata* (see below).

10.8.   Queen Success and Worker-Brood Genetic Relatedness

Recall the system of serial polygyny in *R. marginata*, described in section 9.2. The data we obtained in that study could be used to examine the role of genetic relatedness in yet another way. Perhaps the most striking feature of that study was the enormous variation in queen success (Table 9.2).

Among the 21 queens studied, their tenures ranged from 0 to 394 days, the number of offspring produced per day of tenure ranged from 0 to 2.3, and the proportion of eggs laid that successfully developed into adults ranged from 0 to 0.37. What makes some queens much more successful than others? Surely worker-brood genetic relatedness during the tenure of a queen is one of the factors we must examine in the effort to understand the determinants of queen success. My students and I considered the variables mentioned above—queen tenure, total number of offspring produced during the tenure, number of offspring produced per day of the tenure and proportion of eggs laid that successfully became adults—indicators of queen success. The first three are obvious indicators of queen success but the fourth needs some explanation. A substantial fraction of the eggs laid in a colony fail to reach adulthood, perhaps because of cannibalism by the workers, perhaps because of neglect of the larvae. The proportion of eggs that successfully reach adulthood is therefore not only an important indicator of queen success but probably a more direct measure of the extent of cooperation that the queen is able to obtain from the workers. We also considered three possible determinants of queen success. These are age in days at the time of takeover, the mean worker-brood ratio during the tenure, and the mean worker-brood genetic relatedness during the tenure. We considered age at takeover as a possible determinant because young queens may have a long tenure ahead of them. A high worker-brood ratio is of obvious advantage in ensuring brood-rearing efficiency, and high worker-brood genetic relatedness can, in principle, ensure better cooperation between queens and workers. As in the case of indicators of queen success, there is also considerable variation from queen to queen in the possible determinants of queen success. Age at takeover varied from 4 to 78 days, worker-brood ratio from 0.04 to 0.25, and worker-brood genetic relatedness from 0.26 to 0.70. This data set is thus sufficiently rich to explore potential determinants of queen success.

Our result was simple and straightforward. For every indicator of queen success, there was a clear statistically significant dependence on worker-brood genetic relatedness but not on worker-brood ratio or age at takeover (Table 10.7) (Gadagkar et al., 1993a). This suggests that queens of *R. marginata* are more successful when the worker-brood genetic relatedness during their tenure is high. But how does this come about? One possibility is that despite everything discussed in this chapter, there may be some sub-

Table 10.7.   Relationships between indicators and possible determinants of queen success. (Data from Gadagkar et al., 1993a.)

| Indicators of queen success | Possible determinants of queen success | | | | | |
| | Age at takeover | | Mean worker-brood ratio | | Mean worker-brood genetic relatedness | |
	r or τ	b	r or τ	b	r or τ	b
Queen tenure (days)	−0.15	−0.76	0.32	336	0.84[c]	442[c]
Total number of offspring produced	0.06	1.41	0.21	403	0.52[c]	641[d]
Number of offspring produced per day of tenure	0.16	0.01	0.13	0.42	0.48[c]	2.99[b]
Proportion of eggs laid that develop into adults	0.44	0.002	0.11	−0.39	0.49[a]	0.48[a]

Note: The body of the table gives the values of the Pearson product moment correlation (r) or Kendal's rank correlation coefficient (τ) between the variable in the row heading and that in the column heading, and Model II regression coefficients (b)(Sokal and Rohlf, 1981) computed with Barlett's three−group method with the variable in the row heading as the dependent variable and that in the column heading as the independent variable. Correlation and regression coefficients significantly different from zero are marked and the p values are given in footnotes. All variables except the total number of offspring produced and the number of offspring produced per day are normally distributed. For these two variables, correlation was measured by the nonparametric Kendall's rank correlation τ and the significance of the regression coefficient was tested by a randomization test (Sokal and Rohlf, 1981).

a. $p < 0.05$.
b. $p < 0.02$.
c. $p < 0.01$.
d. $p < 0.002$.

tle way in which workers can assess the intracolony levels of worker-brood genetic relatedness and offer better cooperation (by more efficient rearing of brood) to queens when such relatedness is high. This idea is quite consistent with inclusive fitness theory. But there is another possibility. Rather than long queen tenures resulting from high worker-brood genetic relatedness values, long queen tenures may instead cause high worker-brood genetic relatedness. Long queen tenures can indeed result in high worker-brood genetic relatedness because of the absence of multiple matrilines. A healthy queen may be capable of having a high tenure as well as producing more offspring, even on a per-day basis; high worker-brood genetic relatedness may thus be just a by-product. But can we explain why a higher proportion of the eggs laid by queens with long tenures successfully develop into eggs? This sounds like a variable that has a very attractive link to worker-brood genetic relatedness. Yet it is entirely reasonable to assume that if queens with long tenures are intrinsically healthy individuals, most workers would be selected to cooperate better with them, because a change in queen is likely to further reduce worker-brood relatedness for all workers that do not become replacement queens.

In summary I think that we are not quite yet in a position to distinguish between the two possible hypotheses: (1) that worker-brood genetic relatedness directly increases worker cooperation and (2) that high worker-brood relatedness is a secondary by-product of an intrinsically healthy queen. The first hypothesis requires successful intracolony kin recognition, for which we have so far consistently found no evidence. The second hypothesis is probably therefore the more likely one. Absence, or at least the unlikeliness, of intracolony kin recognition has also been noted in other species of primitively eusocial wasps and bees (see, e.g., Gamboa, Reeve, and Pfennig, 1986; Queller, Hughes, and Strassmann, 1990; Strassmann, 1996; Strassmann et al., 1997; Solís et al., 1998; Bull and Adams, 1999).

Let me now recapitulate the findings dicussed in this and the previous chapter. *R. marginata* queens can mate with 2–3 males, mix sperm received from them, and produce multiple patrilines among their daughters, reducing their interrelatedness to values approaching 0.5. The system of serial polygyny, involving frequent queen replacements, further reduces levels of intracolony genetic relatedness to values well below 0.5. *R. marginata* workers have an efficient system of nestmate discrimination that is based on recognition labels and templates acquired, from the nest and nestmates, after eclosion of the adult wasps. This makes intracolony kin recognition unlikely. Wasps move from nest to nest in the colony founding stage. In the laboratory, young wasps are accepted onto foreign colonies, where they become well integrated into the social system and go on to become foragers and even queens in their foster colonies. There is no apparent cost to living with unrelated wasps. The only discordant resultis that there is a positive correlation between worker-brood genetic relatedness and queen success, but that may well be due to a mechanism not requiring intracolony kin recognition. On balance, we can conclude that it is extremely unlikely that inequality 8.5 will be satisfied in favor of worker behavior because of asymmetries in the genetic relatedness values. Thus there is no evidence for genetic predisposition for the evolution of eusociality in *R. marginata*. Needless to say, we should now focus on other factors, particularly ecological, physiological, and demographic predispositions, in the search for a solution to the problem of the evolution of eusociality.

11   ECOLOGICAL PREDISPOSITION

$\mathscr{B}$ Y THE criteria laid down in Chapter 8, asking whether *R. marginata* is ecologically predisposed to the evolution of eusociality is equivalent to asking if the productivity of workers (β in inequality 8.5) is greater than the productivity of solitary foundresses (b in inequality 8.5) owing to ecological factors, so that the inclusive fitness of workers is greater than that of solitary foundresses, even if $\rho = $ r and $\sigma = $ s.

11.1.   The Costs and Benefits of Group Life

One way to ask if *R. marginata* is ecologically predisposed to the evolution of eusociality is to compare the success of single-foundress nests with that of multiple-foundress nests. If the success rates are different, one can ask if these differences can be attributed to ecological factors. The study by Shakarad and Gadagkar (1995, 1996) described in Chapter 4 provides an extensive data set with which to do this (see Kojima, 1989 for similar data on *Ropalidia fasciata*). In that study, we tracked 51 single-foundress nests and 94 multiple-foundress nests. Single-foundress nests had a significantly lower rate of success than multiple-foundress nests (Figure 11.1A). Of 145 nests studied, 80 were located in the vespiary where the predator *Vespa tropica* is excluded, and 65 were located in the field, presumably with natural rates of predation. Note that although tachinid flies and ichneumonid wasps sometimes parasitize the brood of *R. marginata,* no parasitization was observed in any of the 145 nests in this study. Hence predation pressure from *V. tropica* is what is of concern here. As might be expected, overall success rate (of both single- and multiple-foundress nests) was higher in the vespiary than in the field (Figure 11.1B). Can the increased rate of

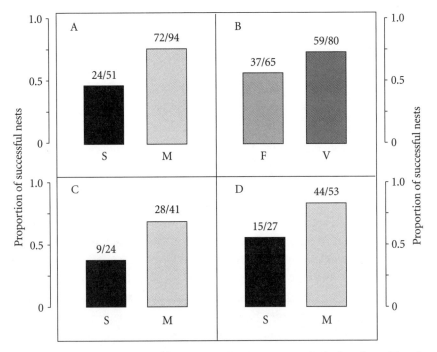

Figure 11.1. Comparison of success rates of pre-emergence single-foundress (S) and multiple-foundress (M) nests. A: Single-foundress nests have significantly lower success than multiple-foundress nests (data pooled from field and vespiary) ($G = 12.7$, $p < 0.001$). B: Nests in the field (F) have significantly lower success than nests in the vespiary (V) (data pooled from single-foundress and multiple-foundress nests) ($G = 4.5$, $p < 0.05$). C and D: When analyzed separately, single-foundress nests had significantly lower productivity both in the field (panel C, $G = 5.9$, $p < 0.025$) and in the vespiary (panel D, $G = 6.7$, $p < 0.01$). (Data from Shakarad and Gadgkar, 1995.)

failure in the field be attributed entirely to predation pressure, which was lacking in the vespiary? In the vespiary 21/80 (26.2%) nests failed, obviously owing to causes other than predation. In the field, although 28/65 (43.1%) nests failed, only 16 nests were attacked by the predator. Of the remaining 49 nests not attacked, 16 (32.7%) nevertheless failed. This rate of failure should therefore be attributed to causes other than predation. Now the failure rate of 21/80 in the vespiary and that in the field of 16/49, both due to causes other than predation, are not significantly different from each other ($G = 0.59$, $p > 0.05$). It is reasonable to conclude therefore that the increased failure rate in the field is indeed due to predation by *V. tropica*.

Both in the field and in the vespiary, single-foundress nests had significantly lower rates of success than multiple-foundress nests (Figure 11.1C and D). Why should single-foundress nests have significantly lower rates of success than multiple-foundress nests? There are two reasons why this cannot be due to higher rates of predation on single-foundress nests. The first reason is that single-foundress nests have a significantly lower rate of success even in the vespiary, where there is no predation. Second, there is more direct evidence that *V. tropica* does not prefer single-foundress nests to multiple-foundress nests. Shakarad and I witnessed 21 cases of attack by *V. tropica* (sometimes more than once on the same nest). Of these, 7 instances were directed toward single-foundress nests and 14 instances toward multiple-foundress nests. Of course this does not mean that the predator prefers single-foundress nests to multiple-foundress nests because only 51 single-foundress nests were present in the population and 94 multiple-foundress nests were present. But we also cannot directly compare 7/51 (13.7%) with 14/94 (14.9%) and conclude that the predator had approximately the same preference for both kinds of nests. The seven instances of predation of single-foundress nests can only be compared with the 14 instances of predation of multiple-foundress nests if we also compare the relative proportions of single- and multiple-foundress nests available to the predator during each event of predation. These proportions are known from our data. But since the number of single- and multiple-foundress nests available for each event of predation was different, we could not perform any standard statistical test; we therefore used Monte Carlo simulations.

We simulated each of the 21 instances of predation using the null hypothesis that the predator would simply choose single- and multiple-foundress nests based on their relative abundance, at the appropriate study site on the appropriate day. After each of the 21 instances of predation was thus simulated once, the proportion of times the predator chose single-foundress nests in the simulations was computed. In the same way the 21 simulations were repeated 10,000 times to yield as many values of proportion of times the predator chose to attack single-foundress nests. The distribution of these 10,000 values is shown in the top right-hand panel of Figure 11.2. For comparison, the empirically observed values of the proportion of times the predator preferred single- and multiple-foundress nests is shown in the corresponding left-hand panel of Figure 11.2. What is

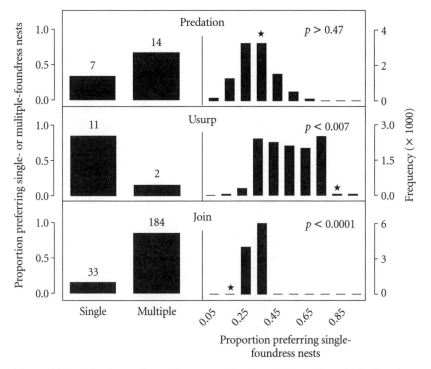

Figure 11.2. Left: Proportions of wasps preferring single- and multiple-foundress nests for joining and usurpation, and proportions of single- and multiple-foundress nests being attacked by *Vespa tropica*. The numbers leading to these proportions are given above the bars. Right: Frequency distribution of the expected proportion of conspecifics choosing single-foundress nests for joining (top), the expected proportion of conspecifics choosing to usurp single-foundress nests (middle), and the expected proportions of single-foundress nests attacked by the predator *V. tropica* (bottom). In each right-hand panel, the star indicates the location of the empirically derived value. (Redrawn from Shakarad and Gadagkar, 1995.)

the probability that a value as high as the empirically observed one for the proportion of times single-foundress nests are attacked ($7/21 = 0.33$) could have been observed by chance alone, if the predator chooses nests simply on the basis of their availability? In 4774 of the 10,000 simulations, the proportion of times single-foundress nests were attacked was greater than 0.33. Thus we accept the null hypothesis that the predator has no particular preference for single-foundress nests and chooses its target nest simply on the basis of availability.

Why then do single-foundress nests fail so much more often than multiple-foundress nests? In single-foundress nests, the lone wasp has to forage for food and building material. This is a high-risk activity, and single-foundress nests failed in our study simply because foundresses did not return from their foraging trips—not so much because the brood was attacked by any predator or parasite. The probability of loss of all foundresses is of course expected to decrease with an increase in the number of foundresses. Multiple-foundress nests have the advantage that even if one or some of the wasps die others can continue to tend the brood. Thus, to paraphrase Kern Reeve (1991), the advantage of multiple-foundress nests seems to be that they have "survivorship insurance." I will discuss this further in Chapter 13, where I will consider such factors under the rubric of demographic predisposition.

There is another difference between single- and multiple-foundress nests. As shown in Chapter 4, pre-emergence nests can be usurped: alien wasps can join and immediately replace the existing queen. We (Shakarad and Gadagkar, 1995) witnessed 13 instances of usurpation, 11 in single-foundress nests and 2 in multiple-foundress nests. By considering the relative abundance of single- and multiple-foundress nests during each event of usurpation and performing a computer simulation, similar to the one described above for predation, we found that usurpers did not choose target nests on the basis of their availability—they specifically targeted single-foundress nests. Of the 10,000 simulations, only in 68 cases was the proportion of single-foundress nests usurped as high or higher than the empirically observed proportion (Figure 11.2, middle panel). Since usurpation is a rare and unpredictable event, we have no behavioral observations of the act of usurpation, but it is easy to imagine that the usurper finds it harder to usurp nests containing more than a single wasp. The significantly lower rate of usurpation is thus a clear benefit of multiple-foundress associations (see Gamboa, 1978, for a similar conclusion with *Polistes metricus*; see also Klahn, 1988; Reeve, 1991; Gamboa et al., 1992; Bagnères et al., 1996).

But two caveats need to be mentioned: (1) The higher rate of usurpation of single-foundress nests cannot account for the observed lower rate of success of single-foundress nests, because usurped nests were not classified as unsuccessful nests in the preceding analysis—only nests that failed to produce a single adult offspring were classified as unsuccessful. (2) While there should be selection pressure on queens to be tolerant toward subordinates to avoid being usurped, this pressure would not necessarily select for joiners to choose multiple-foundress nests. But, for reasons that we do not understand, joiners undoubtedly prefer multiple-foundress nests. Recall from section 9.3 that of 217 instances of alien conspecifics joining pre-emergence nests, 33 joiners preferred single-foundress nests while 184 preferred to join multiple-foundress nests. In 10,000 computer simulations, not even once was such a high preference for multiple-foundress nests observed (Figure 11.1). Thus where the intention is not to usurp the position of the queen but to remain as a subordinate worker (at least for some time), joiners clearly prefer multiple-foundress nests. Why should this be so?

11.2. Productivity and Foundress Number

Is the preference of joiners for multiple-foundress nests related to the higher success rate of multiple-foundress nests? First, let us see if the higher success rate of multiple-foundress nests translates into higher per capita productivity. Addressing this question was in fact one of the aims of the Shakarad and Gadagkar (1995) study of colony founding in *R. marginata*. This of course is an old and controversial topic but one that is continually debated (see Michener, 1964b; Gibo, 1978; Jeanne, 1986a, 1999; Wenzel and Pickering, 1991; Jeanne and Nordheim, 1996; Karsai and Wenzel, 1998). We see from Figure 11.3 (panel A) that the total productivity (defined as the total number of eggs + larvae + pupae at the time of eclosion of the first adult offspring) of the colonies increased significantly with an increase in the number of foundresses. However, it is not the total productivity of the colony that is of interest but the per capita productivity, which does not change significantly with the number of foundresses (Figure 11.3, panel B). It must be emphasized that all failed nests were included (with a value of zero for their productivity) in computing the mean productivity for nests with different numbers of foundresses and in performing the regression analysis. If the failed nests are excluded from the analysis, the slope of per capita productivity as a function of foundress number is significantly negative (not shown). Regression of per capita productivity on foundress number (after we include failed nests, as we should) (Figure 11.3, panel B) shows that, on average, each individual, whether in a single-foundress nest or in a multiple-foundress nest, accounts for about

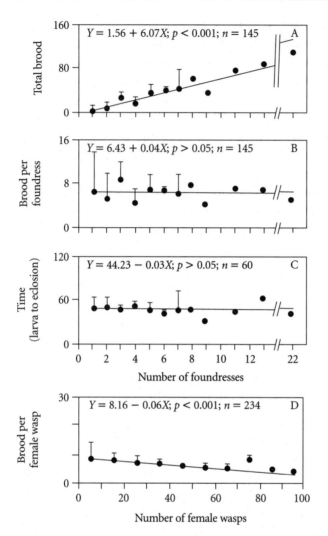

Figure 11.3. Total productivity (A), per capita productivity (B) at the time of the eclosion of the first adult, and time elapsed from the hatching of the first egg to the eclosion of the first adult (C), all as a function of the number of foundresses on pre-emergence nests. Per capita productivity is also shown as a function of the number of adults present on 234 nests at the time of their collection (D). Mean, one standard deviation, linear regression equations, sample size, and *p* values relating to testing deviation of the slopes from zero are shown. (Redrawn from Shakarad and Gadagkar, 1995.).

six items of brood (three eggs, two larvae, and one pupa) at the time of eclosion of the first adult offspring. An important conclusion emerges from this finding of constant per capita productivity as a function of foundress number. The advantage of multiple-foundress nests derived from their significantly higher rate of success (compared to single-foundress nests) has disappeared because, even though we included all failed nests in the analysis, single-foundress nests have the same per capita productivity as multiple-foundress nests. In other words, the higher success rate of multiple-foundress nests does not translate into higher per capita productivity in comparison with single-foundress nests. So we are back to square one in our attempt to understand why subordinate cofoundresses prefer to join multiple-foundress nests.

It is good to bear in mind that we have discovered one advantage of multiple-foundress nests that is not nullified by the demonstration of constant per capita productivity. This is the advantage of the lower rates of usurpation enjoyed by multiple-foundress nests. But leaving that aside, let us dwell a bit more on the observed constant per capita productivity. True, the per capita productivity does not increase as a function of foundress number in the pre-emergence period, but does it increase in post-emergence periods? Perhaps post-emergence nests are qualitatively different, with better cooperation and a better division of labor.

To test this possibility, we regressed the total brood against the number of adult female wasps present at the time of collection for 234 nests used in various experiments in my laboratory during a 12-year period (1983–1994) (Figure 11.3, panel D). These nests are a haphazard sample representing a range of colony sizes in different stages of development. Here we found a significant negative slope, suggesting that in post-emergence colonies per capita productivity may actually decrease with the number of wasps. This might be because not all female wasps present in post-emergence colonies contribute to the colony's labor. In any case there is no evidence of increase in per capita productivity with number of wasps, either in pre-emergence nests or in post-emergence nests. We also looked for and ruled out other possible advantages of large group sizes as compared to small ones. Developmental time (duration from hatching of the first egg to the eclosion of the first adult offspring) did not vary significantly with the number of foundresses (Figure 11.3, panel C). The mean fresh weight of individuals eclosing from single-foundress nests ($40.66 + 8.61$ mg, $n =$

27) was not significantly different from that of those individuals eclosing from multiple-foundress nests (42.10 + 6.80 mg, $n = 86$).

The increase in total productivity with number of foundresses is not surprising, but it does suggest that single-foundress nests in particular and small colonies in general are not limited by the queens' egg-laying ability: rather they are limited by rates of inflow of building material and food and rates of nest building and brood care, all functions performed by the increasing work force available in larger colonies. The lack of increase in per capita productivity is surprising, especially since it is not compensated by faster developmental periods or heavier individuals in larger nests. One might expect that in large nests, owing to a more efficient division of labor, there would have been an increase in the efficiency of brood rearing and hence an increase in the per capita productivity. However, there appears to be a decrease in brood-rearing efficiency as the number of cofoundress increases, even in the pre-emergence period; in spite of relatively lower rates of failures, multiple-foundress colonies have about the same per capita productivity as single-foundress nests do. Shakarad and I have speculated that the apparent decrease in the efficiency of brood rearing with an increase in group size has to do with the primitively eusocial status of *R. marginata* (Shakarad and Gadgkar, 1995). Unlike workers in most highly eusocial species, which rely almost exclusively on the indirect component of inclusive fitness, workers in primitively eusocial species such as *R. marginata* clearly attempt to obtain both direct fitness (by becoming replacement queens at some point) and indirect fitness (by caring for the queen's brood). This is what probably limits the efficiency of brood care in multiple-foundress colonies of primitively eusocial species (Shakarad and Gadgkar, 1995).

A hidden advantage of larger group sizes can perhaps be discerned from Figure 11.3, panel B: there appears to be more variation in the per capita productivity in the smaller group sizes. Wenzel and Pickering (1991) pointed out that the predictability with which the mean per capita productivity is attained is an important and frequently overlooked issue. Because small groups are expected to be more unpredictable and show larger deviation from the expected mean than larger groups, Wenzel and Pickering (1991) claimed correctly that reduction in variance constitutes an "automatic" advantage of group living. To assess evidence for such an advantage for large groups in our data, we regressed the coefficient of variation of the

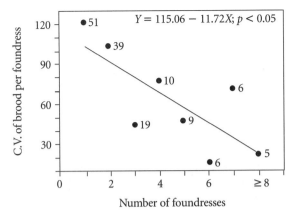

Figure 11.4. Coefficient of variation (C.V.) of the per capita brood (eggs + larvae + pupae) as a function of the number of foundresses. The linear regression and the p value relating to testing of the deviation of the slope from zero and sample sizes (number of nests) are shown for each point. Note that the coefficient of variation decreases significantly with foundress group size despite the drastic reduction in sample sizes for larger groups. Note also that failed nests were assigned a value of zero productivity and included in the data set. (Redrawn from Shakarad and Gadagkar, 1995.)

per capita brood content of colonies on the number of foundresses and found, as expected from the arguments of Wenzel and Pickering (1991), a significant negative slope (Figure 11.4). The reduction in coefficient of variation with foundress number could, however, be due either to a decrease in the standard deviation or an increase in the mean. But we already saw in Figure 11.3, panel B, that the mean is nearly constant. Even though the mean per capita productivity does not increase with foundress number, the higher predictability of achieving the mean per capita productivity does appear to be an advantage of multiple-foundress colonies (Shakarad and Gadagkar, 1995).

11.3.　Do Wasps Choose Their Nesting Strategies?

So far I have demonstrated two advantages of multiple-foundress associations over single-foundress associations. One is that single-foundress associations have a higher rate of nest usurpation than multiple-foundress nests, and the other is that multiple-foundress nests achieve their mean per capita productivity with greater certainty than single-foundress nests do.

Although these advantages must contribute to the ecological predisposition of *R. marginata* for the evolution of eusociality, we cannot easily assess their influence on the relative magnitudes of β and b (in inequality
8.5). Yet the finding of constant mean per capita productivity at different
group sizes appears, at first sight, to have a simple consequence, namely,
$\beta = b$. This makes the altruism on the part of the subordinate cofoundress
even more paradoxical because it appears that a cofoundress gains no particular advantage from joining another individual's nest rather than initiating her own. But this argument makes the assumption that all wasps are
equal. More specifically, it makes the assumption that the subordinate
cofoundresses would have achieved the same productivity as the solitary
foundresses do had they themselves chosen the solitary nesting strategy.

But imagine for a moment that a wasp who chose to become a subordinate cofoundress is one who fares very poorly as a solitary foundress. If
this is true, to see if a cofoundress is doing the right thing (in terms of
inclusive fitness), one must compare her productivity as a subordinate
cofoundress in a multiple-foundress nest with the productivity she might
have achieved on her own had she chosen to become a solitary foundress,
rather than with the productivity of those individuals who naturally opt
for the solitary founding strategy. But how does one make such a comparison? Mallikarjun Shakarad and I decided that we would attempt to force
wasps who had chosen to be subordinate cofoundresses to nest alone
(Shakarad and Gadagkar, 1997). We used 77 naturally initiated pre-emergence nests that were in the very early egg stage of the nesting cycle. Of
these, 28 were natural single-foundress nests and we did not manipulate them in any way. The remaining 49 were multiple-foundress nests
and of these we randomly chose 26 nests to force one of the cofoundresses
to nest alone—we removed the queen and all but one randomly chosen
cofoundress on the day the nest was located. To control for the disturbance
caused by such a manipulation, the remaining 23 multiple-foundress nests
were manipulated such that all cofoundresses were removed and the queen
was forced to nest alone. The number of cells and eggs present at the time
of manipulation was not significantly different between the nests subjected
to different manipulations (t-test, $p > 0.05$, df $= 47$ to 52). Each of the 77
nests, now all effectively single-foundress nests, was monitored until the
eclosion of the first adult offspring or until the lone wasp disappeared.

The proportions of successful nests ranged from 0.15 to 0.22 and did not differ significantly between nests with the three categories of wasps—natural solitary foundresses, cofoundresses forced to nest alone, and queens forced to nest alone (Figure 11.5, panel A). This result is not surprising because, as shown above, single-foundress nests fail primarily because of loss of foundresses. Next Shakarad and I looked at the productivity (defined, as before, as the number of eggs + larvae + pupae at the time of eclosion of the first adult offspring) of the successful nests in each category. It is here that the difference between different kinds of nests became evident. The productivity of queens forced to nest alone was not significantly different from that of naturally occurring solitary foundresses. However, the productivity of cofoundresses forced to nest alone was significantly lower than that of queens forced to nest alone and that of natural solitary foundresses (Figure 11.5, panel B). We confirmed that the lower productivity of cofoundresses forced to nest alone was not compensated by any faster brood development: the time taken from the hatching of the first egg to the eclosion of the first adult offspring was the same for all three categories of nests (Figure 11.5, panel C).

Why did cofoundresses forced to nest alone have lower productivity than queens forced to nest alone or natural solitary foundresses? Perhaps subordinate cofoundresses forced to nest alone were incapable of laying as many eggs as queens forced to nest alone or natural solitary foundresses. Somewhat to our surprise, this hypothesis was not supported. Cofoundresses forced to nest alone, queens forced to nest alone, and natural solitary foundresses all showed similar rates of egg laying during the period of observation (Figure 11.5, panel D). Laboratory experiments also show no evidence of differential egg-laying abilities. But first let me describe the relevant laboratory experiments. The 62 cofoundresses and 21 queens removed from the field colonies in order to force cofoundresses and queens to nest alone were isolated in individual laboratory cages (plastic boxes used for the nestmate discrimination experiments described in section 10.2) and provided with an *ad libitum* supply of food and building material. We know that under these conditions, wasps can initiate single-foundress nests and successfully produce offspring (see Chapter 12). By the same criterion as in the field, about 40% of the isolated cofoundresses and about 43% of the isolated queens were successful in the laboratory, that is, they produced at least one adult offspring (Figure 11.5, panel E). Success rate in the laboratory was significantly higher than it was in the field (compare

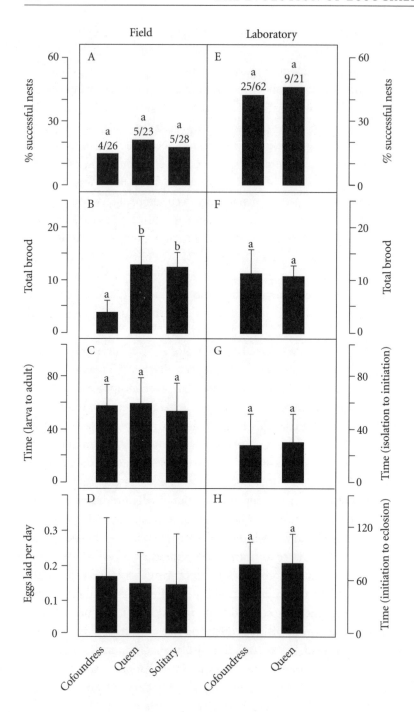

Figure 11.5 (facing page).   Left panel: Comparison between randomly chosen cofoundresses forced to nest alone, queens of multiple-foundress nests forced to nest alone, and naturally occurring solitary foundresses. Right panel: Comparison between cofoundresses and queens from naturally occurring multiple-foundress nests isolated individually in laboratory cages. In both panels, mean values and, where appropriate, standard deviation values are shown. Except in the case of the total brood in the field experiment ($p < 0.01$), all comparisons are insignificant ($p > 0.05$) by the Kruskal-Wallis test. Within each box, bars carrying different letters are significantly different from each other by the Mann-Whitney U test ($p < 0.05$). Brood produced by the cofoundresses forced to nest alone in the field was also significantly less than the brood produced by isolated cofoundresses and queens in the laboratory cages (Mann-Whitney U test, $p < 0.05$). Brood produced by the isolated queens and the solitary foundresses in the field were not significantly different from the brood produced by the cofoundresses and queens isolated in laboratory cages (Mann-Whitney U test, $p > 0.05$). All nests were therefore monitored until the eclosion of the first adult offspring or until they were abandoned, whichever was earlier. Identical results were obtained when cells, eggs, larvae, and pupae were analyzed separately. For brevity, only results with the total brood are shown. Since the time of hatching of the first egg was known more precisely than the time of laying of the first egg under field conditions, brood developmental time was measured as the time between the production of the first larva and the eclosion of the first adult offspring. (Redrawn from Shakarad and Gadagkar, 1997.)

---

panels A and E in Figure 11.5). This was not surprising, if single-foundress nests failed in the field because of loss of foundresses during the risky task of foraging. But contrary to the results in the field there was no difference between the total brood production of isolated queens and isolated cofoundresses in the laboratory (Figure 11.5, panel F). In the laboratory, the isolated queen and isolated cofoundresses also did not differ with respect to the time required for nest construction and eclosion of the first adult offspring (Figure 11.5, panels G and H). In the field queens forced to nest alone and natural solitary foundresses had productivities that were indistinguishable from the productivities of isolated queens and isolated cofoundresses in the laboratory. Only the cofoundresses forced to nest alone in the field had significantly lower productivity when compared to all the others (compare panels B and F in Figure 11.5). Their similar rates of egg laying in the field and identical productivities in the laboratory suggest that cofoundresses are as fertile as queens and solitary foundresses. The significantly lower productivity of cofoundresses forced to nest alone in the field thus seems to arise not from their inability to lay sufficient eggs but perhaps from their inability to forage for as large a quantity of brood and yet have the same probability of survival (which they do) as queens forced to nest alone and natural solitary foundresses.

Thus it appears that *R. marginata* females choose their nesting strategies on the basis of their brood-rearing abilities: the relatively "inferior" subordinate cofoundresses may prefer not to initiate their own nests, as do the "superior" solitary foundresses and queens of multiple-foundress nests. But note that when they find themselves in a multiple-foundress nest, subordinate cofoundresses appear to be able to rear as much brood as can solitary foundresses and queens forced to nest alone. This conclusion arises from the observation that per capita productivity does not change with group size. That the same individuals manage to rear more brood when working in the company of other wasps in multiple-foundress nests seems paradoxical at first glance (Shakarad and Gadagkar, 1997). But there is a possible explanation for the apparently high brood-rearing ability of the same individuals when they find themselves in multiple-foundress nests compared to their brood-rearing ability when forced to nest alone. Recall that when failed nests were excluded in the analysis, there was in fact a significant reduction in per capita productivity with increasing foundress number (see section 11.2), and it was only when failed nests were included in the analysis that there was a nearly constant per capita productivity. Thus the higher probability of survival of multiple-foundress nests appears to compensate for the inferior brood-rearing abilities of subordinate cofoundresses, so that, when averaged over all nests, the per capita productivity remains constant with foundress number. In other words, subordinate cofoundresses forced to nest alone have inferior brood-rearing abilities because their nests fail more often. When the same individuals find themselves in multiple-foundress nests, they manage to rear more brood because their nest do not fail as often.

I believe that the study by Shakarad and Gadagkar (1997) demonstrates that, when forced to nest alone, subordinate cofoundresses have inferior brood-rearing abilities and that their productivity as workers must therefore be compared with their productivities as solitary foundresses and not compared with the productivity of natural solitary foundresses. Before we go on to make such a comparison, let me mention two alternative but unlikely interpretations of our results. We have concluded that, when forced to nest alone, subordinate cofoundresses are inherently incapable of raising as much brood as queens forced to nest alone or natural solitary foundresses. One alternative possibility is that they are unwilling to do so (by "unwilling" of course we mean "will not be selected to, even if they are capable of") rather than incapable of doing so—the unwillingness potentially arises from the expected lower genetic relatedness between the cofoundresses forced to nest alone and the brood, which may belong to the queen that we removed. For several reasons, this interpretation is unlikely to be correct. Not all the brood reared by cofoundresses forced to nest alone belonged to the removed queen; often these wasps destroyed the eggs of the removed queen and laid their own. For the four successful cofoundresses forced to nest alone, 7.1–66.7% (mean ± S.D. = 37.4 ± 33.8%) of the brood they reared belonged to them rather than to the removed queen. Moreover, we have good reason to believe that cofoundresses may not after all be unwilling to rear brood that does not belong to them. In the pre-emergence phase, foundresses move between nests and thus nests may receive joiners from up to four different source nests (section 9.3). Polyandry and serial polygyny lower intracolony genetic relatedness to such an extent that workers in multiple-foundress nests rear brood rather distantly related to them (section 9.2). If wasps are unwilling to rear brood not closely related to them, they should be equally unwilling to do so in multiple-foundress nests, where the brood is distantly related to them. We are only comparing the relative difference between the brood reared by subordinate cofoundresses in the situation where they are forced to nest alone and in the situation where they act as workers in multiple-female nests, perhaps with some level of unwillingness in both situations.

The second potential alternate interpretation of our results is that the subordinate cofoundresses had already been suppressed by their queens into "inferiority" before we removed the queens. This also seems unlikely. We removed the queen very early in the nesting cycle. There is a prevailing impression that all wasps in primitively eusocial species are equivalent at eclosion and that all differences between them are of post-imaginal origin. But we have good evidence, in another context, that there can be strong pre-imaginal effects even in a species such as *R. marginata* (see Chapter 12). Besides, why would queens suppress their workers and make them inferior at brood rearing? The most likely explanation for our results is therefore that subordinate cofoundresses forced to nest alone are incapable of rearing as much brood as queens forced to nest alone or natural solitary foundresses. This constitutes a substantial predisposition to the evolution of eusociality. It is reasonable to think that it arises owing to ecological factors because of the argument that subordinate cofoundresses forced to nest solitarily lower their brood-rearing efficiency in order to have the same probability of surviving as their "superior" counterparts that found their own nests (natural solitary foundresses and queens of multiple-foundress nests) rather than join nests founded by other wasps. Thus we can approximate b (the productivity of cofoundresses forced to nest alone) as 4.2 and β (the per capita productivity of multiple-foundress nests) as 12.3 and infer that β is substantially greater than b.

11.4.   Compensation for Being a Worker

We saw above that a subordinate cofoundress forced to nest alone had a productivity of about 4.2 (total brood at the time of eclosion of the first adult wasp), while a natural solitary foundress had a corresponding productivity of about 12.3 items of brood (Figure 11.5, panel B). We also saw in section 11.2 that the mean per capita productivity remained almost constant even when the number of foundresses varied from 1 to 22. We may approximately assume from this that while a solitary female has a productivity of 12.3, each additional cofoundress contributes another 12.3 items of brood to the colony's productivity. Thus a subordinate cofoundress gains by a factor of $12.3/4.2 = 2.9$ by choosing not to found her own nest but to work in another queen's nest. A wasp that can have a productivity of only 4.2 as a solitary foundress can thus trade this advantage of a 2.9-fold increase in her productivity, by rearing brood less related to her than her own offspring would be. She would break even by working for brood related to her by 2.9 times less than her own brood would be. Given that her own brood would be related to her by 0.5 (in an outbred population), she should thus be willing to rear brood related to her by $0.5/2.9 = 0.17$. We saw in sections 9.1 and 9.2 that polyandry and serial polygyny reduce worker-brood genetic relatedness to values ranging from 0.20 to 0.38 (Table 9.4). Thus the higher efficiency with which subordinate cofoundresses rear brood in multiple-foundress nests, as compared with what they can achieve on their own as solitary foundresses, appears to more than compensate for the cost of not producing their own offspring and instead rearing brood of low genetic relatedness in multiple-foundress nests. The higher efficiency with which workers seem to be able to rear brood in multiple-foundress nests (compared to their efficiency when they are forced to be solitary) may thus be thought of as adequate compensation for functioning as sterile workers.

To sum up, we saw three positive results in this chapter: (1) Multiple-foundress nests have significantly higher rates of usurpation than single-foundress nests. (2) Multiple-foundress nests achieve their expected mean per capita productivity with greater certainty than do single-foundress nests, (3) Wasps that choose to be subordinate cofoundresses in multiple-foundress nests have substantially inferior brood-rearing abilities as solitary foundresses. It is not easy to translate the first two results into quantitative inclusive fitness consequences. But the third result is clear evidence of ecological predisposition, making $\beta > b$ and thus tilting the inclusive fitness inequality in favor of workers. In Chapter 9, in an attempt to explore possible genetic predisposition for the evolution of eusociality, we found instead that there was a predisposition to be solitary: worker-brood relatedness values were below the relatedness expected between a solitary foundress and her offspring. This chapter shows that the ecological predisposition to eusociality compensates for the genetic predisposition to solitary life (see Chapter 10). I refer the reader to Reeve and Nonacs (1992, 1997), Nonacs and Reeve (1993, 1995), Bull and Scharwz (1996, 1997), and Field et al. (1998, 1999) for ways of investigating the ecology of cooperation in primitively eusocial wasps and bees not pursued here.

12 PHYSIOLOGICAL PREDISPOSITION

꒐N THIS chapter I will ask the same question that I asked in Chapter 8: Is $\beta > b$? But now I will focus on physiological reasons for the difference between the expected productivity of an individual that adopts the single-foundress strategy and the productivity expected if the same individual adopts a worker strategy. The motivation for the experiments described in this chapter also came from a realization that one should not compare the productivity of individuals that naturally choose the solitary nesting strategy with the productivity of other individuals that naturally choose the worker strategy, because these two kinds of individuals may differ from each other in their reproductive potential. The question then is: Can we demonstrate that there are differences in the potential productivities of different wasps? The discussion in Chapter 11 showed that there are such differences. But in that context my speculation was that subordinate cofoundresses forced to nest alone were lowering their efforts at brood rearing in order to have the same probability of being alive as the natural solitary foundresses. But perhaps there are more fundamental differences unrelated to survival probabilities. In the context of primitively eusocial species such as *R. marginata*, this possibility needs to be explicitly investigated because it may well be that all individuals are equivalent at eclosion and that only post-imaginal processes, such as suppression of some individuals by others, lead to inequalities among the adult wasps. In the highly eusocial species, we know that pre-imaginal processes leading to differences in nutritional and hormonal status of different larvae result in adults with very different potentials—indeed, caste determination is often completed during the larval stage in many highly eusocial species (Wilson, 1971). To achieve this requires specific, highly evolved behavioral and physiological processes, the production and differential feeding of royal jelly in honey bees being an excellent example (Winston, 1987). It may be argued, however, that primitively eusocial wasps have not developed that level of sophistication. The absence of morphological differentiation between queens and workers in primitively eusocial species strongly suggests that caste determination takes place in the adult stage.

12.1.   Are All Eclosing Females Capable of Laying Eggs?

Thus my students and I set out to test the simple null hypothesis that all eclosing female *R. marginata* are capable of laying eggs (Gadagkar et al., 1988). To do so we isolated wasps immediately upon their eclosion and prevented them from being manipulated or suppressed in any way by their nestmates, and provided them with the wherewithal for nest building and egg laying. Twenty-two naturally occurring nests were collected from several localities in and around Bangalore. An attempt was made to collect nests of different sizes with the hope that they would represent different phases in the colony cycle. Nests were also collected at different times of the year so that any seasonal effects on egg-laying potential could also be detected.

Upon collection, the nests were brought to the laboratory, cleared of adults, larvae, and eggs and maintained in ventilated plastic jars. Eclosing female wasps were isolated in ventilated plastic jars, provided with a piece of soft wood as a source of building material, and given an *ad libitum* diet of final instar *Corcyra cephalonica* larvae, honey, and water. Each wasp was observed daily for signs of nest building and egg laying. Wasps were fixed in Dietrich's solution on the day they laid their first egg or on the day of their natural death. The number of larvae consumed by the wasps was estimated as the difference between the number provided and the number remaining uneaten. Honey consumption was not estimated. One hundred and ninety-seven female wasps were thus isolated and tested for their ability to lay eggs. Of these, 97 individuals successfully built nests and laid at least one egg. We referred to them as egg layers. The remaining 100 wasps died without doing so and we referred to them as non–egg layers.

12.2.   Correlates of Egg Layers and Non–Egg Layers

The short answer to the question raised above is that not all eclosing female wasps are capable of laying eggs (Table 12.1). Of the 97 individuals
that built nests and laid eggs, 95 built apparently normal nests with stalks
and only 2 built nests without stalks. Another telling observation was that
none of the 100 wasps that died without laying eggs attempted to initiate a
nest or even to build a stalk. The correlation between nest building and egg
laying was thus perfect. Why did the non–egg layers die without laying
eggs? Because of their inability to build a nest or their inability to lay eggs?
The latter is much more likely because workers that die without laying eggs
in natural colonies routinely participate in nest building. Why then were
the non–egg layers incapable of laying eggs? Unfortunately, we could not
determine the state of the ovaries of the non–egg layers because we had to
wait until they died a natural death and they were therefore quite dried up
by the time they were discovered dead and thus could not be properly dissected. But we do know that the reason for the non–egg layers dying without building a nest and laying eggs was not a shorter lifespan than the egg
layers had. Non–egg layers survived on average significantly longer than
the time taken by the egg layers to build nests and lay their first eggs (Table
12.2). Thus there appears to be genuine heterogeneity at eclosion in the
egg-laying abilities of the wasps. In spite of the fact that all of them were
treated identically and none was allowed to interact in any way with her
conspecifics, only about 50% of them built nests and laid eggs. It is therefore of obvious interest to understand what makes some individuals egg

Table 12.1.   Evidence for pre-imaginal caste bias. Numbers (and percentages) of egg
layers and non–egg layers among eclosing wasps that were isolated from
all adult conspecifics in three independent experiments.(Data from
Gadagkar et al., 1988, 1990, 1991b.)

Expt. no.	No. of wasps tested	No. of egg layers	No. of non–egg layers
1	197	97 (49%)	100 (51%)
2	102	53 (52%)	49 (48%)
3	87	47 (54%)	40 (46%)
Total	386	197 (51%)	189 (49%)

Table 12.2   Characteristics of egg layers and non–egg layers (all values given are
mean ± standard deviations). (Data from Gadagkar et al. 1988)

	Egg layers (sample size = 97)	Non–egg layers (sample size = 100)
Days before egg laying or life span[a]	62 ± 38	87 ± 66
Feeding rate[b]	0.24 ± 0.27	0.16 ± 0.12
Interocular distance[c]	0.31 ± 0.02	0.32 ± 0.02
Ocello-ocular distance	0.52 ± 0.03	0.52 ± 0.03
Head width	3.20 ± 0.12	3.19 ± 0.16
Head length	2.82 ± 0.13	2.82 ± 0.15
Mesoscutellum width	2.25 ± 0.15	2.26 ± 0.16
Mesoscutellum length	2.34 ± 0.18	2.34 ± 0.18
Wing length	9.96 ± 0.67	10.09 ± 0.73

a. Number of days taken to lay the first egg for egg layers and number of days alive for non–egg layers.

b. Number of final instar *Corcyra cephalonica* larvae eaten per day.

c. This and all subsequent measurements are in mm.

layers and others non–egg layers. The time of the year when a wasp eclosed was not one of the reasons; the numbers of egg layers and non–egg layers among wasps eclosing at different times of the year were not significantly different from each other.

To examine the possible influence of other variables, my students and I used logistic regression analysis, in collaboration with Anil P. Gore and Ashok Shanubhogue (Cox and Snell, 1989; Shanubhogue and Gore, 1987). We did this because each wasp in our experiment either lays eggs or she does not, and the dependent variable is therefore a binary one, which of course may be influenced by a variety of independent variables. In the first model the following properties of the nest from which a wasp eclosed were considered independent variables that might potentially be correlated with an individual's probability of laying eggs: numbers of eggs, larvae, pupae, parasitized cells, empty cells, and males and females. In the second model the following variables associated with the individual wasps themselves were considered independent variables that might potentially be correlated with the probability of laying eggs: rate of food intake during adult life, interocular distance, head width, head length, mesoscutellum width, mesoscutellum length, and wing length, all but the first being indices of body size. The mean and standard deviation of all independent variables listed, computed separately for egg layers and non–egg layers, are given in Table 12.2. Independent variables were modeled to influence the probability of egg laying.

In the first model, where nest properties were tested for their potential correlation with the probability of wasps becoming egg layers, only one regression coefficient was significantly different from zero: the one associated with the number of empty cells (Table 12.3). In the second model, where we tested variables associated with the wasps themselves for their potential correlation with the probability of becoming egg layers, the regression coefficient associated with adult feeding rate was highly significantly different from zero. (The coefficient associated with ocello-

Table 12.3. Logistic regression analysis of determinants of probability of egg laying by females of *R. marginata*. (Data from Gadagkar et al., 1988.)

Variable	Estimated coefficient(β)	S.E.	Z
Model 1: nest properties as determinants of the probability of egg laying by eclosing females			
Intercept	−0.1854	0.3335	−0.5558
No. of eggs	−0.0072	0.0155	−0.4641
No. of larvae	0.0060	0.0150	0.4020
No. of pupae	0.0101	0.0225	0.4472
No. of parasitized cells	−0.1536	0.1489	−1.0315
No. of empty cells	0.0519	0.0218	2.3845[a]
No. of males	0.1753	0.1771	0.9896
No. of females	−0.0111	0.0194	−0.5700
Model 2: feeding rate and body size as determinants of the probability of egg laying			
Intercept	1.5631	4.1547	0.3762
Feeding rate	3.4993	1.2871	2.7188[b]
Interocular distance	−6.4072	7.1170	−0.9003
Ocello-ocular distance	−11.7276	5.9464	−1.9722[c]
Head width	1.2445	1.4517	0.8573
Head length	1.3271	1.4336	0.9257
Mesoscutellum width	−1.0160	1.5312	−0.6635
Mesoscutellum length	1.1687	1.3327	0.8770
Wing length	−0.2291	0.3216	−0.7125

a. $p < 0.02$.
b. $p < 0.007$.
c. $p < 0.05$.

ocular distance was also mildly significant, but see below). There is good reason to believe that the logistic regression models tested here are not unreasonable. The expected numbers of wasps in each probability class calculated from the models are indistinguishable from the corresponding observed numbers. We conclude therefore that wasps eclosing on nests with large numbers of empty cells and those that feed well during their early adult life have a high probability of becoming egg layers. Conversely, wasps eclosing on nests with small numbers of empty cells and those that feed poorly during their early adult life have a low probability of becoming egg layers.

In the beginning of this chapter I referred to behavioral and physiological processes that lead to pre-imaginal caste determination in highly eusocial species and admitted the possibility that a primitively eusocial species such as R. marginata may not have attained such a level of sophistication. But the finding that nearly 50% of eclosing wasps die without laying eggs, under conditions that permit the remaining 50% to build nests and lay eggs, suggests that we must be prepared for the possibility that similar behavioral and physiological processes might exist in R. marginata also. My colleagues and I have therefore postulated a phenomenon in R. marginata that we call pre-imaginal caste biasing. Using an additional set of 102 wasps from 17 independent nests and employing additional statistical methods we have since obtained a complete reconfirmation of the results reported above (Gadagkar et al. 1990).

12.3. The Mechanism of Pre-Imaginal Biasing of Caste

The influence of adult feeding rate is easy to understand. Wasps that feed well become egg layers and those that feed poorly do not become egg layers. Egg laying requires extra resources, which are obtained by the egg layers through better feeding rates. But which is the cause and which is the effect? Does the extra feeding cause egg laying or does the egg laying cause extra feeding? Such cause-effect relationships are usually difficult to tease apart, but there are two lines of evidence which suggest that high rates of feeding may be the cause of the high probability of laying eggs. First, it must be emphasized that since the experiment was terminated as soon as wasps laid their first eggs, our estimation of feeding rates pertains only to the period before the laying of the first egg. It must also be emphasized

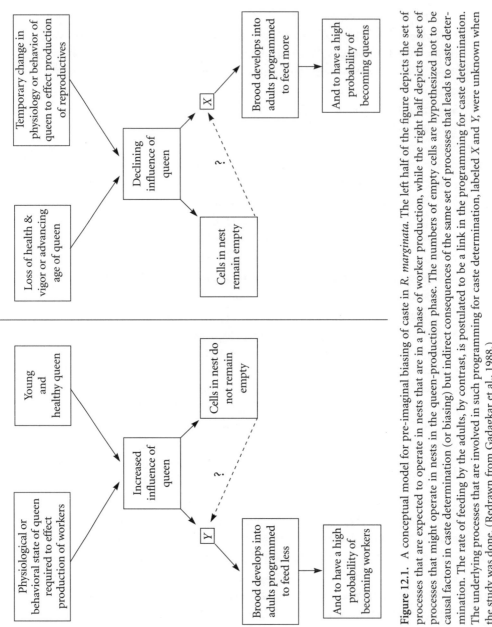

Figure 12.1. A conceptual model for pre-imaginal biasing of caste in *R. marginata*. The left half of the figure depicts the set of processes that are expected to operate in nests that are in a phase of worker production, while the right half depicts the set of processes that might operate in nests in the queen-production phase. The numbers of empty cells are hypothesized not to be causal factors in caste determination (or biasing) but indirect consequences of the same set of processes that leads to caste determination. The rate of feeding by the adults, by contrast, is postulated to be a link in the programming for caste determination. The underlying processes that are involved in such programming for caste determination, labeled X and Y, were unknown when the study was done. (Redrawn from Gadagkar et al., 1988.)

that all wasps—egg layers and non–egg layers—were given an *ad libitum* supply of food. Some chose to eat less and others chose to eat more. That high rates of feeding result in egg laying is therefore a reasonable hypothesis. Perhaps we may conclude that some wasps feel more hungry, eat well, and become egg layers while others don't feel so hungry, eat less, and become non–egg layers.

The correlation of a high number of empty cells in the nest from which egg layers are more likely to eclose is not so easy to interpret. But taking a cue from highly eusocial species and speculating that the queen may be involved in some way in biasing the caste of her brood, we constructed a conceptual model of caste biasing (Figure 12.1) (Gadagkar et al. 1988). In a normal, healthy nest, cells almost never remain empty. Queens usually lay eggs in cells vacated by pupae within hours of their becoming vacant. When new cells are being constructed, an egg is usually laid while the new cell is barely large enough to hold an egg; proper shaping and elongation of the cell take place only after the cell has an egg in it. It is therefore reasonable to assume that the accumulation of empty cells in a nest, especially more than the one or two cells that may conceivably be left empty by chance alone, is a strong indicator of the queen's declining influence. The queen's influence could decline owing to old age or poor health. But if *R. marginata* has indeed attained a certain level of sophistication in its mechanism of caste determination, then the queen's influence may also decline temporarily and in a programmed fashion during a certain phase of the colony cycle that might be devoted to the production of future reproductives. Whatever the cause of the queen's declining influence, our model postulates two consequences. One is the accumulation of empty cells and the other is the production of adult wasps that are programmed to feed more and have a high probability of becoming egg layers. Conversely, there would be other phases of the colony cycle that might be devoted to the production of a large worker force. During this phase we expect the queen to exert a strong influence, leading to few or no empty cells and to the production of adult wasps programmed to feed less and have a low probability of becoming egg layers.

When we suggested this model in 1988, we had no idea how the queen's weak influence could lead to the production of egg layers and the queen's strong influence could lead to the production of non–egg layers. We therefore postulated two hypothetical processes *X* and *Y* resulting from the

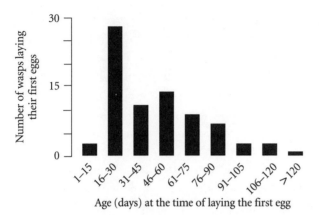

Age (days) at the time of laying the first egg

Figure 12.2. Frequency distribution of wasps in different age classes at the time of laying their first eggs. Two hundred ninety-nine freshly eclosed female *R. marginata* were isolated in individual laboratory cages. Of these, 147 died without laying eggs. Of the remaining 152 that built a nest and laid eggs, 73 were "caught" in winter, which is unfavorable for the initiation of egg laying by isolated females in the laboratory. The remaining 79 individuals that initiated egg laying before getting caught in winter were used to construct this frequency distribution, which has a mean ± S.D. of 48 ± 31 days and a range of 14–191 days. (Redrawn from Gadagkar et al., 1990.)

queen's weak influence and strong influence respectively, and hoped eventually to unravel X and Y. Today, I think we know what X and Y probably are, but before I get to that I should mention another intriguing result of the 1988 study. Among the egg layers there is considerable variation in the time taken to lay eggs. About 25% of the wasps laid their first eggs within the first 4 weeks of their adult life, another 25% did so between 4 and 8 weeks after eclosion, a third quartile laid their first eggs between 8 and 12 weeks of their adult life, and the rest took even longer (in one case, up to 217 days) (Figure 12.2). In that study we failed to detect any correlates of this variability in time required to initiate egg laying, and it was some years before I began to appreciate its profound significance (see Chapter 13) for the evolution of eusociality. For another kind of example of how the presence or absence of brood can influence caste determination, see Solis and Strassmann (1990).

12.4. The Role of Larval Nutrition

At about this time, I came increasingly under the influence of the persuasive arguments of James H. Hunt that his data, that of many others, and in-

deed my own can be interpreted to infer a role for larval nutrition in caste
determination and hence in the evolution of eusociality (see Hunt, 1991,
and references therein; see also O'Donnell, 1998d). The next course of action was obvious: another experiment made even more tedious by the requirement that observations be made on the nests from which the experimental wasps eclose so that information could be obtained about the
feeding rates of the experimental wasps when they were larvae.

Seetha Bhagavan, K. Chandrashekara, and C. Vinutha readily participated in this study (Gadagkar et al., 1991b). We used six post-emergence
nests of R. marginata, made 4–6 days of behavioral observations on each
of them, using methods similar to the ones described in Chapter 5, and
then collected these nests, cleared them of adult wasps, eggs, and larvae,
and isolated the eclosing female wasps in the manner described above. We
managed to obtain 87 wasps and, once again, close to half of them (47) became egg layers, while the remaining 40 became non–egg layers. From the
data obtained during behavioral obervations, we computed for each nest
the number of times an average larva was fed per hour and used this value
as the independent variable in the logistic regression model. The result was
clear: the logistic regression coefficient associated with larval feeding rate
was significant and positive (Table 12.4). Thus individuals that are well fed
as larvae appear to grow into adults that have a high probability of becoming egg layers. But individuals that are poorly fed as larvae seem to grow
into adults that aren't hungry, eat less as adults also, and become non–egg
layers.

In this study we were also successful in discovering a correlate of the
time taken by adults to start laying eggs. There was a weak but statistically
significant negative correlation between the time taken by an adult wasp to
attain reproductive maturity and start laying eggs and the rate at which the

Table 12.4. Logistic regression analysis of larval nutrition as a determinant of
the probability of egg laying by eclosing females of R. marginata.
(Data from Gadagkar et al., 1991b.)

Variables	Estimated Coefficient(β)	S.E.	Test statistic	p
Intercept	−0.5612	0.3963	−1.4160	0.1556
No. of times an average larva is fed per hour	3.5462	1.6525	2.1459	0.0316

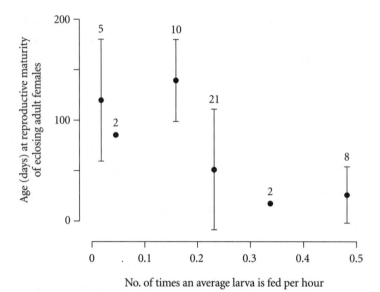

Figure 12.3. The weak but statistically significant negative correlation between time taken by an animal to start laying eggs (= age at reproductive maturity) (shown as mean ± S.D.) and the rate at which larvae are fed in the nest from which they eclose. (Pearson product moment correlation, $r = -0.42$; $df = 45$; $p < 0.01$.) The numbers on top of each vertical line are sample sizes. (Redrawn from Gadagkar et al., 1991b.)

larvae are fed in the nest from which they eclose (Figure 12.3). Larval nutrition thus seems to influence both forms of pre-imaginal caste bias: the differentiation into egg layers and non–egg layers and the further differentiation of the egg layers into early egg layers and late egg layers.

Thus we deciphered the nature of the postulated processes X and Y in our model for pre-imaginal caste bias (Figure 12.1); $X =$ high rate of larval feeding and $Y =$ low rate of larval feeding. Now I can tell the complete story: —individuals well fed as larvae develop into adults that feel hungry, eat well, and become egg layers, and individuals poorly fed as larvae develop into adults that do not feel hungry, eat less, and become non–egg layers. But what is the correlation between the queen's influence and larval feeding rate, and, similarly, what is the correlation, if any, between the number of empty cells and larval feeding rate? An attractive hypothesis can potentially link all these variables (Gadagkar et al., 1991b).

Imagine the following reasonable sequence of events in the life of a colony. New nests begin with relatively young and healthy queens that may leave few or no empty cells. Such new nests may also have a modest work force that will be busy expanding the nest to match the queen's egg-laying rate. The food that is brought to the nest at this time may therefore be limited because few workers are foraging. Besides, the food brought has to be shared with the relatively large number of larvae. Thus there will be relatively less nourishment for the larvae, which will program them to become non–egg layers. In the course of time the colony acquires a large work force and also expends less effort in expanding the nest. In addition, the queen, either on account of old age and weakness or for the purpose of producing reproductives, may slow down her rate of egg laying, causing an accumulation of empty cells and soon a reduction in the number of larvae to be fed, relative to the amount of food available. Larvae at this time may therefore be better nourished and hence be programmed to develop into egg layers. Such a scenario is consistent with what is known in other eusocial insects about the colony cycle (Jeanne, 1972), variation in larval and adult nutrition through colony development (West-Eberhard, 1969; Wilson, 1971; Wheeler, 1986; Hunt, 1988), and the postulated role of nutrition in social evolution (Pardi and Marino Piccioli, 1970, 1981; Hunt, 1984, 1990, 1991; Hunt et al., 1987, 1996; Rossi and Hunt, 1988). It must be cautioned however that we do not yet have direct evidence for this appealing but nevertheless hypothetical scenario.

I should touch upon four additional points before concluding this chapter.

1. One might have expected the egg layers to be larger in body size than the non–egg layers, but this does not seem to be so. We have tested seven indices of body size, and none of them appears to show any strong correlation with the probability of an individual becoming an egg layer. The regression coefficient associated with one of them, the ocello-ocular distance, was significant (though at a lower level than the other significant variables) in the first experiment (Gadagkar et al., 1988) but not in the second experiment and not when data from both experiments were combined (Gadagkar et al., 1990). It is necessary therefore to conclude that there is really no demonstrable influence of body size on the probability of wasps becoming egg layers. Larval nutrition has a remarkable influence on the reproductive potential of adult wasps, but apparently it acts by some route other than making the egg layers larger in size. Unfortunately we did

not have data on such variables as size of fatbodies and hormone titers in this study.

2. That we used unmated females to test whether they are capable of initiating nests and laying eggs may, at first sight, be considered a weakness of our experimental design. *R. marginata* females do not mate readily in the laboratory and hence it is almost impossible to use only mated females for experiments such as we have done. In most social insect species, unmated females are potentially capable of developing their ovaries and laying unfertilized, haploid eggs that develop into males (Wilson, 1971). Thus why some females are capable of at least producing male offspring while others are incapable of doing even that much is also a significant issue. More important, as discussed in Chapter 4, mating is unnecessary for a female *R. marginata* to develop her ovaries, suppress egg laying by all her nestmates, and completely monopolize egg laying in her colony. Moreover, an unmated female can first establish herself as the sole egg layer of a colony, lay some haploid eggs, and then mate later. I therefore do not consider the use of unmated females a weakness of our experimental design.

3. We conducted our studies under artificial laboratory conditions, but all conditions were identical for all the wasps, and also those conditions may not be particularly unnatural for the wasps. When all the members of a colony are transferred to a plastic box of the kind used in these experiments, and indeed even to much smaller boxes without proper ventilation, a new nest, with the original queen in control, is often established in less than a day (see Chapter 3).

4. What is the significance of egg layers and non–egg layers in our experimental set-up, in relation to queens and workers in natural colonies? I do not directly equate the egg layers in our experiments with queens in nature or the non–egg layers in our experiments with workers in nature. The proportion of queens and workers is by no means equal in natural populations: the proportion of queens is rather small compared to that of workers, about 35% in the founding population of the Shakarad and Gadagkar (1995) study. Instead, I expect that the egg layers in our experiment will have a high probability of capitalizing on opportunities to become queens. Similarly, I expect the non–egg layers in our experiment to have a low probability of capitalizing on opportunities to become queens. In other words, I do not expect all the egg layers in our experiment to become queens or all the non–egg layers in our experiment to become workers un-

der natural conditions. Instead, I expect that most queens in nature come from among the individuals we classified as egg layers and most of the individuals we classified as non egg–layers end up as workers in nature.

In conclusion, it is clear that there is considerable physiological predisposition in *R. marginata* for the productivity of some individuals to be high when they behave as workers (their β would be high because they don't have to lay their own eggs) compared to the productivity these same individuals could achieve if they acted as solitary foundresses (their b would be low because they have to lay their own eggs) (but see Field and Foster, 1999, for evidence against the role of such "subfertility" in a stenogastrine wasp). However, we are unfortunately not in a position to quantify the effect of such a physiological predisposition. We can assert that pre-imaginal caste bias is bound to make $\beta > b$, but we cannot yet say by how much.

13 DEMOGRAPHIC PREDISPOSITION

*I*s *R. MARGINATA* demographically predisposed to the evolution of eusociality? In inequality 8.5, is $\sigma > s$, so that even if $\rho = r$ and $\beta = b$, the inclusive fitness of workers can be greater than that of solitary foundresses? Recall that σ and s are the demographic correction factors for workers and solitary foundresses, respectively. I argued in Chapter 8 that in order to compare the inclusive fitness of workers and solitary foundresses, we should devalue the intrinsic productivity factors of workers and solitary foundresses (β and b, respectively) not only by the genetic relatedness between the adult wasps and the brood they rear (ρ and r, respectively) but also by the respective demographic correction factors (σ and s, respectively). In this chapter I will illustrate how one can estimate the appropriate demographic correction factors.

Demography has generally received little attention in either theoretical or empirical studies of the evolution of eusociality (but see Queller, 1989, 1994, 1996; Strassmann and Queller, 1989; Gadagkar, 1990a, 1991b, 1996a; Nonacs, 1991; and Field et al., 2000). Mortality is a crucial aspect of demography and $\sigma > s$ may imply that workers have a lower rate of mortality than solitary foundresses. This is probably true but quantitative data on survivorship of solitary foundresses are hard to obtain. In any case, I will try to show that even if workers and solitary foundresses have identical mortality rates, the consequences of such mortality for the magnitude of the demographic correction factors and thus for inclusive fitness can be quite different for workers and solitary foundresses. Thus I will use data on survivorship of workers obtained from field colonies of *R. marginata* (Figure 13.1). In using survivorship data on workers, rather than solitary foundresses, I will perhaps overestimate the inclusive fitness of solitary

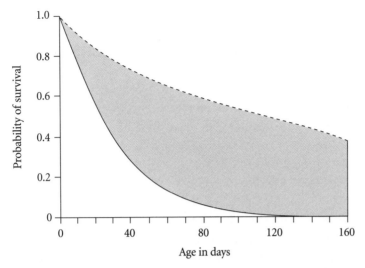

Figure 13.1. Survivorship curves for *R. marginata* females under field (solid line) and laboratory (broken line) conditions. The field survivorship curve was obtained by fitting the hazard function given by Hjorth (1980) to census data so that $l_t = [\exp(-\delta t^2/2)]/(1 + \beta t)^{\theta/\beta}$ where l_t is the probability of surviving to the age t and β, θ, and δ are constants with values of 1.0×10^{-5}, 2.8281×10^{-2}, and 1.8480×10^{-4}, respectively. The laboratory survivorship curve was obtained by fitting a quadratic model given by Bain (1978) to census data so that $l_t = \exp[-(at + bt^2/2 + ct^3/3)]$ where a, b, and c are constants with values of 9.8973×10^{-3}, -1.2204×10^{-4}, and 6.984×10^{-7}, respectively. (Redrawn from Gadagkar, 1990a.)

foundresses, but to the extent that I do so, my estimate of the advantage of the worker strategy will be conservative. This is safer than overestimating the inclusive fitness of workers, because here I am attempting to explore the possible role of hitherto unexplored factors that might help promote the evolution of eusociality. I discuss below a hierarchy of models that illustrate how demographic inequalities between workers and solitary foundresses can strongly select for worker behavior in *R. marginata* (Gadagkar, 1990a, 1991b).

13.1.　Delayed Reproductive Maturation

We saw in section 10.3 that there was a great deal of variation in the time taken by wasps isolated at eclosion to attain reproductive maturity and begin to lay eggs. The distribution of such "waiting times" has a mean and standard deviation of 48 ± 31 days (Figure 12.2). As I mentioned in Chapter 10, when those data were first obtained, I did not appreciate their significance. In the context of a possible demographic predisposition to eusociality, the significance of delayed reproductive maturation becomes clear. Delayed reproductive maturation will make the worker strategy more attractive because solitary foundresses have to survive, first until they become reproductively mature, and then until they bring their offspring to adulthood. Workers, by contrast, are unaffected by their own rate of reproductive maturation because their work (and fitness) depends on the queen's supplying them with brood. Using the mean delay in reproductive maturation of 48 days and an estimate of egg-adult brood developmental period of 62 days (Gadagkar, 1990a), I have computed the asymmetrical fitness consequences of such delayed reproductive maturation for the inclusive fitness of solitary foundresses and workers. The demographic correction factor s for solitary foundress is simply the probability of survival for $48 + 62 = 110$ days, which is 0.015 (from Figure 13.1). The demographic correction factor σ for workers is the probability of their survival merely for the duration of the brood developmental period of 62 days, which is 0.12 (also from Figure 13.1).

These rather different values for the demographic correction factors obtained for a solitary foundress on the one hand and a worker on the other hand illustrate the disadvantage of delayed reproductive maturation for a solitary foundress compared to a worker. If we assume that $b = \beta$ (in the spirit of all other things being equal), the threshold ρ value required for satisfying inequality 8.5 is given by the equation

(13.1)   threshold $\rho = s/2\sigma = 0.06$

This means that, other things being equal, workers would break even with solitary foundresses, in spite of rearing brood related to them by a mere 0.06. Alternatively, if we assume that $\rho = r = 0.5$, the threshold b/β value required for satisfying inequality 8.3 is given by the equation

(13.2)   threshold $b/\beta = \sigma/s = 8.0$

This means that if $\rho = r$, then workers would break even with solitary foundresses in spite of solitary foundresses being capable of performing eight times more work per unit of time. Following Queller (1989), I compare this advantage of delayed reproductive maturation with the maximum advantage possible under the haplodiploidy hypothesis (when the brood reared by workers consists entirely of full sisters). The maximum value for b/β under the haplodiploidy hypothesis (if we assume $\sigma = s$) is given by the equation

(13.3) $b/\beta = \rho/r = 0.75/0.5 = 1.5$

Compared to the maximum threshold of b/β of 1.5 under the haplodiploidy hypothesis, delayed reproductive maturation should thus be 8.0/1.5 = 5.3 times more effective in promoting the evolution of a worker caste.

13.2.   Variation in Age at Reproductive Maturity

Thus delayed reproductive maturation appears to be capable of providing a substantial advantage for workers over their solitary nesting counterparts (Table 13.1). Indeed, the advantage for workers is so great that one begins to wonder why the solitary nesting strategy persists at all. But the fact is that a mixture of single-foundress and multiple-foundress nests occurs in all populations of R. marginata that I have seen and indeed in a variety of primitively eusocial species (for reviews see Michener, 1974; Gadagkar, 1991a; Reeve, 1991). Any reasonable hypothesis should therefore explain not only the existence of multiple-foundress nests with their sterile worker

Table 13.1.   The advantage of assured fitness returns, delayed reproduction, and both acting in concert in R. marginata. (Data from Gadagkar, 1991b.)

Parameter	Delayed reproduction	Assured fitness returns	Both acting in concert
Demographic correction factor for solitary nest foundresses (s)	0.015	0.12	0.015
Demographic correction factor for workers (σ)	0.12	0.43	0.43
Threshold ρ	0.06	0.14	0.017
Threshold b/β	8.0	3.6	28.7
Relative strength compared to haplodiploidy	5.3	2.4	19.1

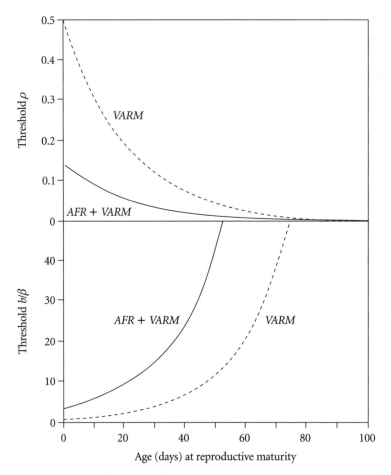

Figure 13.2. The advantage of variation in age at reproductive maturity and assured fitness returns. The figure shows threshold ρ (above) and threshold $b/\text{ß}$ (below) required to satisfy inequality 8.5 as a function of age at reproductive maturity for the individual action of variation in age at reproductive maturity (*VARM*; dashed line) and for assured fitness returns *(AFR)* and variation in age at reproductive maturity acting in concert (solid line). (Redrawn from Gadagkar, 1991b.)

castes but also why single-foundress nests and multiple-foundress nests coexist.

The variation in age at reproductive maturity, with a range of 14–191 days (Figure 12.2), may provide the basis for such a hypothesis. With increasing delay in attaining reproductive maturity, the threshold ρ required for satisfying inequality 9.2 decreases while the threshold b/β increases (Figure 13.2). Since the advantage of the worker strategy is small for early reproducers and large for late reproducers, it is reasonable to conclude that individuals that experience a small delay in attaining reproductive maturity will be relatively easily selected to become solitary foundresses and that those that experience large delays in attaining reproductive maturity will be relatively easily selected to become workers. Comparing this model with the haplodiploidy hypothesis again, I should emphasize that the latter does not provide any explicit explanation for the coexistence of single- and multiple-foundress nests.

13.3. Mixed Reproductive Strategies

The models and data discussed above also suggest the possibility of selection for the adoption of a mixture of worker and queen strategies. If there is likely to be a delay in attaining reproductive maturity, an individual would maximize her inclusive fitness by first being a worker and then, after attaining reproductive maturity, switching to the role of foundress or queen. The data in Figure 12.2 on the time required for wasps to start laying eggs show that 28% of the wasps could have completed the rearing of one entire brood (by working for 62 days) before they became reproductively mature (Figure 13.3). Since *R. marginata* follows a perennial, indeterminate nesting cycle where queen supercedure leading to serial polygyny is quite common, many individuals first act as workers and later become queens in their natal colonies. And since there is always a substantial advantage of becoming the queen of a large well-established colony, there should be selection on individuals to attempt to take over the role of queen. But if there is a delay in attaining reproductive maturity, individuals should be selected to work for their colonies before becoming queens. I believe that this explains why social organization in *R. marginata* is not wrecked by potential queens. Individuals that replace queens indeed appear to behave like all other workers until they actually become queens. Recall from Chapter 6 that my students and I are unable to predict the identity of a potential queen until after the event of queen replacement and that we have dubbed potential queens unspecialized intermediates. In addition, we also occasionally see wasps working for some time in their natal colonies and later leaving to found single- or multiple-foundress colonies.

It should be pointed out that if working for the colony constitutes a significant drain on the potential queen's physiology and leads to a further

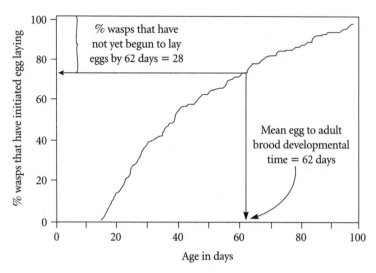

Figure 13.3. Scope for mixed reproductive strategies. The data in Figure 12.2 are here rearranged to show that 28% of the wasps had not yet begun to lay eggs by the age of 62 days, which is the egg to adult brood developmental period for the species. These wasps could thus have reared one entire brood as workers before switching over to the role of queen or foundress. (Redrawn from Gadagkar, 1991b.)

delay in her attainment of reproductive maturity, natural selection will work against the mixed reproductive strategy. We saw in Chapter 5 that all workers in *R. marginata* colonies do not perform equally risky tasks and that the system of behavioral caste differentiation into sitters, fighters, and foragers makes possible a highly nonrandom distribution of the risky tasks associated with foraging. We also saw some preliminary evidence that potential queens are probably drawn preferentially from among the sitters, which perform the least risky tasks but nevertheless contribute to the welfare of the colony; they perform intranidal tasks such as feeding larvae and extending the walls of the cells. I suggest that the system of behavioral caste differentiation that allows individuals to contribute to the welfare of the colony without necessarily taking on the most risky tasks creates a favorable background for the adoption of mixed reproductive strategies.

13.4.   Assured Fitness Returns

The asymmetry in the demographic correction factors for solitary foundresses and workers seen so far was due to the requirement that solitary foundresses have to survive longer than workers to rear the same amount of brood. I will now consider a factor that will cause differences in the magnitude of the demographic correction factors for solitary foundresses and workers even when they survive for the same length of time. While solitary foundresses must survive until the end of the developmental period of the brood to recoup their investment in it, workers have a special advantage. If a worker cares for some larvae for a part of her developmental period and dies before bringing them to independence, there is a good chance that another worker from the colony will continue to care for the same larvae. Thus workers in multifemale nests are relatively more assured of fitness returns for their labor even if they work only for a fraction of the brood developmental period. I have called this phenomenon "assured fitness returns" (Gadagkar, 1990a, 1991b).

13.4.1. The Head-Start Hypothesis

My idea of assured fitness returns was inspired by Queller's (1989) head-start hypothesis. Queller argued that a worker has a "reproductive head-start" because "there are already young of various ages present on her natal nest" so that "her efforts can immediately result in some of these reaching the age of independence." For this reason Queller concluded that $\sigma = 1$. (Queller [1989] uses a different system of notations, but for the sake of uniformity, I will discuss his work using the corresponding notations I have used throughout this book.) In the case of solitary foundresses, Queller correctly computed s as their probability of survival for the entire duration of the brood developmental period. As I am doing, Queller (1989) used survivorship data on workers to estimate s to be equal to 0.057, 0.187, 0.097, and 0.167 for four polistine wasps, *Polistes exclamans*, *Polistes chinensis*, *Polistes gallicus*, and *Mischocyttarus drewseni*. Assuming that $\sigma = 1$, as discussed above, he computed threshold r values ranging from 0.03 to 0.09 and threshold b/β values ranging from 6.0 to 17.4 (Table 13.2). In other words, under Queller's model, workers would break even with their solitary counterparts in spite of rearing brood with an average genetic relatedness to themselves as low as 0.03 to 0.09 or even if solitary foundresses are 6 to 17.4 times more efficient at rearing brood than they are.

Given my experience with *R. marginata*, which suggested repeatedly that I should look beyond the role of haplodiploidy, I was immediately attracted to Queller's head-start hypothesis. Such a powerful force favoring

Table 13.2. Comparison of head-start hypothesis and the assured fitness returns model for polistine wasps. (Modified from Gadagkar, 1990a.)

Species	Polistes exclamans[a]	Polistes chinensis[a]	Polistes gallicus[a]	Mischocyttarus drewseni[a]	Ropalidia marginata survivorship data from	
					Natural colonies	Laboratory colonies
Brood developmental period (days)[b]	38	52	38	46	62	62
s	0.057	0.187	0.097	0.167	0.1215	0.6475
Head-start hypothesis[c]						
Threshold σ	0.03	0.09	0.05	0.08	—	—
Threshold b/β	17.4	5.3	10.3	6.0	—	—
Relative strength[d]	11.6	3.5	6.9	4.0	—	—
Assured fitness returns model (assuming constant age-specific mortality)						
m	0.9274	0.9683	0.9404	0.9619	0.9666	0.9930
σ	0.3170	0.4774	0.3750	0.4569	0.4100	0.8078
Threshold ρ	0.0899	0.1958	0.1293	0.1828	0.1482	0.4008
Threshold b/β	5.5614	2.5529	3.8660	2.7359	3.3745	1.2476
Relative strength[d]	3.7076	1.7019	2.5773	1.8239	2.2497	0.8317

a. Brood developmental period and s values taken from Queller (1989).
b. Values rounded off to the nearest day.
c. Values in this part of the table are taken from Queller (1989) for comparison and the corresponding computations are not made for *R. marginata*.
d. Compared with the advantage of haplodiploidy.

sociality as Queller's head-start hypothesis had seldom been suggested. And it was obvious that the head-start hypothesis was ideal for a comprehensive exploration of the role of demography in the evolution of eusociality. But it was equally obvious that there was a problem with the head-start hypothesis as it was formulated (Gadagkar, 1990a). The problem relates to Queller's assumption that $\sigma = 1$ because "there are already young of various ages present" on the natal nest of a worker. This assumption amounts to giving full credit for the rearing of one offspring from egg to adulthood (independence) to a worker that may have eclosed only one or a few days before the completion of development of the offspring and thus cared for it only for that one or those few days. This assumption overlooks the fact the queen or other workers performed all the duties of rearing that offspring from egg to the stage at which the worker in question found it. The credit (contribution to fitness) for this part of the work should go to the queen or the other workers, not to the worker that eclosed later. The contribution to any worker's fitness should clearly be in proportion to her contribution to the rearing of each offspring; otherwise the full fitness benefit for rearing a given larva gets assigned to several workers. I have argued therefore that the magnitude of the advantage provided by Queller's head-start hypothesis is partly due to the unfair advantage that he gives to workers in formulating his model (Gadagkar, 1990a). The reader is encouraged to consult Gadagkar (1990a) for a more detailed discussion of my interpretation of the problems associated with the head-start hypothesis and Queller (1996) for his more recent formulation of the model.

13.4.2.   Computing the Demographic Correction Factor for Workers

There is, however, an advantage that a worker has over a solitary foundress, which I have called the advantage of assured fitness returns. Instead of thinking of the head start of a worker that ecloses after the brood in her nest have already reached an advanced stage of development, leaving only a small amount of work to be done, I am thinking of a more general situation. If any worker works for part of the developmental period of the brood and dies before bringing the brood to independence, and if other workers can continue the work left unfinished, the dead worker should be assigned credit (fitness) for her fractional contribution to the survival and growth of the brood, irrespective of whether she cared for the brood at the beginning, middle, or later part of the brood developmental period.

The problem is how to quantify the advantage of the assured fitness returns that accrue to a worker but not to a solitary foundress. The parameters of interest in inequality 8.5 are the demographic correction factors s and σ. The advantage of assured fitness returns is best built into the asymmetry between s, the demographic correction factor for solitary foundresses, and σ, the demographic correction factor for workers. Factor s can simply continue to be the probability of survival of the solitary foundress up until the time that the brood under her care complete development and thus became independent of her care. That is how I computed s in the previous models of delayed reproductive maturation and variation in age at reproductive maturity. In the assured fitness returns model solitary foundresses gain no extra advantage, so s remains unchanged. It is σ that should be computed so as to incorporate the advantage of assured fitness returns.

Before I proceed, I must remind the reader that here I am attempting to compute the advantage of assured fitness returns in the absence of delayed reproductive maturation, variation in age at reproductive strategies, and mixed reproductive strategies. Assured fitness returns can act in concert with the other models and I will return to such concerted action later. Here I consider a pure assured fitness returns model. The essential point about the assured fitness returns model is that if a worker cares for some brood but dies before the brood reach independence, that worker should get some credit for her work. I therefore assign fitness to workers in proportion to the fraction of their contribution. The difference in the procedure for assigning fitness to solitary foundresses, which must survive for the entire brood developmental period, and workers, which need not, is illustrated graphically in Figure 13.4. Let $b = \beta$ and the brood developmental period be equal to n days. If a solitary foundress has a lifespan of x days (when $x < n$), I give her zero credit because she rears no offspring. If a worker has a lifespan of x days, I give her credit for rearing β (x/n) individuals. With a lifespan of y days (when $y > n$) a solitary foundress gets credit for rearing b offspring and a worker gets credit for rearing β offspring. No individual gets more credit than b or β despite surviving for more than n days because for the sake of simplicity I am considering only one synchronously produced batch of brood.

Thus σ for an average individual in the population is given by

$$(13.4) \qquad \sigma = \sum_{i=1}^{n-1} p_i (i/n) + \sum_{i=n}^{\infty} p_i$$

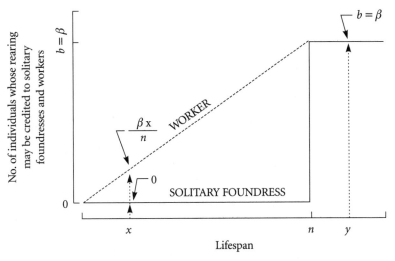

Figure 13.4. A procedure for assigning credit for rearing brood to solitary foundresses and workers with different lifespans. The expression $b = ß$ is the number of individuals that solitary foundresses and workers, respectively, can rear provided that they survive for the entire brood developmental period, and n is the brood developmental period in days. Solitary foundresses get zero credit if they survive for less than n days and b if they survive for n or more days. Workers, by contrast, get credit in proportion to the fraction of brood developmental period that they survive with a maximum of β if they survive for n or more days. Thus a solitary foundress that survives for x days (where $x < n$) gets zero credit but a worker that survives for x days gets a credit of $(ßx)/n$. A solitary foundress that survives for y days (where $y > n$) gets a credit of b and a worker who survives for y days gets a credit of $ß$. No individual gets more credit than b or $ß$ even if she survives for longer than n days, because I am considering only one synchronously produced batch of brood. That is the reason for the flat portion of the curves beyond n. (Redrawn from Gadagkar, 1990a.)

where p_i is the proportion of workers that have a lifespan of i days (that is, those that die on day $i + 1$) and n is the brood developmental period in days. When β is multiplied by σ, the first term in equation 13.4 is equivalent to giving credit for rearing β (i/n) individuals to those workers that survive for i days (where $i < n$) and giving credit for rearing β individuals to all workers that survive for n or more days. As in the example in Figure 13.4, workers that survive for more than n days get no more credit than β because of the assumption of only one synchronously produced batch of brood. Thus $\beta\sigma$ (in the left-hand side of inequality 8.5) is equivalent to a weighted average of the credit that accrues to workers with different lifespans.

13.4.3. Assured Fitness Returns in Ropalidia marginata

Let us now explicitly apply the assured fitness returns model to R. marginata. In the context of the delayed reproductive maturation model, I have already computed $s = 0.12$, which remains valid here. Given the survivorship curve in Figure 13.4, equation 13.4 yields a value of 0.43 for σ. The ratio between σ and s is thus the asymmetry between the fitness of a worker which has assured fitness returns, and a solitary foundress, which does not. To put it more formally, as I did in the case of delayed reproductive maturation, if we assume that $b = \beta$, the threshold value of ρ required for the maintenance of eusociality is given by the equation

(13.5)   threshold $\rho = s/2\sigma = 0.14$

Alternatively, if we assume that $\rho = r = 0.5$, the threshold b/β value required for the maintenance of eusociality is given by the equation

(13.6)   threshold $b/\beta = \sigma/s = 3.6$

This means that, owing to the advantage of assured fitness returns, workers would break even with solitary foundresses in spite of rearing brood related to them by a mere 0.14 or in spite of solitary foundresses being able to perform 3.6 times more work per unit of time. The assured fitness returns model is thus 2.4 times more effective than the haplodiploidy hypothesis $(3.6/1.5 = 2.4)$ in driving the evolution of eusociality (Table 13.1).

13.4.4. Survivorship in Laboratory Experiments

Survivorship curves estimated from natural colonies almost certainly overestimate mortality, because a wasp's disappearance from a colony is taken as death but some wasps leave their nests to join other nests or to found their own. This phenomenon is particularly significant in R. marginata because new nests are initiated throughout the year and because wasps eclosing at all times of the year have all options open to them (Chapter 4). It is therefore necessary to correct for this bias in the estimated probability of survival. And furthermore, it is desirable to have some estimate of the range of values that the parameters in the model, such as s, σ, threshold ρ, and threshold b/β, may take. All this, however, is easier said than done. It is virtually impossible to differentiate between actual death and emigration of the wasps. The next best option is to repeat the foregoing analysis using survivorship data from experiments in which freshly eclosed wasps were isolated and maintained in the laboratory. I must emphasize that under these conditions mortality is almost certainly underestimated because of the absence of predation and other risks of foraging that would be present in nature. Thus I have no choice but to swing from one extreme (underestimating survival in natural colonies) to another (overestimating survival in the laboratory).

But, for a first approximation, I liken survivorship values obtained from natural colonies and those obtained from laboratory experiments to the lower and upper bounds, respectively, of the range of survival probabilities for *R. marginata*. The survivorship curve obtained from wasps isolated in the laboratory (Figure 13.1) yields values of $s = 0.65$, $\sigma = 0.79$, threshold $p = 0.41$, and threshold $b/\beta = 1.22$. Notice that even these values are sufficient to satisfy inequality 8.5, although the potential advantage of the assured fitness returns here is marginally less than the possible maximum advantage given in the haplodiploidy hypothesis, the relative strength of the former being $1.22/1.5 = 0.8$.

13.4.5.   Other Polistine Wasps

It is obviously interesting to compare the relative efficiency of Queller's head-start hypothesis and my assured fitness returns hypothesis. But Queller applied his model to four species of polistine wasps other than *R. marginata* and I have applied my model only to *R. marginata*. I shall therefore attempt to apply the assured fitness returns model to the four species used by Queller (1989). There is one stumbling block however; I do not have the survivorship curves for the species used by Queller; he provides only the probability of survival to the end of the brood developmental period. Hence I make the simplifying assumption of constant age-specific mortality (for which there is evidence, at least in *R. marginata*; see Chapter 4) and estimate the day-to-day probability of survival m from the equation

(13.7) $m^n = s$

where m is the probability of survival for 1 day, n is the brood developmental period, and s is the probability of survival for n days. I thus compute the proportion of individuals surviving to i days as m^i. The proportion of individuals expected to have a lifespan of i days is the difference between the proportion of individuals surviving up to i days and the proportion surviving up to $i + 1$ days. Threshold σ may now be computed directly from equation 13.3. The results of this exercise (Table 13.2) show that the assured fitness returns model, as expected, gives a smaller advantage to workers than the presumed advantage of the head-start hypothesis. Nevertheless, these values are sufficient to satisfy inequality 8.5 and thus to derive the evolution of eusociality, in all four species.

To ascertain that the results of this analysis are not sensitive to my assumption of constant age-specific mortality, I repeat the analysis for *R. marginata* without using the entire survivorship curve. Instead I use, as I have done in the case of Queller's species, the probability of survival until the end of the brood developmental period to compute the day-to-day probability of survival (m) from equation 13.7. From this I compute the proportion of individuals having a lifespan of i days and then apply equation 13.4 to compute σ. These results are remarkably similar to those obtained with the explicit use of the survivorship curve (see Table 13.2). Hence I conclude that my results are different from those of Queller (1989) because of a genuine difference between his head-start hypothesis and my assured fitness returns model rather than because of my assumption of constant age-specific mortality.

13.4.6.   The mn Plot

The lack of the complete survivorship data for the species used by Queller and the need to apply the simplifying assumption of constant age-specific mortality is a blessing in disguise. Because the assumption of constant age-specific mortality appears to be reasonable, the number of critical parameters needed to investigate the assured fitness returns model are now reduced to two: the day-to-day probability of survival (m) and the brood developmental period (n). The advantage of assured fitness returns is expected to be highest when the day-to-day probability of survival is low and the brood developmental period is long. Both these quantities are likely to vary from species to species, thus varying the efficacy of assured fitness returns in selecting for worker behavior. I have therefore plotted the day-to-day probability of survival (m) versus the brood developmental period (n) (Figure 13.5).

This is a very useful plot because it allows us to search through the

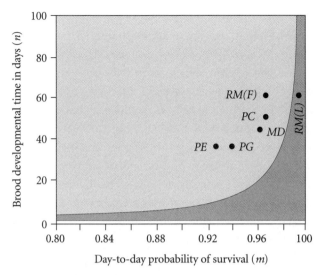

Figure 13.5. The *mn* plot. Contour lines are shown in the parameter space of day-to-day probability of survival (*m*) and the brood developmental period (*n*) that separate the regions: (1) where there is no advantage of assured fitness returns (b/ß < 1.0) (the narrow band of white close to the X axis where *n* < 2); (2) where the advantage of assured fitness returns promotes worker behavior but is less effective in doing so than haplodiploidy (1.5 > b/ß > 1.0) (dark gray region); and (3) the region where the advantage of assured fitness returns not only promotes worker behavior but is more effective in doing so than haplodiploidy (b/ß > 1.5) (light gray region). The positions of all wasp species discussed in the text are also shown. *PE = Polistes exclamans, PG = Polistes gallicus, MD = Mischocyttarus drewseni, PC = Polistes chinensis, RM(F) = Ropalidia marginata* (survivorship data from natural colonies), and *RM(L) = Ropalidia marginata* (survivorship data from a laboratory experiment). (Redrawn from Gadagkar, 1990a.)

entire *mn* parameter space (by assuming ρ = *r* and β = *b*) and delineate regions where workers get more, equal, or less fitness than do solitary foundresses. There are three qualitatively different regions in the *mn* plot. These are regions where

1. the assured fitness returns model does not promote worker behavior (*b*/β < 1.0);
2. the assured fitness returns model promotes the evolution of worker behavior, but is less effective in doing so than haplodiploidy (1.5 > *b*/β < 1.0); and

3. the assured fitness returns model is more effective than
   haplodiploidy in selecting for worker behavior ($b/\beta < 1.5$).

What is the use of the *mn* plot? It allows us to apply the assured fitness returns model to a variety of species (and not necessarily just of insects) by
plotting their position in the *mn* parameter space. I illustrate the use of the
mn plot by showing the position of all the wasp species considered in this
chapter. Notice that all five species of wasps lie in that region of the *mn*
plot where assured fitness returns is more powerful in selecting for worker
behavior than is haplodiploidy. Only the point pertaining to R. *marginata*
laboratory survivorship data is in the region where haplodiploidy can, in
principle, be more effective in selecting for worker behavior.

   Let us now see how assured fitness returns can act in concert with the
other three demographic models considered at the beginning of this chapter—delayed reproductive maturation, variation in age at reproductive
maturity, and mixed reproductive strategies—to provide an even more
powerful force in selecting for worker behavior.

13.5. Assured Fitness Returns and Delayed Reproductive Maturation Acting in Concert

Assured fitness returns and delayed reproductive maturation can act in
concert because although delayed reproductive maturation can lower *s* and
reduce the fitness of workers, assured fitness returns can raise σ and increase the fitness of workers. To quantify such concerted action of delayed
reproductive maturation and assured fitness returns, I simply have to use
the value of the demographic correction factor for solitary foundresses
that I used in the delayed reproductive maturation model, namely, $s =$
0.015. This value of *s* now has to be contrasted not with $\sigma = 0.12$, as in section 13.1, but with $\sigma = 0.43$, as in section 13.4. This is reasonable because
when delayed reproductive maturation and assured fitness returns act in
concert, a solitary foundress must survive not only until she becomes reproductively mature but also until she lays eggs and brings her offspring to
adulthood. By contrast, a worker need not wait until she becomes reproductively mature and also need not survive for the entire developmental
period of her brood. Now the threshold ρ assumes a value of 0.015 and the
threshold b/β assumes a value of 28.7. Thus delayed reproductive matura-

tion and assured fitness returns acting in concert are 19.1 times more powerful than the maximum benefit that the haplodiploidy hypothesis can provide (Table 13.1).

13.6.　Assured Fitness Returns and Variation in Age at Reproductive Maturity Acting in Concert

Just as delayed reproductive maturation and assured fitness returns can act in concert and provide a more powerful force for the evolution of worker behavior, assured fitness returns and variation in age at reproductive maturity can also act together and provide, for any given value of delay in attaining reproductive maturity, a more powerful advantage for the worker strategy. The quantitative effects of such action in concert are shown in Figure 13.2.

13.7.　Assured Fitness Returns and Mixed Reproductive Strategies Acting in Concert

In the absence of assured fitness returns, the advantage of mixed reproductive strategies can only be exploited by those individuals who have a delay of 62 or more days in attaining reproductive maturity. When assured fitness returns and mixed reproductive strategies act in concert, however, the advantages of mixed reproductive strategies become available to individuals with a variety of values for the delay in attaining reproductive maturity. Their labor as workers is not wasted if they leave to become solitary foundresses or queens of other multiple-foundress nests, because assured fitness returns will give them fitness returns in proportion to the fraction of brood developmental period during which they worked. By making the advantage of adopting mixed reproductive strategies available to a wide range of individuals, the concerted action of assured fitness returns and mixed reproductive strategies facilitates selection for the worker strategy because conditions suitable for an individual's shift from a worker role to that of a queen can hardly be expected to appear predictably after 62 days or its multiples.

　　The hierarchy of models presented in this chapter show how several factors, acting alone and in concert with one of them (assured fitness returns), can differentially affect the demographic correction factor in the inclusive fitness inequality in 8.5 and thus make the worker strategy more advantageous than the solitary nest-founding strategy. Indeed, these models are consistently more powerful in selecting for worker behavior than the haplodiploidy hypothesis. A particularly satisfying feature of these models is that the advantage of being a worker, rather than being a solitary foundress, can be more pronounced for some individuals than for others. This suggests how natural selection can bring about the coexistence of the single-foundress strategy and the multiple-foundress strategy in the same population. The most relevant conclusion, however, is that *there is* a substantial demographic predisposition for the evolution of eusociality in *R. marginata.*

14   SYNTHESIS

$\mathscr{T}$HE theoretical framework discussed in Chapter 8 permitted the classification of various factors that influence the relative inclusive fitnesses of solitary foundresses and workers into at least four categories: genetic, ecological, physiological, and demographic. In Chapters 9–13, I showed that genetic (relatedness) factors tilt the inclusive fitness inequality in favor of solitary foundresses but that ecological, physiological, and demographic factors all tilt the inequality in favor of workers. We should now progress beyond exploring the direction of selection likely to be promoted by a given factor and consider all factors simultaneously. What is the specific question such an integrated investigation should ask? Can we ask, for example, whether a unified model that takes all factors simultaneously into consideration will tilt the inclusive fitness inequality in favor of solitary foundresses or workers? The answer is no, because solitary foundresses and workers coexist in *R. marginata* populations. Whether the unified model predicts that the inclusive fitness inequality is tilted in favor of solitary foundresses or that it is tilted in favor of workers, that prediction would be wrong. What we need is a unified model that predicts coexistence of solitary foundresses and workers. And to be very precise, we should demand that the unified model make specific predictions for individual wasps—we should be able to input values of the relevant parameters for any wasp and the unified model should be capable of predicting whether that particular wasp should opt for the solitary founding strategy or for the worker strategy. But I think we are very far indeed from devising such a model. Let us then become a little less ambitious and attempt at least to predict what proportion of the population should opt for a solitary founding strategy and what proportion should opt for the worker strategy.

14.1. Toward a Unified Model

Let me attempt to construct such a unified model that integrates expected inequalities between ρ and r, β and b, and σ and s. I considered inequalities in ρ and r in Chapter 9. If we assume outbred populations, r, the genetic relatedness between a solitary foundress and her offspring, can be safely set at 0.5. What about ρ, the worker-brood genetic relatedness? Considering both polyandry and serial polygyny, I estimated worker-female brood genetic relatedness to range from 0.22 to 0.46. These values would be relevant if workers raise an all-female brood. If workers can skew investment in the ratio of their relatedness to the male and female brood, then the weighted mean relatedness to the brood they rear would be relevant, and these ranged from 0.20 to 0.38 (Table 9.4). As a first approximation, let me therefore assume that ρ ranges from 0.20 (the low limit of the two ranges) to 0.46 (the upper limit of the two ranges). In Chapter 11, I considered inequalities between β and b and found that a natural solitary foundress raised 12.3 items of brood while a cofoundress forced to nest alone raised only 4.2 items of brood. If all wasps initiated solitary nests, b, their productivity, would range from 4.2 to 12.3. Since the per capita productivity does not change significantly with the number of foundresses (section 11.2), I assumed that each worker contributes to the rearing of an additional 12.3 items of brood. Thus I can set β at 12.3. In Chapter 13, I considered inequalities in σ and s, the demographic correction factors for workers and solitary foundresses, respectively. I found inequalities due to delayed reproductive maturation and variation in age at reproductive maturity, but my calculations were based on data on delayed reproductive maturation in wasps kept in isolation in laboratory cages. Although those calculations were useful to illustrate how delayed reproductive maturation can affect the demographic correction factors of workers, they cannot be used to compute absolute values of σ. But the assured fitness returns model, which also concerns values of s and σ, is more useful. There s and σ were calculated by using survivorship curves derived both from field colonies and from wasps reared in the laboratory. The survivorship values derived from field colonies overestimated mortality while laboratory derived values underestimated mortality, and hence I likened the field and laboratory derived survivorship values to the possible lower and upper limits of survival probabilities. Using the field derived survivorship curve (Figure 13.1, solid line), I estimated $s = 0.015$ and $\sigma = 0.43$ (Table 13.3), and using the laboratory derived survivorship curve (Figure 13.1, broken line), I estimated $s = 0.015$ and $\sigma = 0.79$. The strength of the assured fitness returns model in promoting the evolution of worker behavior depends on the ratio σ/s. Thus I will set the field derived $\sigma/s = 0.43/0.12 = 3.6$ as one limit and the laboratory derived $\sigma/s = 0.79/0.65 = 1.2$ as the other limit.

Armed with values for all the parameters, I can now explore the entire parameter space to see where the fitness of solitary foundresses is greater than that of workers and where the fitness of workers is greater than that of solitary foundresses. I do this by using the condition that worker behavior will be favored when

$$(14.1) \qquad \frac{\sigma}{s} > \frac{rb}{\rho\beta}$$

It turns out that inequality 14.1 is satisfied in 94.9% of the parameter space while it is not satisfied in 5.1% of the parameter space. This can be interpreted to mean that 5.1% of the wasps should be selected to become solitary foundresses while 94.9% of them should be selected to become workers (Figure 14.1). And what do the wasps actually do? We saw in Chapter 4 that in the Shakarad and Gadagkar (1995) study of 145 pre-emergence nests, 4.6–7.5% of the wasps nested solitarily and 92.5–95.4% of the wasps nested in groups (most of these wasps of course were workers, but see below). By any standards, this is a remarkable fit between predictions of a model and the relevant empirical data. But I must emphasize that this is by no means a definitive model and that it represents no more than an illustration of how one might proceed in constructing a unified model that integrates genetic, ecological, physiological, and demographic factors that might influence the inclusive fitness of wasps choosing solitary nest-founding strategies and those choosing worker strategies. For this reason I think we should not attribute too much significance to the close quantitative agreement between model and data. That there is a qualitative agreement is sufficient encouragement—both model and data show that the solitary nesting strategy should be the preferred option for a small minority of the individuals and that the worker strategy should be the preferred option for the vast majority of individuals in the population.

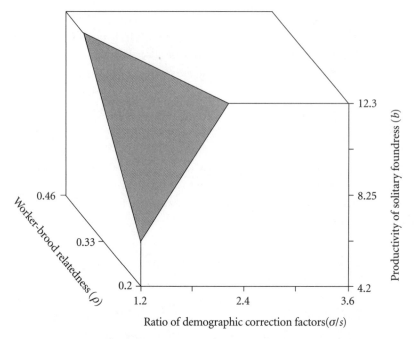

Figure 14.1. A graphic illustration of the unified model showing the parameter space where worker behavior is selected (unshaded) and the missing chip of the block where solitary nesting behavior is favored.

14.2.   Some Simplifying Assumptions of the Unified Model

It is instructive to focus on some of the simplifying assumptions made in constructing the unified model and those made in generating the relevant empirical data. In constructing the model, I used a single mean value for β and, although I used a range of values for σ/s, ρ, and b, I made the simplifying assumption that these variables are uniformly distributed within those ranges. The model provides predictions of the expected proportion of solitary foundresses and workers, but these predictions are compared with the observed proportion of solitary foundresses and group-nesting wasps; the latter include workers as well as queens. The model ignores the fact that some of the wasps in the multiple-foundress nests will become queens and have much greater fitness than the rest of the workers and even than the solitary foundresses. If that possibility were taken into consideration, more wasps should be expected to opt for the worker strategy. In other words, the model underestimates the proportion of workers. But this is compensated by the fact that the empirical data also probably underestimate the proportion of workers. My reasoning is as follows. The Shakarad and Gadagkar (1995) study considered only pre-emergence nests, that is, those of the founding population. This is justified for annual colonies such as the many North American *Polistes* species, where all workers and queens die during winter and only the next generation of female wasps hibernates and initiates new nests in the next spring (for a review, see Reeve, 1991). In the tropical *R. marginata*, with its perennial, indeterminate nesting cycle, new colonies are founded throughout the year by wasps leaving their natal nests, although many wasps remain in the natal nest, functioning as workers and occasionally becoming replacement queens. There are no hard quantitative data, but it is very likely that among the wasps that remain in their natal nests, a greater proportion die as workers than the proportion of wasps that end up as workers among the wasps that leave their natal nests to found new nests. Thus the data, being exclusively drawn from newly initiated nests, also probably underestimate the proportion of wasps that opt for the worker strategy. To what extent these two errors cancel each other out is of course not known.

While the empirical estimation of the proportions of wasps preferring single-foundress and multiple-foundress nests focused on the founding population only and ignored the wasps remaining in post-emergence nests, worker-brood relatedness values were estimated only on the basis of post-emergence nests. This may be a serious error if one is dealing with annual colonies, of North American *Polistes* for example. In annual colonies, pre-emergence multiple-foundress colonies consist of sisters nesting together, so that workers rear their nieces (and nephews, if they also rear male brood). Post-emergence colonies consist of a mother queen and daughter workers that rear siblings. So one could not use estimates of worker-brood relatedness in post-emergence colonies to estimate the corresponding values for single-foundress nests. But with their perennial, indeterminate nesting cycle and frequent queen replacements, post-emergence colonies of *R. marginata* consist of a group of female wasps of varying degrees of relatedness to each other, one of which functions as the queen. Thus workers are not necessarily daughters of the queen. If a group of wasps leaves a post-emergence colony and founds a new multiple-foundress nest, with one of them functioning as the queen, there is no reason why worker-brood relatedness in pre-emergence colonies would be any different from that in post-emergence colonies. So it is probably all right that I have used worker-brood relatedness derived from post-emergence nests to estimate the corresponding values for pre-emergence nests.

Despite its shortcomings, I think the unified model is a good beginning, especially because, to the best of my knowledge, this is the first empirical study that attempts to integrate genetic, ecological, and demographic factors to understand the evolution of eusociality. It is also, I believe, the first attempt to predict the expected proportions of individuals in a population that should opt for the selfish, solitary nest-founding strategy and the proportion that should opt for the altruistic, group-founding strategy with its associated risk that most individuals in multiple-foundress nests end up as sterile workers. I hope that future work in my laboratory will help refine both the model and the empirical estimation of parameter values.

14.3.   Multiple Origins of Eusociality in the Hymenoptera

In Chapters 9–10 we saw that the asymmetries in genetic relatedness expected to be created by haplodiploidy are broken down by polyandry and serial polygyny so that, if anything, the selfish solitary nesting strategy would be favored on the basis of genetic considerations alone. But several other factors—ecological, physiological, and demographic—tilt the inclusive fitness in favor of the sterile worker strategy for an overwhelming proportion of the wasps in the population (Chapters 11–13). This finding reopens the important question of why there have been a disproportionately large number of independent origins of eusociality in the Hymenoptera—a question that was thought to have been satisfactorily answered. For example, Wilson (1971) used the fact that eusociality is estimated to have independently originated at least 11 times in the Hymenoptera but only once in all diploid animals (termites were then the only known case of eusociality in diploid organisms) as a strong argument in favor of the role of haplodiploidy in promoting the evolution of eusociality. If, as I have argued here, relatedness asymmetries potentially created by haplodiploidy are not the driving force for the evolution of eusociality and it is other ecological, physiological, and demographic factors that are largely responsible for the evolution of eusociality, then why have there been a disproportionately large number of origins of eusociality in the Hymenoptera? There are fours possible kinds of answers to this reopened question (see also Bourke and Franks, 1995, for an excellent critic of the haplodiploidy hypothesis).

1. One possibility is that the paradox of the unique taxonomic distribution of eusociality may eventually disappear with the steady increase in the number of discoveries of independent origins of eusociality in diploid animals—aphids, shrimps, beetles, and mole rats have now been added to the lone example of termites known at the time of the writing of Wilson's 1971 book (see Chapter 1). I must confess that I rather doubt if this paradox can ever be so wished away, especially in light of Michener's (1990) suspicion that during the evolution of the bee family Halictidae, "eusocial behavior has arisen repeatedly dozens or hundreds of times."

2. A second possibility is that haplodiploidy has indeed been involved in promoting the evolution of eusociality disproportionately more often in the Hymenoptera, but not through its effects on relatedness. There are other ways by which haplodiploidy can potentially promote the evolution of eusociality:

2A. Haplodiploid females can produce at least some offspring (males) without having to mate, and this makes it possible for them to forgo mating, function as workers, and yet recover some direct fitness (Seger, 1991).

2B. Haplodiploidy permits females to create an extreme sex ratio bias and produce an all-female brood. In the absence of males, their daughters may have no option but to function as unmated workers (Yanega, 1988, 1989).

2C. Reeve (1993) showed that a hypothetical, dominant (but not recessive) gene that causes its bearers to rear sibs instead of offspring has a lower chance of being lost by drift, and thus a higher chance of going to fixation, in haplodiploid populations than in diploid ones.

2D. Because the males are haploid, recessive alleles get exposed and eliminated, which leads to an overall reduction in recessive lethals. This can permit haplodiploids to withstand higher levels of inbreeding, which in turn can cause relatedness asymmetries and potentially promote eusociality (Saito, 1994). It must be emphasized, however, that this is a theoretical possibility (and not a particularly rigorously demonstrated one at that) that has little or no empirical support.

3. The third possibility is that no matter how much reduction there is in intracolony genetic relatedness due to polyandry, there will always persist a

higher value of intracolony genetic relatedness in haplodiploid populations than in diploid populations with corresponding levels of polyandry
(Page and Metcalf, 1982; Gadagkar, 1985b). This is best seen by computing
relatedness among the daughters of a queen as a function of the number of
males that the queen mates with. At any given level of multiple mating, relatedness in the Hymenoptera is always slightly higher than it is in the corresponding diploid populations. Notice that this will also apply to reduction in worker-brood relatedness caused by serial polygyny—in nearly all
examples of distantly related brood reared by *R. marginata* workers (Tables
9.3 and 9.4), the values of worker-brood relatedness would have been even
lower had *R. marginata* been a diploid species. Thus haplodiploid species
may be said to be always genetically more predisposed to the evolution of
eusociality than diploid species. The only problem, and a serious one in
my opinion, is that this hypothetical role of haplodiploidy in promoting
the origin of eusociality can never be put to a rigorous empirical test.
However low a value of worker-brood relatedness I demonstrate, someone
always responds that it would have been even lower if *R. marginata* had
been diploid.

4. The fourth possibility is that the order Hymenoptera has features, independent of haplodiploidy, that can promote eusociality:

4A. Extended parental care is a conspicuous feature of hymenopterans,
including solitary species, and this may be an important preadaptation for
the evolution of eusociality (Wilson, 1971; Alexander, 1974; Andersson,
1984; Eickwort, 1981; Hansell, 1987; Myles, 1988; Seger, 1991).

4B. The presence of an often formidable sting that makes it possible to
protect the large quantity of commonly reared brood has often been considered a factor that has enabled hymenopterans to evolve social life (Starr,
1985, 1989; Kukuk et al., 1989; Seger, 1991).

4C. The apparently high rate of speciation in the Hymenoptera has led
to a very large number of solitary species, providing many starting points
for the evolution of eusociality (Andersson, 1984; Alexander, Noonan, and
Crespi, 1991).

4D. Sherman (1979) pointed out that hymenopterans generally have
high chromosome numbers, and this, he argued, facilitates social evolution
through parental manipulation. High chromosome numbers increase variance in the genes shared by sibs. Such variance hinders discrimination of
close from distant kin and renders nepotism difficult. This should make it
easier for mothers to manipulate their offspring into helping each other.

4E. The ecological, physiological, and demographic factors that I have found to predispose *R. marginata* to the evolution of eusociality may be more common among solitary Hymenoptera than among other groups of solitary insects—we know little about the distribution of these factors in solitary or presocial species (Gadagkar, 1991b; but see Cowan, 1991).

The idea that features within the order Hymenoptera other then asymmetries in genetic relatedness created by haplodiploidy could promote the evolution of eusociality has recently received a significant boost from a particularly elegant study that combines assessment of the salience of various traits for eusocial evolution with phylogenetics (Hunt, 1999). In spite of the multiple origins of eusociality in the order Hymenoptera, eusocial species are clustered in only 4 closely related families out of the total of 80 families. Hunt therefore correctly argues that, rather than focus exclusively on traits characteristic of the entire order (ordinal-level characteristics), we should focus on traits characteristic of those infraordinal taxa that actually contain eusocial species. Using cladograms of the superfamilies and families in the order Hymenoptera and of the family Vespidae, Hunt has attempted to identify traits that are associated with the appearance of the entirely eusocial family Vespidae. He finds that "high salience traits that closely subtend vespid eusociality include nesting, oviposition into an empty nest cell, progressive provisioning of larvae, adult nourishment during larval provision malaxation and inequitable food distribution among nestmates." But what about haplodiploidy itself? Hunt's analysis suggests no role for genetic asymmetries created by haplodiploidy but does suggest a significant role for the ability of haplodiploidy to help produce an all-female brood (see point 2B above).

15 FACTORS THAT REMAIN TO BE EXPLORED

$\mathcal{T}$HE unified model presented in Chapter 14 does not include some factors that must surely influence the inclusive fitness of solitary foundresses and workers in *R. marginata* (see, for example, Gadagkar et al., 1991c). Because I had no good data on delayed reproductive maturation under natural conditions, I assumed that neither solitary foundresses nor workers experienced any delay in attaining reproductive maturity. Because I did not know how to assess the consequences of the observed pre-imaginal caste bias, leading to 50% non–egg layers under laboratory conditions, I did not incorporate the effects of physiological predisposition in computing inclusive fitness. Also, the fact that foundresses often drift from one nest to another during the founding of a colony should lead to a reduction in worker-brood relatedness in pre-emergence colonies; but because I had no way to assess this effect quantitatively, it was not incorporated in the model.

15.1. Mutualism, "Hopeful" Queens, and the Gambling Hypothesis

I believe that another neglected factor is mutualism, made famous in the context of social evolution in insects by Lin and Michener (1972). The presence of permanently sterile worker castes is the most prominent and seemingly paradoxical feature of highly eusocial insects. Not surprisingly, most theories of the evolution of eusociality attempt to explain the attainment of this evolutionary peak. Somewhat unfortunately though, such an emphasis has sometimes obscured the rather obvious fact that the many social insect species that do not possess permanently sterile workers could nevertheless have preadaptations necessary for the evolution of the highly eusocial state. Lin and Michener (1972) lamented that because of the excessive focus on kin selection, it is commonly assumed that most primitively eusocial insects live in mother-daughter colonies and that transfer of individuals from one colony to another is abnormal. They argued that these assumptions are erroneous, that "there is considerable evidence that social behavior in insects is in part mutualistic," and therefore that social colonies can exist without altruism. Similarly, West-Eberhard (1978b) says that "whether among relatives or not, as long as a female has 'hope' of laying eggs—at least some small probability of future reproduction—her participation in the worker tasks of the colony can be viewed as possibly or partially an investment in her own reproductive future."

I have always thought that these arguments are particularly relevant to understanding the evolution of worker behavior in *R. marginata* (Gadagkar, 1985a), because of several characteristics of the species: the low levels of intracolony genetic relatedness; the drifting of foundresses between pre-emergence nests; the lack of intracolony kin discrimination; thecooperation among unrelated individuals; and the availability of opportunities for queen replacements during the perennial, indeterminate nesting cycle in a tropical environment (see especially Chapters 4, 9, and 10). It has been argued correctly that any mutualistic benefits get automatically accounted for in the benefit and cost terms of Hamilton's rule and that mutualism is therefore subsumed under kin selection (see, for example, Bourke and Franks, 1995). But the theory of mutualism has sometimes been dismissed as incapable of explaining the evolution of a sterile worker caste because the term mutualism suggests that all participants benefit (see, for example, Crozier, 1977). This argument deserves reexamination. Consider a situation where an individual that nests with a group obtains more fitness, on average, than it would as a solitary individual. If we replace the concept of alleles programming individuals into workers with the idea that alleles program individuals to take the risk of being part of the group, then under certain ecological conditions, such "gamblers" will be fitter than the risk-averse solitary individuals. Let us take a specific example. Consider two wasps that come together and nest jointly. If their joint productivity (say, 21 offspring) is greater than the sum of their individual productivities (say, $10 + 10 = 20$) in the solitary mode, and if the roles of fertile queen and sterile worker are assigned randomly, these wasps who take the risk of joint nesting will, on average, produce 10.5 offspring and thereby do better than those who shy away from the risk of joint nesting and produce only 10 offspring. This gambling hypothesis does not require that the wasps discriminate between close and distant kin but does require the ability to detect those potential partners with which joint nesting is more risky than it is with others (Gadagkar, 1990e, 1991a).

It appears to me, therefore, that maximizing inclusive fitness is not just a matter of identifying the closest available relatives, staying close to them, and working hard to allow them to reproduce. Instead, it probably involves a great deal of strategic decision making, based on the assessment and comparison of many variables, internal and external to the wasps. Rather than a well-developed ability to discriminate kin, highly complex, flexible, and cognitive behaviors may be the required preadaptation for the origin of eusociality. Much study of such behaviors is needed to determine their possible role in social evolution.

15.2.   Complex, Flexible, and Cognitive Behaviors in *Ropalidia marginata* and *Ropalidia cyathiformis*

What is the evidence for complex, flexible, and cognitive behavior in *R. marginata*? Recall the following facts: An *R. marginata* female has several options, one of them being to leave her natal nest to found or join another colony (Chapter 4). Whether a female founds her own solitary nest or joins another colony as a worker appears to depend on her brood-rearing abilities in the solitary founding mode (Chapter 11). While attempting to join other nests, females may move from one nest to another, perhaps looking for a less risky partner to nest with (Chapter 10). Another option is to remain in the natal nest, work for some time, and then attempt to replace the original queen (Chapters 4 and 6). During queen succession, no individual seems guaranteed to become the next queen; potential queens are best described as unspecialized intermediates (Chapter 6). The system of behavioral caste differentiation appears to have evolved to permit the wasps to achieve an optimum balance between maximizing the chances of their direct reproduction and the welfare of their colonies (Chapter 5). Mature colonies allow unrelated intruders that are young to remain in the vicinity, without displaying much aggression. But intruders that may pose a reproductive threat are singled out for intense aggression, often resulting in the death of the intruders. While most of the colony members are busy sorting out different kinds of intruders and dealing with them appropriately, some of the wasps from the same colony defect and team up with those intruders that are not killed and form new colonies with them, even if that means accepting one of the unrelated intruders as their new queen (Chapter 10). The young intruders who do get accepted into alien colonies appear to become so well integrated into their foster colonies that they are behaviorally indistinguishable from the natal wasps. The accepted aliens may go on to become foragers and also have a fair chance of becoming replacement queens in their foster colonies (Chapter 10).

Surely the great variety of behavior exhibited by *R. marginata* females requires a fairly complex and flexible behavioral repertoire. The related species *R. cyathiformis* appears to have an even richer repertoire of complex, flexible, and cognitive behaviors. I describe below three very suggestive examples.

15.2.1.   The Case of a Colony Fission

A colony of *R. cyathiformis* under observation in April-May 1981 began to show a steep decline in both the number of adults present on the colony and the number of brood being reared. I feared that, as often happens, the wasps would abandon the colony, bringing a premature end to my study. What happened instead was very interesting. On the evening of 31 May, I had left the colony with 11 adult females, all individually marked with unique spots of colored paint. On my arrival on the morning of 1 June, I noticed with dismay that only 6 of the 11 females remained on the nest. It is not unusual for 1 or 2 wasps at a time to disappear from such colonies. But the disappearance of 5 wasps (nearly half the population) overnight disturbed me. I started looking for the missing wasps, and amazingly only a few minutes later I found all 5 of them. But even more amazingly the wasps were not just sitting there; they had a small nest of their own.

Clearly, these 5 wasps had deserted their original colony, perhaps revolting against the queen, and had started building their own nest. One individual, OTBAA, one of the particularly aggressive individuals on the original nest, had established herself as the queen in the new nest. My disappointment at the temporary loss of half my wasps turned into great excitement. Clearly, half the population had left their declining colony and ventured out on their own. Perhaps the aggressive OTBAA had led the revolt and left with her followers.

I could easily imagine that OTBAA became dissatisfied with the state of

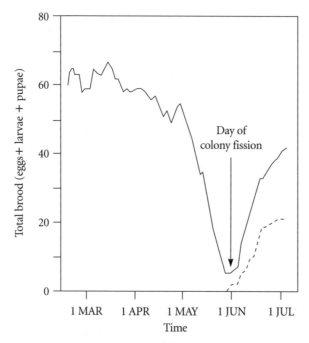

Figure 15.1. Total brood on the three combs of the parent nest until 30 May 1981, the day of colony fission. After the fission, total brood on the parent nest and the new nest put together (solid line) and brood on the new nest alone (broken line) are shown. (Redrawn from Gadagkar and Joshi, 1985.)

the original colony, and not being able to dislodge the original queen, BLATA, she simply left. But, I wondered, what would be the consequences of her departure for the rebels that left with her and indeed for the loyalists that remained in the original colony? The easiest way to find out was of course to include the new nest in my observations. Remarkably, the colony fission turned out to be good for both the rebels and the loyalists. The rebels did very well; their colony grew rapidly and they began to rear brood quite successfully. The loyalists in the original colony also benefited. Before the fission, the total number of brood had been declining precipitously; after the fission, the loyalists began to rear brood quite successfully (Figure 15.1). Clearly, the fission increased the fitness of both the rebels and the loyalists (Gadagkar and Joshi, 1985). But why was there such a difference in the state of the original nest before and after fission? It was my impression that there was a great deal of aggression among the wasps before fission. A quantitative analysis of the behavior of the wasps before and after fission confirmed this suspicion. There were significantly higher rates of dominance behavior per individual on the original colony before the fission than there were per individual among either the loyalists or the rebels after fission.

An analysis of the patterns of aggression before the fission was even more instructive. Having identified the loyalists and the rebels, I could go back to the behavioral data on these particular wasps in my computer files and compare the behavior of the loyalists and the rebels before the fission occurred. It turned out that the loyalists were the real aggressors; they showed much more aggression toward the rebels than the rebels did toward the loyalists (Figure 15.2). Indeed, the loyalists appeared to have driven away a number of other individuals during April and May 1981, although I do not know the fate of these wasps (Figure 15.3). It is reasonable to conclude therefore that high rates of aggression reduced the efficiency of brood rearing before colony fission and that low rates of aggression, in both colonies after fission, allowed efficient brood rearing (Gadagkar and Joshi, 1985).

An examination of the caste and age composition of the loyalists and rebels before the fission was equally interesting. We (Gadagkar and Joshi, 1985) used multivariate statistical analysis of time-activity budgets to assess behavioral caste differentiation before and after colony fission. Recall that in both *R. marginata* and *R. cyathiformis,* wasps can be classified into three behavioral castes: sitters, fighters, and foragers. Recall also that queens in *R. marginata* are sitters and queens in *R. cyathiformis* are fighters (see Chapter 5). At the time of the colony fission, the original queen, BLATA (herself a fighter), was 72 days old; she managed to retain with her STNBA, a 58-day-old forager; OTYA, a 24-day-old forager; BLTDA, a 10-day-old sitter; and BATRA and BATYA, sitters aged 5 and 3 days, respectively. The rebel queen, OTBAA, also herself a fighter in the colony before fission, although only 19 days old on the day of the fission, managed to take away 4 foragers, STDGA, RTDGA, OTLBA, and BATSA, 58, 37, 17, and 8 days old, respectively, on the day of the fission. It appears therefore that the group of 11 wasps split about as evenly as possible under the circumstances, an interpretation that is strengthened by the fact that the behavioral castes of the wasps, among both the loyalists and the rebels did not change after the fission (Table 15.1). I imagine that no sitter went with the rebels because at least 2 of the 3 sitters were probably too young to leave.

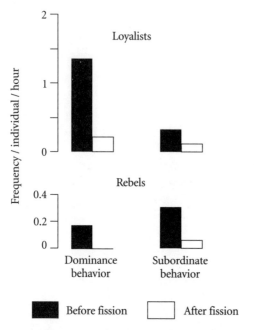

Figure 15.2. Mean frequencies of dominance behavior and subordinate behavior per individual per hour shown by loyalists and rebels before and after the colony fission. (Redrawn from Gadagkar and Joshi, 1985).

But how did the rebels manage to get together and leave at the same time and reach the same site to start a new nest? Was it a "snap" decision made on the night of May 31 or was it the result of something "brewing" for some time? Was there some form of "groupism" well before the actual fission? To investigate these questions, or at least to begin to do so, we (Gadagkar and Joshi, 1985) measured behavioral coordination within and between subgroups by computing Yule's association coefficients between pairs of wasps. We then asked the question whether there was more co-ordination within subgroups than between subgroups. For instance, did wasps within a subgroup synchronize their trips away from the nest and did rebels and loyalists avoid each other? The rebels had high association coefficients among themselves. Of the 10 possible pairs of individuals among the 5 rebels (excluding 5 self-pairs and 10 repeats), 5 pairs had the highest possible association of $+1$, while the mean for the 10 pairs was 0.69. Monte Carlo simulations show that this value could not have been obtained by chance alone ($p < 0.01$). Similarly, the loyalists among themselves also had positive association coefficients, although the results were

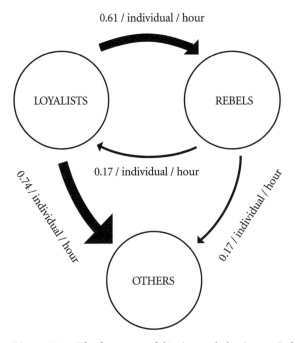

Figure 15.3. The frequency of dominance behavior per individual per hour shown by loyalists and rebels toward each other and toward others. Here "others" refers to animals that were present at some time before colony fission but had disappeared by the day of the fission and therefore could not be classed as either loyalists or rebels. (Redrawn from Gadagkar and Joshi, 1985.)

not statistically significant ($p > 0.05$). In contrast, when we chose pairs of wasps such that one was a loyalist and the other a rebel, of the 15 possible pairs with 6 loyalists and 5 rebels, 4 pairs had the lowest possible value of -1 and the mean of the 30 pairs was -0.26. This result also could not have been obtained by chance ($p < 0.05$). Together, these results demonstrate that the wasps had differentiated into two subgroups well before the fission, with the loyalists and rebels behaving as two coordinated groups and avoiding each other (Gadagkar and Joshi, 1985). These results also suggest that the wasps are capable of individual recognition and had some way of deciding when to leave and where to go.

15.2.2.   Do Wasps Form Alliances?

In early 1985 I had another nest of *R. cyathiformis* under observation for the purpose of removing the queen to see which wasp would be the next queen. The behavior of two of the wasps was particularly interesting. RT

Table 15.1.   Caste and age composition of loyalists and rebels. (Data from Gadagkar and Joshi, 1985.)

Wasp	Loyalist/rebel	Behavioral caste[c]	Age in days at the time of fission
BLATA[a]	Loyalist	Fighter	72
STNBA	Loyalist	Forager	58
OTYA	Loyalist	Forager	24
BLTDA	Loyalist	Sitter	10
BATRA	Loyalist	Sitter	5
BATYA	Loyalist	Sitter	3
STDGA	Rebel	Forager	58
RTDGA	Rebel	Forager	37
OTBAA[b]	Rebel	Fighter	19
OTLBA	Rebel	Forager	17
BATSA	Rebel	Forager	8

a. Queen in the original colony as well as in the loyalist group.

b. Queen after fission, in the rebel group.

c. The same caste was maintained after fission, for all animals.

was very aggressive and particularly so toward DBA. She harassed DBA so often and for such prolonged periods of time that on several occasions I noticed that the queen intervened. The queen actually climbed on the grappling mass of RT and DBA and separated them. This intervention was clearly of great help to DBA, which was no match for RT. I got the distinct impression that DBA in turn was trying not only to avoid RT but also to appease the queen.

The most dramatic example of this behavior occurred one day when DBA returned with food but before she could land on the nest RT noticed her and poised herself to snatch the food from DBA. It appeared to me that DBA did not want to give the food to RT. It also appeared that she wanted to give the food to the queen. But the queen was looking the other way and did not see DBA arrive. DBA's response was very interesting. She landed on the leaf on which the nest was built about 2 cm away from the nest, something that returning foragers seldom do—they usually alight on the nest. With DBA sitting on the leaf and RT sitting on the nest, they went through what might be described as a war of attrition for over 5 min; DBA would attempt to get on the nest but RT would block her way and try to grab the food. Having succeeded neither in attracting the attention of the queen nor in climbing onto the nest without losing the food load to RT, DBA finally simply walked around the nest until she was in full view of the queen. The queen seemed to sense immediately what was going on. She let DBA climb onto the nest and took the food load from her, but at the same time RT pounced on DBA and bit her. Before too long, DBA managed to escape from RT and fly away.

The goings-on in this nest became even more intriguing after I removed the queen. Clearly, RT was the next most dominant individual and I had little doubt she would be the next queen. But to my surprise, it was DBA who become the next queen, in spite of the presence of RT. Indeed, RT stayed on for over a month after DBA took over, although I cannot help describing her behavior as sulking: she would do nothing at all except occasionally take some food from one of the foragers. She did not participate in any nest activity. Even sitters do a fair amount of intranidal work, but not RT, an individual that therefore could not be classified as either a sitter, a fighter, or a forager by the multivariate techniques that I normally use— she was a clear outlier.

Why was RT so much more aggressive toward DBA than toward other individuals? Why was the queen so "considerate" to DBA? Was there some kind of alliance between DBA and the queen? If so, did this in any way influence DBA's becoming the next queen when I removed the original queen, even though RT was higher in the dominance hierarchy? Since the events I described above were witnessed only once, I will not pretend to answer these questions, but clearly they are pointers to the potential flexibility and complexity of the behavior of the wasps and to the possibility of alliances between wasps.

15.2.3.   Do Workers Choose Their Queens?

During a similar queen-removal experiment with *R. cyathiformis*, I once had a situation when there were two contenders, as it were, to replace the existing queen. These were DBT and OT, both more or less equally dominant. When I removed the queen on 9 March 1985, DBT became the replacement queen and OT promptly left the colony. However, DBT apparently was not a very "good" queen. All the other wasps stopped foraging and began to simply sit on the nest. Even when they did go out, they always returned with nothing. Clearly DBT had eggs to lay because she began to cannibalize existing eggs to make room for her to lay her own, since no wasp would bring building material or build new cells. Eventually, other

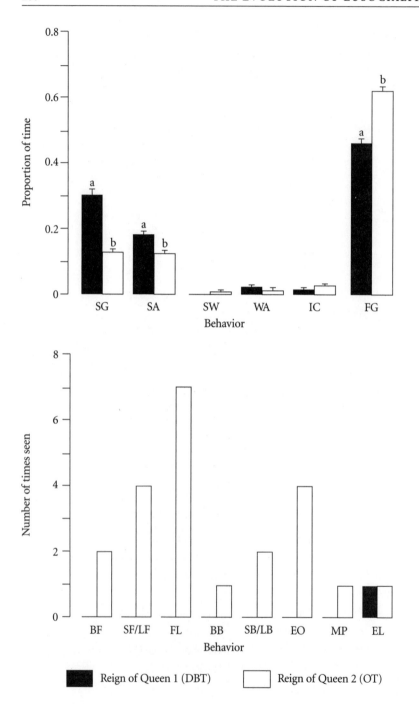

Figure 15.4 (facing page). Behavioral variation in a group of *R. cyathiformis* females during the reigns of two different queens. SG = sit and grom; SA = sit with raised antennae; SW = sit with raised antennae and wings; WA = walk; IC = inspect cells; FG = forage; BF = bring food; SF/LF = snatch/lose food; FL = feed larvae; BB = bring building material; SB/LB = snatch/lose building material; EO = extend walls of existing cells; MP = mouthing pedicel; and EL = egg laying. Above: Mean proportions of time spent in six common behaviors by workers. Below: Mean rates of performance of eight behaviors by workers. Comparisons are by the Mann-Whitney U test. (Redrawn from Gadagkar, 1995.)

---

wasps began cannibalizing brood too, and I was afraid that the colony would be abandoned. I was amazed to notice, however, that OT had not quite given up. She occasionally came back, as if to check on how DBT was doing. She never spent the night on the nest, merely visiting occasionally. By about 20 March, OT returned for good and DBT left. A pity that I was not there to witness their meeting! Now the behavior of the rest of the wasps changed dramatically. They began to work—they foraged, brought food, fed larvae, extended the walls of the cells of the growing larvae, and even brought building material and built new cells for their new queen to lay eggs in.

The story does not quite end there. DBT also, it turned out, had not quite left the nest. From time to time, she also came to visit, as if to see how her rival was doing. After a few days DBT rejoined the nest, but not before experiencing a great deal of hostility from the resident wasps. DBT had to spend nearly a whole day waiting and be subordinated by several residents before she was accepted back. It is the striking difference in the behavior of the same set of wasps during the reign of two different queens that is most suggestive (Figure 15.4). This variation in behavior once again not only points to these wasps' capacity for individual recognition but also suggests that they are able to modify their behavior on the basis of such recognition. Why did the wasps not cooperate with DBT when she first took over as the queen? If she was simply not good enough to be a queen, why did she succeed in the first place, especially in the presence of OT? Again, I will not pretend to be able to answer these questions, but even without answers, the questions themselves provide evidence of the complexity and flexibility of the behavior of these wasps.

15.3.   Comparison of *R. marginata* and *R. cyathiformis*

Since *R. marginata* and *R. cyathiformis* are two congeneric primitively eusocial species that coexist in the same habitats, it might be appropriate to ignore any differences between them and draw upon the information available on either of them to explore the question of the origin and evolution of social life in insects. But that is not why I have turned to *R. cyathiformis* from time to time. Whether it was in Chapter 5, where I compared behavioral caste differentiation in the two species, in Chapter 6, where I compared the position of the queen in the dominance hierarchies of the two species, or in this chapter, where I am contrasting the apparent levels of complex, flexible, and cognitive behaviors seen in the two species, my emphasis is always on the differences between the two species. Focusing on the differences between closely related species is often far more helpful than ignoring the differences and assuming that one can simply enlarge one's empirical database by pooling information on both of them.

Let me first recapitulate the behavioral differences between *R. marginata* and *R. cyathiformis*. *R. marginata* is strictly monogynous, but the queen is a docile sitter, seldom at the top of the behavioral dominance hierarchy of a mature colony. Most of the time, she appears to use a nonbehavioral (probably pheromonal) means of regulating the reproductive activities of her workers, and seems to do nothing to regulate the nonreproductive activities of her workers. But during the founding of new colonies or during the period when an individual is establishing herself as the new queen, she is strikingly different—very active and aggressive and at the top of the dominance hierarchy. In contrast, *R. cyathiformis* colonies are often polygynous, even though the queen is always one of the most aggressive and the most behaviorally dominant individual and therefore gets classified as a fighter in my system of behavioral caste differentiation.

There are other striking differences between the two species that I mentioned briefly in Chapter 3. Unlike *R. marginata* colonies, *R. cyathiformis* colonies are rather hard to work with. The reason is not that the wasps are aggressive. On the contrary, they are extremely timid—they prefer to fly away and even abandon their colonies, brood and all, at the slightest disturbance. They are smaller than *R. marginata* and although they have a sting it can hardly pierce my skin. Their timidity makes marking the wasps difficult and transplanting the colonies into the vespiary almost impossible. They do not even accept a wire-mesh hood around their nest as a protection from the predator *Vespa tropica*, which incidentally has no hesitation in attacking their rather small nests. My student Sujata Kardile is finding it very difficult to do queen-removal experiments, and even when she succeeds in removing the queen, it is hard to put her back—as soon as the queen is released onto the nest, she usually flies away, and seldom returns. A possible underlying explanation of this behavior is that in *R. cyathiformis*, the wasps have relatively little "commitment" to their nest and their brood; they seem much more concerned about their own safety than about that of their brood. This possibility, coupled with the finding that *R. cyathiformis* queens use overt physical means (rather than pheromones) to suppress their workers and that they are not very successful in doing so, permits the speculation that social life in *R. cyathiformis* is more primitive that of *R. marginata*.

There are yet other differences between the two species that are consistent with such an interpretation. Dominance behavior in *R. marginata* is almost always ritualized and appears not to cause any physical harm to the participants (except sometimes when a new queen is establishing herself). In *R. cyathiformis*, by contrast, dominance behavior is much less ritualized and much more diverse in its form. Dominant individuals not only attack, peck, chase, and nibble subordinates, as in *R. marginata*, but also frequently hold another individual's legs, wings, or antennae in their mandibles or sit on them and thus immobilize them for several minutes at a time. Never in *R. marginata*, but often in *R. cyathiformis*, I have seen dominance interactions between more than two wasps at a time. Sometimes one individual simultaneously dominates more than one opponent, but quite often, and much more interestingly, one individual, usually the queen, separates two fighting wasps, thus, temporarily at least, ending their interaction. Because of the nonritualized nature of dominance behavior, such intervention by the queen may be of considerable help to the weaker of the two that she is separating. Simultaneous, nonritualized aggressive interactions between several wasps at a time and the possibility of some of them intervening and helping some interactants more than others—such behavior surely sets the stage for the formation of coalitions and the evolution of the other complex behaviors I have described above. My students and I have observed *R. marginata* colonies for so long and so intensely that we could not have missed seeing these behaviors in that species; indeed, as my students confirm, the contrasts between the two species become evident after just a few days of observation.

I propose that the distinctly greater degree of complex, flexible, and cognitive behaviors displayed by *R. cyathiformis*, compared to *R. marginata*, is linked to the primitive level of its social organization relative to the level of social organization of *R. marginata*. At first one might imagine that social life, rather than solitary life, would require a greater capacity for complex, flexible, and cognitive behaviors. But as I have argued before, maximizing inclusive fitness during the origin or early evolution of eusociality may involve a great deal of strategic decision making, which requires highly complex, flexible, and cognitive behaviors as a preadaptation. Thus it might well be that as social life becomes more advanced and interactions become more ritualized and communication and control begin to involve pheromones, the behavioral repertoire of the wasps actually degenerates into less variable and more stereotyped components. Turning to *R. cyathiformis* in the search for better examples of the role of complex, flexible, and cognitive behaviors in the origin of eusociality is therefore entirely reasonable. For this reason I consider the role of complex, flexible, and cognitive behaviors among the most important factors not included in my quantitative modeling of factors responsible for the evolution of sociality. There is no denying, however, that the road to assessing and quantifying the import of mutualism in general, and of what might be called "intelligent" behaviors in particular, is bound to be long and arduous.

PART IV

$\mathscr{B}$eyond *Ropalidia marginata:* Social Evolution, Forward and Reverse

16   A ROUTE TO SOCIALITY

$\mathcal{U}$NTIL now, my main focus has always been *R. marginata*, although I have occasionally turned to *R. cyathiformis* in the effort to better understand *R. marginata*. In this last part, however, I propose to go beyond *R. marginata* and consider three general issues in social evolution: (1) the possible sequence of events that might have led to evolution from the solitary to the highly eusocial state, in this chapter; (2) the possible evolutionary forces responsible for the extreme differentiation of castes in social insects, in Chapter 17; and (3) the possibility of reversal of social evolution, in Chapter 18.

Many people have considered the question of the evolutionary route to eusociality (Wilson, 1971). William Morton Wheeler (Wheeler, 1923, 1928), who was greatly influenced by the early studies on wasps, formally labeled and championed the so-called subsocial route earlier proposed by Roubaud (1916). This hypothesis envisages that colonies with extended maternal care are starting points for the ultimate evolution of the situation where the daughters stay back and help their mother produce more offspring. Realizing the obvious inappropriateness of this hypothesis for many bee species, Michener (1958, 1974; see also Lin and Michener, 1972) proposed the semisocial route, arguing that associations of adults of the same generation, rather than a family unit containing two generations, is a more likely starting point. More recently, West-Eberhard (1978a; see also Carpenter, 1989, 1991; Itô, 1993b) has modified the semisocial hypothesis and proposed the polygynous family hypothesis, which argues that a rudimentary caste containing a polygynous stage may have preceded the permanently monogynous stage. I wish to focus here not on the kinds of individuals or colonies that occur in different stages in evolution (subsocial

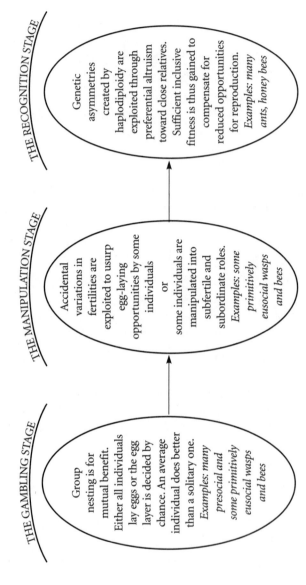

Figure 16.1. The route to the highly eusocial state from the solitary state through the gambling, manipulation, and recognition stages. The examples given for each stage are tentative, since our knowledge of the causes and consequences of group living in most social insect groups is sketchy. (Redrawn from Gadagkar, 1991a.)

THE GAMBLING STAGE

Group nesting is for mutual benefit. Either all individuals lay eggs or the egg layer is decided by chance. An average individual does better than a solitary one. *Examples: many presocial and some primitively eusocial wasps and bees*

THE MANIPULATION STAGE

Accidental variations in fertilities are exploited to usurp egg-laying opportunities by some individuals
or
some individuals are manipulated into subfertile and subordinate roles. *Examples: some primitively eusocial wasps and bees*

THE RECOGNITION STAGE

Genetic asymmetries created by haplodiploidy are exploited through preferential altruism toward close relatives. Sufficient inclusive fitness is thus gained to compensate for reduced opportunities for reproduction. *Examples: many ants, honey bees* or semisocial, polygynous or monogynous), but on the selective advantages of group living that must have sustained social evolution at different stages. Such a focus is important because knowledge of the kinds of colonies or associations does not necessarily tell us the selective advantages that promote them. For example, both subsocial and semisocial colonies may persist either because of opportunities for the workers to reproduce at some future time and thus gain direct fitness or because of opportunities for them to rear very closely related brood and thus maximize their indirect fitness.

When I first proposed the following hypothetical route to eusociality (Gadagkar, 1990d, 1991a), there were three main classes of theories to explain the evolution of eusociality (for an excellent review of the theoretical discussions during the 1970s, see Starr, 1979): (1) the haplodiploidy hypothesis (Hamilton, 1964a, 1964b); (2) the theory of parental manipulation (Alexander, 1974), or (for my purpose here) the equivalent subfertility hypothesis (West, 1967; West-Eberhard, 1975); and (3) the theory of mutualism (Lin and Michener, 1972). It often used to be said that the three ideas are not mutually exclusive and that they may all have played a role in shaping the evolution of eusociality. I cannot disagree with this statement but I think that it is of little heuristic value unless we can assign more specific roles to each theory. Thus my motivation for hypothesizing a route to eusociality was in part to make such assignments. I therefore proposed three sequential stages in the evolution of eusociality (Figure 16.1).

16.1. The Gambling Stage

It seems reasonable to assume that the initial incentive for group living comes from mutualism, reciprocal altruism, and the benefits of "gambling." I call this the gambling stage. As previously discussed, this stage requires no preadaptation for intragroup manipulation or for recognition of genetic relatedness, although group living among kin (individuals that might come together merely by virtue of being neighbors) will evolve more easily (West-Eberhard, 1978a; Schwarz, 1988). The only prerequisite for the evolution of incipient societies by mutualism is a sufficiently complex behavioral repertoire (what I called complex, flexible, and cognitive behaviors in Chapter 15) to permit the necessary interactions. Solitary wasps seem to possess an appropriately diverse array of behaviors (Rau and Rau, 1918; Rau, 1933; Tinbergen, 1932, 1935; Tinbergen and Kruyt, 1938; Alcock, 1974; Evans and O'Neill, 1978; Brockmann and Dawkins, 1979; Brockmann, Grafen, and Dawkins, 1979; Field, 1989). It is easy to imagine that many presocial wasps, such as those listed in West-Eberhard (1978a) and described by West-Eberhard (1987) and Wcislo, West-Eberhard, and Eberhard (1988), are either already at this stage or resemble immediate forerunners of this stage. I would argue that *R. cyathiformis* either is still at this stage or still retains the critical adaptations required for this stage (Chapter 15).

16.2.   The Manipulation Stage

Once group living is established, the situation is ripe for the second stage, which I call the manipulation stage. Accidental variations in food supply leading to subfertility can now be exploited and the ability to manipulate offspring or other nestmates can be selected for. As manipulation becomes increasingly effective, the benefits of group living become increasingly unavailable to some individuals, and they begin to lose reproductive options and get trapped into worker roles. All that we have seen of *R. marginata* so far suggests that it is the perfect example of a species at this manipulation stage of social evolution: plenty of inequalities between individual wasps; a sufficiently complex behavioral repertoire to permit exploitation of those inequalities, but perhaps no longer the extent of behavioral flexibility characteristic of *R. cyathiformis;* and insufficient ability to discriminate between different levels of genetic relatedness to exploit the benefits of haplodiploidy (see below).

16.3.   The Recognition Stage

Now it is precisely at the next and final stage, which I call the recognition stage, that the ability to recognize and give preferential aid to closer relatives will begin to have selective value. In other words, I propose that the benefits of haplodiploidy for social evolution become available at this final stage. This hypothesis helps explain one otherwise curious fact. Although kin recognition and nestmate discrimination studies have a long way to go, there is a growing impression that primitively eusocial species may not have the ability to discriminate between different levels of genetic relatedness once all the individuals are part of the same colony (Chapter 10; see also Gadagkar, 1985b; Gamboa, Reeve, and Pfennig, 1986; Venkataraman et al., 1988; Queller, Hughes, and Strassmann, 1990). Whether highly euso-cial species have the ability to discriminate levels of genetic relatedness within their colonies is debatable. But the highly eusocial species do seem to be better equipped to do so than the primitively eusocial species. There is some evidence that honey bee workers can discriminate between closely and distantly related larvae and adults, although the role of such kin dis-crimination in natural situations is debatable (Getz and Smith, 1983; Breed, Velthuis, and Robinson, 1984; Breed, Butler, and Stiller, 1985; Noonan, 1986; Visscher, 1986; Breed, Williams, and Fewell, 1988; Page, Robinson, and Fondrk, 1989; Carlin and Frumhoff, 1990; Oldroyd, Rinderer, and Buco, 1990; 1994; Page, Breed, and Getz, 1990; Page and Robinson, 1990, 1991; Arnold et al., 1996). There is strong evidence that, as predicted by the worker policing hypothesis, honey bee workers selectively destroy worker-laid eggs (related to them as sons of half-sisters) rather than queen-laid eggs (related to them as brothers) (Ratnieks, 1988; Rat-nieks and Visscher, 1989).

But perhaps the most compelling evidence for intracolony kin recogni-tion comes from the recent discovery of a green beard gene in the fire ant *Solenopsis invicta*. Hamilton (1964b) postulated, and Dawkins (1976) made famous, the idea of green beard genes—genes that recognize each other. For many years, most theorists were skeptical of green beard genes and there was a lack of empirical evidence for them. Recently Keller and Ross (1998) found a genetic locus, *Gp-9*, that has two alleles, *B* and *b*. The *bb* homozygotes, both queens and workers, appear to die early from intrin-sic causes. *Bb* heterozygotes are found among both queens and workers. *BB* homozygotes are found among workers and nonreproducing queens but not among egg-laying queens. Behavioral observations have shown that *BB* workers (those possessing at least one copy of the green beard gene, *b*) se-lectively killed nearly all those *BB* queens (those not possessing at least one copy of the green beard gene, *b*) that attempted to reproduce. Keller and Ross (1998) interpret *b* as the green beard gene itself, or as a gene that is closely linked to a green beard gene that confers (1) an externally percepti-ble label (smell), (2) the ability to recognize the presence and absence of this label on other individuals, and (3) the behavioral repertoire required to behave differently toward those who possess the label (not to kill) and those that lack it (to kill) (see also Gadagkar, 1998a, 1998b; Grafen, 1998; Hurst and McVean, 1998). There is, however, no such evidence for effective worker policing or for green beard genes in any primitively eusocial species.

If haplodiploidy and intracolony kin recognition were important for the origins of insect sociality, one would expect workers in primitively eusocial species to have the capacity to exploit the genetic asymmetries created by discriminating between close and distant relatives. But if the origin of group living lies in mutualistic benefits, as argued here, and if haplo-diploidy and intra-colony kin discrimination play a role in its subsequent maintenance, then the observed distribution of kinship discrimination abilities is no longer a paradox. Indeed, it is difficult to imagine how the ability to manipulate conspecifics or to discriminate between different levels of genetic relatedness among conspecifics (of the same sex) would have had selective value before the advent of group living. This view is consistent with the argument, made frequently in the literature (Lin and Michener, 1972; Alexander, 1974; Wilson, 1975; West-Eberhard, 1975, 1978a; Evans, 1977; Andersson, 1984; Gadagkar, 1990c, 1991c; Hughes et al., 1993), that haplodiploidy has little to do with the origin of eusociality.

The route to eusociality I propose above of course remains to be tested. One way to test it is to assess, in the spirit of what my students and I have done for *R. marginata*, the relative roles of various potential factors in tilt-ing the inclusive fitness inequality, for a variety of species, caught at vari-ous stages in the progress from the solitary to the highly eusocial stage.

17   THE EVOLUTION OF CASTE
POLYMORPHISM

$\mathcal{W}$ITH very few exceptions, highly eusocial insects, especially ants and termites, are unparalleled in the extent of intraspecies, intrasexual size variations and allometry. The most widespread differentiation is between fertile reproductives (queens) and sterile workers (Wilson, 1971). In the ants and termites, subgroups among workers may also be morphologically differentiated among themselves into soldiers and/or major, minor, and media workers (Wilson, 1953; Oster and Wilson, 1978; and Hölldobler and Wilson, 1990). Such morphological caste differentiation can be so extreme that different castes, if encountered separately, may get classified as different species (Wheeler, 1910). The greatest size variation has been recorded in the Asian marauder ant *Pheidologeton diversus* (Moffett, 1987), where major workers weigh 500 times as much as minor workers and have a head width 10 times as large. How such polymorphism has evolved and why, in its extreme form, it is essentially restricted to social insects are questions that have not yet received satisfactory answers.

17.1.   Evolutionary Novelty

The analogous problem of the appearance of major novel structures and novel lifestyles during evolution has often concerned evolutionary biologists. The solutions suggested appear to vary widely but I will argue that they have a common underlying theme: *genetic release followed by diversifying evolution.* When an organism becomes adapted to an environmental condition, stabilizing selection will normally prevent it from drifting away from the optimally adapted condition. In such a situation, a new factor can come along and release the organism from the constraints of stabilizing se-

lection and permit new directional selection, thereby permitting diversifying evolution and the appearance of evolutionary novelty. The mechanisms of genetic release and of the subsequent diversification differ from
case to case, but the action of stabilizing selection prior to the genetic release and the action of directional selection after the release are recurring
features in the appearance of evolutionary novelty (Gadagker, 1997a). Let
us consider some examples in the plant and animal worlds.

17.2.   Escape and Radiation Coevolution

The coevolution of escape and radiation was first proposed by Ehrlich and
Raven (1964) as a hypothesis for the coevolution of families of flowering
plants and butterflies, and has since been christened "escape and radiation
coevolution" and applied to interactions between parasites and their hosts
in general by Thompson (1989, 1994; see also Gilbert, 1990; Berenbaum,
1990). When plants produce novel secondary metabolites by mutation and
recombination, they become free of their usual insect predators, and thus
enter new niches and undergo evolutionary radiation. Mutations and recombination in some of the insect populations permit them to overcome
the plant resistance. This allows them to escape from competition with
unmodified insect populations, and thus to enter new niches and to undergo new evolutionary radiation.

17.3.   Enemy-Free Space and the Evolution of Host Shifts

Selection of host plants by herbivorous insects is dependent not only on
the allelochemical profiles of the host species but on a variety of other
host-related factors. The presence or absence of ants on the host plants is
an unusual cue that butterflies of the family Lycaenidae use to land as well
as to oviposit on particular host plants (Fukuda et al., 1978; Atsatt, 1981a;
1981b; Henning, 1983; Pierce and Elgar, 1985). Because the ants protect
the butterfly larvae from their predators and parasites, this cue has important fitness consequences for the ovipositing female. The use of such "enemy-free" space for oviposition also has other important consequences for
the butterfly species. In many cases this behavior has led, on the part of the
butterfly larvae and pupae, to the evolution of the habit of secreting drop-
lets of carbohydrates and amino acids as food for the ants. Even more im-

portant, it has also led to significant changes in the host ranges of the butterfly species. The ability to exploit such enemy-free space permits the butterflies to utilize host species not normally available to other species that have not coevolved with ants in this manner. A comparative study of 285 species of lycaenid butterflies led Pierce and Elgar (1985) to conclude that "major trends in host plant selection by phytophagous lycaenids may thus be explained in terms of freedoms and constraints that are ultimately the result of pressures exerted by the parasitoids and predators of these butterflies."

17.4. Amelioration of Trade-offs between Resistance and Competitive Ability

Quite often the extent of directional selection on a character is constrained by negative correlation of the character under directional selection with other characters. Perhaps the best known examples are trade-offs between resistance to insecticides or parasites and general competitive ability. Bacterial and insect strains resistant to antibiotics or insecticides are thought to be kept at very low frequencies in the absence of the antibiotic or insecticide because the resistant phenotype confers no advantage in this environment and indeed may be disadvantageous.

An important reason for such negative correlation between the effects of a gene on two or more characters is antagonistic pleiotropy (Caspari, 1952; Wright, 1968; Dykhuizen and Davies, 1980; Falconer, 1981; Rose, 1982; Service and Lenski, 1982; Via, 1984; Lenski and Levin, 1985; Maynard Smith et al., 1985; Roff, 1990; Rose, Graves, and Hutchinson, 1990; Scott and Dingle, 1990). There are, however, ways in which such antagonistic pleiotropy can be broken. Two mechanisms for the amelioration of antagonistic pleiotropy have been suggested in the context of trade-offs between insecticide/parasite resistance and general competitive ability. One is somewhat trivial and involves selection among alternate alleles conferring resistance and fixation of the one that has the least antagonistic pleiotropic effects (Hall, 1983; Cohan and Hoffmann, 1986; Lenski, 1988a, 1988b). The other way, of more interest here, is the selection for epistatic modifiers that ameliorate maladaptive pleiotropic effects (Caspari, 1952; Fisher, 1958; Uyenoyama, 1986). Studies on the development of resistance to the insecticide Diazinon in the sheep blowfly *Lucilia cuprina* (McKenzie,

Whitten, and Adena, 1982; Clarke and McKenzie, 1987) and on the development of resistance to the bacteriophage T4 in the bacterium *E. coli* (Lenski, 1988a, 1988b) provide evidence for both mechanisms. What is most interesting for my purposes here is of course that such amelioration of antagonistic pleiotropy and the consequent amelioration of trade-offs result in genetic release from the constraints of stabilizing selection and permit new directional selection and hence diversifying evolution.

17.5. Evolution by Gene Duplication

First suggested by Haldane (1932) and Muller (1935), the idea of evolution by gene dupliction is that redundant, duplicate copies of genes can accumulate potentially lethal mutations without killing the organism and can eventually give rise to novel genes coding for novel structures via pathways that would be inaccessible to an individual with a single copy of the gene. Thus gene duplication once again releases previously existing constraints due to stabilizing selection and permits new directional selection and hence diversifying evolution. Ohno (1970), the most ardent champion of such a mechanism, has gone to the extent of suggesting that evolution by gene duplication is the only mechanism for the evolution of new genes. Today there is considerable evidence for evolution by gene duplication, which is recognized as an important mechanism for the evolution of genome size and of gene families such as the globin gene superfamily (Hardison, 1991; Li and Graur, 1991).

17.6. Inclusive Fitness

In social insects, especially in the highly eusocial species, one or a small number of colony members (queens) monopolize reproduction while the remaining function as sterile workers and spend their entire life working for the welfare of the colony and its queen(s). The evolution of such altruistic sterile worker castes in the social insects was considered paradoxical until Hamilton proposed the theory of inclusive fitness (Hamilton, 1964a, 1964b). As discussed in Chapter 8, today it is common practice to recognize inclusive fitness as having two components, a direct component, gained through production of offspring, and an indirect component, gained through aiding close genetic relatives. Sterile worker castes in social insects are expected to gain fitness exclusively through the indirect component, and in no other group is there a comparable level of dependence on the indirect component of inclusive fitness (see Chapter 8).

I argue that genetic release followed by diversifying evolution, similar to that discussed above, can occur when some individuals in a species begin to rely on the indirect component of inclusive fitness while others continue to rely on the direct component, as workers and queens in social insects are expected to do. In workers, genes concerned with mating and reproduction (which I will henceforth call queen-trait genes) can be thought of as duplicate, redundant copies, as can genes in queens concerned with foraging, nest building, and brood care (which I will henceforth call worker-trait genes). Despite the fact that queens and workers have the same genes, there is convincing evidence—although there are a few exceptions—that caste determination is due not to the presence of different genes in different individuals but to different nutritional, hormonal, and other environmental conditions experienced by different individuals (Wilson, 1971; Hölldobler and Wilson, 1990; Wheeler, 1986). There must therefore be different subsets or combinations of genes, even if they come from the same overall set, with different temporal patterns and/or intensities of activity that characterize different morphological castes (Figure 17.1). It is hard to imagine how natural selection can perfect the adaptations of different morphological castes for their specific roles without the presence of such caste-specific subsets of genes. The adaptation of castes for their caste-specific functions is indeed most impressive (Wilson, 1971; Hölldobler and Wilson, 1990; Bourke and Franks, 1995).

17.7. Genetic Release

In solitary ancestors of eusocial species, directional selection on queen-trait genes as well as worker-trait genes might be expected to have been constrained by the requirement that both queen and worker traits had to be optimized in the same individual, that is, under a single developmental program. In other words, antagonistic pleiotropy between queen-trait genes and worker-trait genes would have led, in solitary species, to epistasis for fitness among these genes such that the queen phenotype could not improve beyond a certain point without reducing the efficacy of the worker phenotype and vice versa. However, during the evolution of euso-

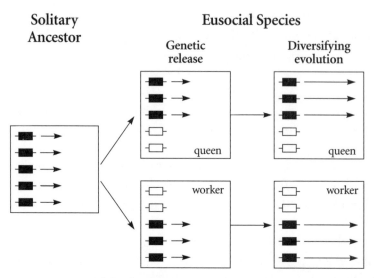

Figure 17.1. Schematic illustration of the evolution of caste polymorphism in social insects through the process of genetic release followed by diversifying evolution. Shown here are five genes (black boxes) in the solitary ancestor with intermediate levels of gene activity (represented by short arrows). In social insects, queens and workers go through different developmental programs, possibly involving activation (black boxes) and inactivation (open boxes) of different subsets or combinations of the same overall set of genes present in the solitary ancestor. Note that gene number three is shown to be active in both queens and workers to illustrate the point that it is the activation of a particular combination of genes that could characterize a caste-specific developmental pathway rather than a different subset of genes. The inactivation of some genes while the others remain active results in a genetic release for the active genes, whose epistatic interactions with the formerly active but presently inactive genes are now broken so that they may undergo new directional selection (represented by the longer arrows emanating from the active genes) leading to diversifying evolution in the final stage. (Redrawn from Gadagkar, 1997a.)

ciality, because of the evolution of a control switch that enables queen and worker developmental programs to be turned on in different individuals, workers begin to rely on queens to mate and provide all or most of the eggs and queens begin to rely on workers to do all or most of the foraging, nest building, and brood care. Hence the antagonistic pleiotropy between queen-trait genes and worker-trait genes breaks down, leading to a relaxation of stabilizing selection on both sets of genes. This in turn leads to divergence between workers and queens, because the structures previously subjected to stabilizing selection will begin to stray away from their previously perfected optima.

17.8. Diversifying Evolution

Further divergence between queens and workers then becomes possible through a process of diversifying evolution made possible by the genetic release. As a result of genetic release from antagonistic pleiotropy, the epistasis for fitness among queen-trait genes and worker-trait genes is broken, so that both sets of genes can now be subjected to new and more extreme directional selection (in contrasting directions), leading to diversifying evolution and hence to further differentiation of the castes. I postulate that continued feedback between genetic release due to relaxed stabilizing selection and diversifying selection due to new directional selection on the appropriate genes can give rise to the levels of queen-worker dimorphism seen in today's highly social insects (Figure 17.2).

17.9. Queen-Trait and Worker-Trait Genes

The idea that different subsets or combinations of genes are active in different castes in social insect species is hardly radical. This must also be true of males and females in all sexually reproducing organisms because, barring the few genes that reside on sex-specific chromosomes such as the Y chromosome, the genetic constitution is the same in males and females. And yet the two sexes can be phenotypically rather different, which must of course involve activation and inactivation of different subsets or combinations of genes (or a difference in the timing of activation and inactivation) from the same overall set characteristic of the species. Recent molecular-biological investigations of sex determination and dosage compensation in *Drosophila melanogaster* have uncovered intricate details of the mechanisms of such sex-specific gene expression, which include alternative splicing, transcriptional control, subcellular compartmentalization, and intracellular signal transduction (Burtis, 1993; Gorman, Kuroda, and Baker, 1993; Gorman, Franke, and Baker, 1995; Henikoff and Meneely, 1993; Kelley et al., 1995). Although no social insect species has been investigated with any comparable level of sophistication, the idea that different

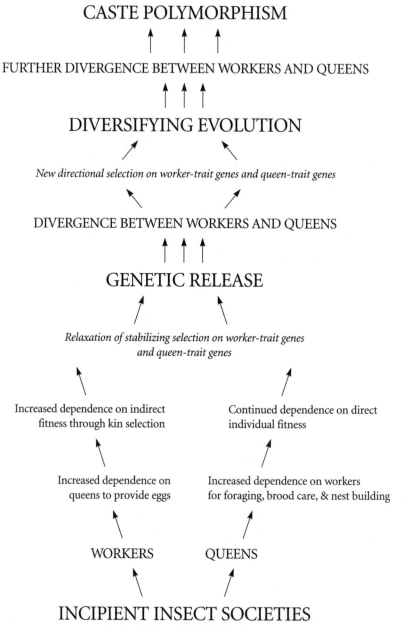

Figure 17.2. A flow chart for the evolution of caste and subcaste polymorphism in social insects, based on the idea of genetic release followed by diversifying evolution. (Redrawn from Gadagkar, 1997a.)

subsets or combinations of genes are active in different castes in social insect species is easily amenable to experimental verification. There is already some evidence of caste-specific gene expression in ants with reference to malate dehydrogenase (Hung and Vinson, 1977), α-glycerophosphate dehydrogenase (Hung, Dowler, and Vinson, 1977), and adenylate kinase (Craig and Crozier, 1978). More recently, Evans and Wheeler (1999) have obtained evidence of both queen-specific and worker-specific gene expression in the honey bee *Apis mellifera*. Indeed, Craig and Crozier (1978) have suggested that "patterns of differences probably arose via a gene duplication followed by functional divergence leading to one isozyme being superior in workers and the other in queens and males, as determined by the isozymes' kinetics and the requirements of the castes of the species studied."

17.10. The Power of Gene Duplication in Social Insects

During evolution by gene duplication, one copy of the gene remains structurally and functionally intact while the other copy is released from the constraints of stabilizing selection and subjected to new directional selection (albeit after a period of drift and accumulation of mutations), both copies being present in the same cell. During genetic release followed by diversifying evolution postulated for the social insects, two previously existing sets or combinations of genes are simultaneously released from the constraints of stabilizing selection and subjected to new directional selection in opposite directions. Such differential selection is made possible even though the two sets or combinations of genes are in the same individual because they are turned on in different individuals during alternate (queen and worker) developmental pathways. Nevertheless, genetic release followed by diversifying evolution, expected in social insects because some individuals begin to rely on the indirect component of inclusive fitness while others continue to rely on the direct component, has much in common with evolution by gene duplication. In addition to this similarity, traditional evolution by gene duplication also acquires a new power in social insects, where indirect fitness is available.

As we saw in Chapter 1, the honey bee queen produces a host of chemical substances that influence the behavior and physiology of the workers in her colony. Perhaps the most important component of the queen's pheromone blend is 9-keto-(E)2-decenoic acid (9 ODA). Also, workers produce mandibular gland secretions that are added to the brood food and may serve as preservatives and nutrients. Instead of the queen's 9 ODA, workers secrete a diacid, hydroxylated at the 10th or ω carbon atom, rather than the 9th or $\omega - 1$ carbon atom, as in the case of the queen's acids. In other words, queens and workers differ essentially only in the position of the carbon atom that is hydroxylated. But how does this difference arise? As a result of a series of experiments involving analysis of the fate of deuterated test compounds applied to excised queen and worker mandibular glands, Plettner and his colleagues (1996), using gas chromatography-mass spectrometry (GC-MS), have proposed a caste-specific, bifurcated, three-step biosynthetic pathway for the production of these compounds.

The starting point is stearic acid, a very common, 18-carbon, straight chain, saturated intermediate of lipid oxidation. In the first step of the proposed pheromone biosynthetic pathway, functionalization is achieved by the addition of a hydroxyl group on either the 18th (ω) or the 17th ($\omega - 1$) carbon atom in stearic acid. This functionalization, which foreshadows the queen-worker differences depending on whether it happens at the ω or the $\omega - 1$ carbon atom, is, however, itself not caste-specific; both ω and $\omega - 1$ functionalizations occur in both castes to about the same extent. In the second step, the 18-carbon hydroxy acids are shortened to give 10-HDA and 9-HDA by the standard chain-shortening cycles of β oxidation that normally occur during fatty acid metabolism. It is the β oxidation step that is caste-specific—queens preferentially channel the $\omega - 1$ compounds and workers channel the ω compounds into the oxidation pathway. In the final step, oxidation of the ω or $\omega - 1$ hydroxy group that was added in the first step results in the formation of diacid in the case of workers and keto acid in the case of queens.

I have hypothesized (Gadagkar, 1996b) that the pheromone biosynthetic pathway employed by the workers deviates relatively little from the typical lipid metabolism pathway and is perhaps simply adopted from there. The diacid they make can be relatively easily channeled into an energy-generating role, and its degradation products can be profitably fed into the Krebs cycle. I speculate, by contrast, that the pheromone biosynthetic pathway of the queens is quite a deviation from the standard lipid metabolism pathway. In particular, keto acid is not what one would expect if energy generation is the immediate goal. The expense involved in further breaking down the keto acid makes it a poor candidate to be fed into the Krebs cycle. I therefore speculate that in the course of making their pheromones, the workers are doing more or less what any solitary insect would do to generate energy from lipids, and that their pheromone biosynthetic pathway is therefore the more ancestral one. Conversely, queens have considerably modified the ancestral lipid metabolism pathway in order to make a pheromone that has only lately (relatively speaking) become necessary. Thus the function of the worker pheromone and the biochemical pathway involved in its production may be thought to be relatively more ancestral, and the function of the queen pheromone and the biochemical pathway involved in its production may be thought to be relatively more derived. If this is true, then it is not difficult to see the tremendous advantage of conventional gene duplication in developing the derived condition from the ancestral one. It seems likely that the enzymes involved in the β oxidation step give rise to specificity for substrates hydroxylated at the ω or $\omega - 1$ positions.

Imagine that the ancestor of the social insect species had a gene that coded for an enzyme that could deal only with the substrate that was hydroxylated at the ω position. The workers in the descendant social species can continue to use this gene and this enzyme to make worker pheromones, which may perhaps even have been made by the ancestor. A duplication of the gene involved can permit the evolution of an alternate enzyme that can handle the substrate hydroxylated at the $\omega - 1$ position. We know that such a substrate must have been available because both kinds of hydroxylations occur to an equal extent in both queens and workers (Plettner et al., 1996). The duplicated gene would now be free to evolve in new directions without reduced fitness owing to the reduction in the efficiency of energy production through lipid metabolism. Thus new directional evolution can sometimes give rise to substances with remarkable properties, such as the queen pheromone. A similar chance occurrence of such a mutation could hardly have been utilized effectively by a solitary species. Because social insects set aside some individuals for the sole purpose of monopolizing reproduction and inhibiting and manipulating all others, they are in a special position to exploit such a consequence of conventional gene duplication and evolve in directions that are not open to solitary species. In other words, conventional gene duplication and the habit of obtaining indirect fitness by aiding relatives can act in concert to

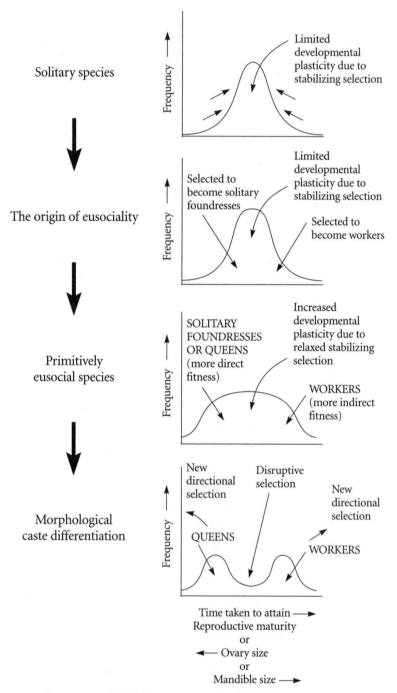

Figure 17.3. The origin and evolution of eusociality through selection for developmental plasticity. See the text for details. (Redrawn from Gadagkar, 1996a.)

provide a more powerful mechanism of genetic release followed by diversifying selection. This is what I was attempting to foreshadow in Chapter 1,
when I discussed the details of the biochemistry and physiology of the
honey bee queen pheromone.

17.11. From the Solitary to the Highly Eusocial

Demographic factors, such as initial variations in time taken to attain reproductive maturity for example, will make it worthwhile for some individuals to adopt worker-like roles and for others to assume queenlike roles,
leading to the origin of eusociality (West-Eberhard, 1978a; Craig, 1983;
Gadagkar, 1991b, 1996a). Such demographic factors can coevolve with
eusociality and become more pronounced because late reproducers will,
for example, have a smaller selective disadvantage in a eusocial species
than in a solitary species because of the possibility of gaining indirect
fitness in the former (Chapter 13). I now suggest genetic release followed
by diversifying evolution as a mechanism for such coevolution that can
potentially lead to sufficient divergence between queens and workers to
give rise to caste polymorphism.

What this means is that we can envisage the evolution of highly eusocial
species starting from completely solitary ancestors through selection for
developmental plasticity (Figure 17.3). In solitary species, any character,
such as time taken to attain reproductive maturity, ovary size, mandible
size, would have limited developmental plasticity on account of stabilizing
selection, because, as explained above, both queen and worker functions
would have to be optimized under a single developmental program. But
even here, under the right ecological, genetic, or demographic conditions
(Chapters 8–11), individuals at the extremes of the distribution of values
for these characters would be selected to take up predominantly or exclusively queenlike or worker-like roles. For example, individuals that have to
wait only a short time to attain reproductive maturity or have larger than
average ovaries would be fitter as queens than individuals at the opposite
ends of the distributions. Conversely, individuals with delayed attainment
of reproductive maturity, smaller than average ovaries, or larger than average mandibles (useful say, in transporting food back to the nest) would be
fitter as workers than individuals at the opposite ends of the distributions.
This stage may thus be thought of as the origin of eusociality. It should be

emphasized that genetic release followed by diversifying evolution is not being postulated as being involved in the origin of eusociality but only as being involved with the subsequent morphological, physiological, and behavioral differentiation of queens and workers. Thus as the worker-like individuals begin to rely increasingly on the indirect component of inclusive fitness and the queenlike individuals continue to depend on the direct component, there would be relaxed stabilizing selection and increased directional selection on appropriate genes, as described above.

As West-Eberhard (1979) has argued, such a process can go far enough to make intermediate individuals to be good neither at being queens nor at being workers and thus reinforce the process of morphological caste differentiation. Indeed, I suggest that genetic release followed by diversifying evolution as discussed and interpreted here provides a mechanism for West-Eberhard's (1979, 1987, 1989, 1992) model of the evolution of insect sociality through "alternative phenotypes." Although I have referred to West-Eberhard several times in this section, it is quite possible that I have not done full justice to her many works on this subject. I therefore refer the reader to West-Eberhard (1986a, 1987, 1989, 1996) for a firsthand account of her arguments.

18   REVERSE SOCIAL EVOLUTION

_I_N contrast to the evolution of altruism and of eusociality, the possibility of reversal of social evolution—the transition from the eusocial to the solitary—has received surprisingly little attention. During a brief discussion, Wilson (1971) concluded that although the solitary habit may sometimes have evolved from primitively eusocial ancestors, highly eusocial species have probably reached a point of no return. In halictine bees, however, which have a great diversity of social organizations, frequent alternations occur between solitary and eusocial states, both between different habitats and between different species within a genus. Halictine bee researchers have therefore long wondered about the transition from the primitively eusocial to the solitary state (Michener, 1964a, 1965, 1990; Packer, 1990; Eickwort et al., 1996; Wcislo, 1996).

What can the analysis of the evolutionary forces favoring worker behavior in _R. marginata_ tell us about the possibility of reversal of social evolution? Can it help determine whether social insect species can retrace their evolutionary pathway and eventually become solitary? It is obvious that Hamilton's rule itself provides a theoretical framework for exploring the emergence of selfishness. It follows, therefore, that inequality 8.5 can again be a useful tool if we reverse the inequality sign: we could say that selfishness would be favored over altruism if

(18.1) $\rho \beta \sigma < r b s$

Indeed, this is exactly why solitary nesting is favored over worker behavior in _R. marginata_ in a small portion of the parameter space explored in Chapter 15. Thus the inclusive fitness model developed for understanding the evolution of altruism is symmetrical with reference to direction and may favor the evolution of altruism or selfishness, depending merely on the numerical values of the parameters. It is this symmetry in the model that aroused in me an interest in the possibility of reverse social evolution (Gadagkar, 1997c). That the solitary nesting strategy is favored in some portion of the parameter space in the case of *R. marginata* is not itself, however, sufficient evidence of reverse social evolution. To demonstrate reverse social evolution, we must find the solitary nesting strategy in the descendants of an obligatorily eusocial ancestor or the complete loss of eusociality in the descendants of an ancestor that was at least facultatively eusocial.

18.1.   Can Solitary Behavior Reemerge in Highly Eusocial Species?

Despite admitting that "social evolution can be reversed and a eusocial species can revert to a solitary condition," Wilson (1971) concluded that "there are reasons for believing that a point of no return [in social evolution] does indeed exist." The reason for this dichotomy is that all suspected examples of possible reversion from the eusocial to the solitary state are primitively eusocial species, and as Wilson (1971) said, "the highest insect societies have lost elements of behavior that would be very difficult to reattain in evolution." Is it pointless then to persist in exploring the possibility of reversal of the highly eusocial state? I believe that there is merit in exploring this possibility (Gadagkar, 1997c), and I suggest a three-pronged approach: (1) intensify investigations of suspected reversals from the primitively eusocial state; (2) identify and investigate possible intermediate states in potential reversals from the highly eusocial state; and (3) identify and investigate "preventing mechanisms" or factors responsible for the rarity of reverse social evolution both at the primitively eusocial and the highly eusocial level. But let us proceed in the reverse order.

18.2.   Preventing Mechanisms

In discussions of the evolution of eusociality, it is customary to talk about additional enabling mechanisms that make it easier for sociality to evolve if inequality 8.5 is satisfied. In the context of eusociality in the Hymenoptera, the sting, the nest, and advanced parental care are the three most commonly discussed enabling mechanisms (see Chapter 15). It is reasonable therefore to consider corresponding "preventing mechanisms" that might make it difficult for solitary behavior to reemerge even when inequality 18.1 is satisfied. On the basis of the discussion of this question in Wilson (1971) and what we know about social insects today, there appear to be at least five major preventing mechanisms that make it difficult or impossible for solitary behavior to reemerge from the highly eusocial state: queen control of workers, loss of spermatheca and of the ability to mate, morphological specialization, caste polyethism, and homeostasis. Of course, all five factors are not likely to be equally powerful—morphology would be relatively hardest to reverse and behavior would be the easiest with chemistry somewhere in between.

18.2.1.   Queen Control of Workers

A fundamental property of all social insect colonies is the control of worker behavior and especially worker reproduction by the queens (Wilson, 1971; Bourke and Franks, 1995; Winston, 1987). This control is achieved by queen pheromones in the advanced social insects, while in the primitively eusocial species it may be achieved entirely by direct physical interactions. In the case of the honey bee, for example, the queen produces a host of chemical substances that influence the behavior and physiology of the workers in her colony. Owing to the fact that each colony consists of a single queen and many thousand workers, communication between the queen and her workers is, as expected, primarily mediated by chemicals. The well-known effects of queen pheromones on workers include rapid detection of the presence or absence of the queen. A retinue of some 8 to 10 workers, the composition of which changes every few minutes, feeds and licks the queen and thereby acquires the queen pheromones and passes them on to other workers. The pheromones also inhibit the development of workers' ovaries and stimulate building and foraging activities. Workers of a queenright colony almost never lay eggs. Instead, they build combs, feed the larvae, groom and feed the queen, protect the hive from intruders, forage, and store honey and pollen (Winston and Slessor, 1992).

Similar chemical control of workers by the queens probably occurs in all highly eusocial species, although the nature of the chemicals and the extent of their effects remain to be studied in most cases. The alternative point of view—that queen pheromones in highly eusocial insects are not agents of control but are signals that workers use to forgo reproduction, which is best left to the queens (Keller and Nonacs, 1993; Gadagkar, 1997b)—does not alter the argument that queen pheromones can act as a preventing mechanism for reverse social evolution. In situations where it is to the advantage of the workers to remain sterile and let the queens lay eggs, workers can hardly be expected to revolt against the hegemony of the queen.

In the primitively eusocial wasp *Polistes fuscatus,* by contrast, the queen regulates the behavior of workers through her own extreme activity. The fact that worker activity is stimulated by queen activity was most strikingly demonstrated in an experiment where introduction of a cooled, inactive queen into a colony depressed worker activity even more than the mere absence of the queen (Reeve and Gamboa, 1983; see also Chapter 6). Queen control of workers must surely serve to prevent reversal of social evolution, and the strength of that prevention must depend on the strength of the queen's control. It is reasonable to suppose that workers in primitively eusocial species can more easily escape from the nudging and biting of the queen than could honey bee workers, whose behavior and physiology are strongly influenced by the queen's pheromones.

18.2.2.   *Loss of Spermatheca and of the Ability to Mate*

In most highly eusocial species workers cannot mate and store sperm. In the queens, the spermatheca acts as a reservoir for sperm and has associated glands to secrete nutrients and keep the sperm viable. In most highly eusocial species workers have reduced and nonfunctional spermathecae; and with one or two exceptions (see below), this completely prevents them from producing female offspring. In many species workers retain rudimentary ovaries and can lay a few haploid eggs upon the death of the queen (Bourke, 1988). But this ability does not give them complete freedom from the queen, and their continuing reliance on the queen has probably been a major preventing factor for the reversal of social evolution.

18.2.3.   *Morphological Specialization*

With very few exceptions, highly eusocial insects, especially ants and termites, are unparalleled in the extent of intraspecies and intrasexual variations in size and allometry, which I discussed at length in section 16.2. Such morphological specialization of workers for nonreproductive and non–colony-founding roles has also probably been a significant factor in making reverse social evolution difficult.

18.2.4.   Caste Polyethism

Division of labor is a striking feature of colony organization in most social insects. The most fundamental division of labor is between the reproductive and the worker castes, but further division of labor within the workers is accomplished either by physical or by temporal means or by a combination of both (Wilson, 1971; Oster and Wilson, 1978). When castes are physically differentiated, behavior is usually correlated with size, and when castes are temporally differentiated, behavior is usually correlated with age. Physical castes are most strongly developed in termites (Noirot, 1989) and ants (Hölldobler and Wilson, 1990), although temporal castes are also seen. Temporal castes are most strongly developed in stingless bees (Sommeijer, 1984) and honey bees (Seeley and Kolmes, 1991). In halictine bees (Michener, 1974), bumble bees (Cameron, 1989), and wasps (Jeanne, 1991a), the correlation of behavior is less pronounced with both size and age. There is now evidence that age polyethism can precede the evolution of morphological differentiation between queens and workers (see Chapter 7). Although such caste polyethism is relatively flexible and allows readjustment of behavior in response to an unusual age or size distribution of workers and unusual levels of demand for work, behavioral specialization based on size and/or age is expected to place some restrictions on the ability of workers to revert to a solitary mode of existence.

18.2.5.   Homeostasis

The ability of social insects to regulate the environment of their nest, sometimes called social homeostasis, is most impressive, especially in the highly eusocial species (Wilson, 1971). While such homeostasis has permitted social insects to gain a considerable measure of independence from the environment, it has also made them dependent on the nearly constant conditions prevailing in their nests. Individual bees and ants that lose their way to the nest die very quickly. This dependence on the homeostatic conditions provided by the nest is also likely to have been a preventing factor in the reversal of social evolution.

18.3.   Intermediate States in Reverse Social Evolution

As a consequence of the unlikelihood of complete reversal from the highly eusocial state to the solitary state, it is useful to contemplate possible intermediate states during potential reversals. Consider three successive steps in such a hypothetical process:

1. Workers revolt against the hegemony of the queen (whether morphologically differentiated queens or merely functional queens) and challenge her status as the only reproductive individual.
2. Workers stop producing queens and one or more of them function as egg layers capable of producing both haploid and diploid offspring (functional queens).
3. Social evolution reverses completely so that the species becomes solitary.

18.3.1.   Thelytoky

There are at least two good examples of workers in highly eusocial species successfully revolting against the hegemony of the queen and challenging her status as the only producer of diploid offspring. This appears to have been accomplished by thelytokous parthenogenesis, which permits unmated workers to lay diploid, viable eggs that will develop into fertile adult females that, under the right conditions, will develop into morphologically differentiated queens. This has happened in the Cape honey bee, *Apis mellifera capensis,* a unique race of honey bees found at the tip of South Africa. Unlike virgin workers in any other race of honey bees, virgin workers in the Cape honey bee lay diploid eggs by thelytokous parthenogenesis, and the laying workers have many queenlike characteristics, including well-developed ovaries with many ovarioles per ovary, well-developed spermathecae, the ability to produce queenlike pheromones, the attendance of a queenlike retinue of workers, high reproductive dominance, and the ability to inhibit production of queen cells (Verma and Ruttner, 1983; Hepburn and Crewe, 1990, 1991; Hepburn et al., 1991; Hepburn, 1992). Another example is the ant *Cataglyphis cursor* (Cagniant, 1973, 1979, 1982; Lenoir and Cagniant, 1986), where, as in the cape honey bee, thelytoky is facultative and has not led to the elimination of the queen.

18.3.2. Queenless Ants

There are several examples among ants where, on an evolutionary time scale, workers have stopped producing morphologically differentiated queens and present-day colonies thus consist of one or more egg-laying workers that can produce both haploid and diploid offspring (functional queens). In many ponerine ants, workers routinely mate and reproduce, and in many such cases there has been a permanent loss of true queens. In many species a single, mated, laying worker, called the gamergate, suppresses ovarian development of all nestmates and functionally behaves like a queen (Peeters, 1987, 1997; Peeters and Crewe, 1984, 1985). The ponerine workers appear to have accomplished this unusual evolutionary feat on account of their morphological primitiveness, which includes retention of a functional spermatheca. Nevertheless, queenlessness (perhaps along with the presence of mated workers) appears, from phylogenetic studies (Baroni Urbani, Bolton, and Ward, 1992) to be a derived condition, and therefore an example of a possible intermediate stage in reverse social evolution.

Despite the loss of spermathecae, in the Japanese myrmicine ant *Pristomyrmex pungens* the same result appears to have been accomplished by the evolution of thelytoky, as in the Cape honey bee and the ant *Cataglyphis cursor*. But because thelytokous parthenogenesis is obligatory in *Pristomyrmex pungens*, I place this species in the second step of reverse social evolution, ahead of the Cape honey bee and of *Cataglyphis cursor*. Queens are entirely absent in *P. pungens* and workers are unusual in that each one of them performs egg laying as well as other nonreproductive tasks through a curious mode of division of labor: young workers remain inside the nests and reproduce while older workers go out of the nest and forage (Itow et al., 1984; Tsuji and Itô, 1986; Tsuji, 1988, 1990).

18.3.3. From the Eusocial to the Solitary

Even though the final step of going from a morphologically well-differentiated highly eusocial state to the solitary state does not appear to have been taken by any social insect, transitions from the primitively eusocial to the solitary are not uncommon in the halictine bees (Wcislo and Danforth, 1997). In the evolution of the Halictinae, Michener (1990) suspects that "eusocial behavior has arisen repeatedly, dozens or hundreds of times, and that reversion to solitary behavior is easy." A particularly clear example emerges from a phylogenetic study of social behavior and nest architecture in the sweat bee subgenus *Evylaeus* (Packer, 1991). When behavioral changes are mapped on to the phylogeny derived from allozymes, it becomes obvious that the solitary state of the species *calceatum* is a condition that is derived from the ancestral eusocial state. Similar reversals from the primitively eusocial to the noneusocial state through the loss of eusociality appear to have occurred in Agochlorini sweat bees (Danforth and Eickwort, 1997), and such reversals through the loss of soldier production appear to have occurred repeatedly in the thrips (Crespi and Mound, 1997; see Chapter 1).

Although highly eusocial species may not have reverted entirely to the solitary state, investigation of suspected intermediate states in the process of reversion may lead to an understanding of why reverse social evolution is not common and might perhaps even lead to the discovery of true reverse social evolution. At the very least, it will lead to a better understanding of why reversal from the highly eusocial state is difficult or impossible. Our understanding of the evolution of eusociality will at best be incomplete if we do not either understand the mechanism of the evolutionary loss of eusociality or understand why eusociality cannot be lost.

SUMMARY

$\mathcal{S}$ocial insects provide a unique opportunity to explore the ecological consequences of sociality and to study the origin and evolution of social life in animals. Social insects exhibit great variety, not only in their life cycles and habits but also in the extent of development of their social organization, social integration, and division of labor. To deal with this social diversity, various levels of sociality are recognized. The most highly evolved stage in social evolution is the eusocial; it is characterized by overlap of generations, cooperative brood care, and reproductive caste differentiation. Among insects, eusociality is seen in the hymenopteran families Formicidae (ants), Vespidae (wasps), Apidae (bees), Sphecidae (in the genus *Microstigmus*), and Halictidae (sweat bees). Outside the Hymenoptera, eusociality is shown by all species of Isoptera (termites) and some species in the orders Hemiptera (aphids), Coleoptera (ambrosia beetles), Thysanoptera (thrips) and class Crustacea (marine shrimp). There is also the now famous eusocial mammal, the naked mole rat.

Social insects are excellent model systems for the study of altruism. Reproductive caste differentiation implies that the worker castes of social insects behave altruistically inasmuch as they remain sterile and work for the welfare of their colonies. Such altruism was considered paradoxical until W. D. Hamilton proposed the theory of inclusive fitness. This idea, most conveniently expressed as Hamilton's rule, provides a powerful theoretical framework for investigating the evolution of eusociality and thus the evolution of altruism.

Among eusocial species, the primitively eusocial ones hold special promise because in these species queens and workers are not morphologically differentiated and workers usually retain some reproductive options.

In many species new nests are founded either by a single female or by groups of females in which only one or a small number reproduces and the rest function as workers. By comparing the solitary and social nesting strategies within the same species researchers can measure the costs and benefits of social life. The social wasp genus *Ropalidia* has long been considered crucial to understanding the evolution of hymenopteran sociality because it contains both independent-founding, primitively eusocial species that build small open nests and exhibit no morphological caste differentiation and swarm-founding, highly eusocial species that often build large, enveloped nests and may exhibit considerable queen-worker dimorphism.

R. marginata, a primitively eusocial species occurring in the tropical climate of southern India and exhibiting a perennial, indeterminate colony-founding cycle, is particularly attractive for studies of the evolution of eusociality. As in many other species, new nests are initiated by one or a small number of female wasps. The nest is made of paper carton whose hexagonal cells are used to rear brood. All colonies are strictly monogynous at any given time. Males eclosing on the nest leave within a few days to lead a nomadic life, mate, and die. Eclosing female wasps, however, have a number of options open to them; they may leave to found new nests by themselves or along with other wasps, they may remain in the natal nest and work for its welfare for the rest of their lives, or they may work for some time and then at an opportune movement drive away the original queen and become replacement queens.

Behavioral observations of individually marked wasps reveal a great deal of interindividual variability. Multivariate statistical analysis of time-activity budgets reveals behavioral caste differentiation of the wasps in a colony into sitters, fighters, and foragers. Sitters are relatively docile individuals that do little or no foraging and participate in little or no dominance interactions, but do perform many intranidal activities. Fighters are aggressive individuals that, while participating in intranidal activities, show the highest levels of dominance behaviors. Foragers are relatively subordinate individuals that are responsible for most of the tasks of foraging for food, water, nectar, and building material. Queens in *R. marginata* always belong to the sitter caste. In a related species, *R. cyathiformis,* queens belong to the fighter caste. It appears that this difference between the behavioral caste of the queens in the two species is related to the fact that *R. marginata* queens face less reproductive threat from their nestmates than do *R. cyathiformis* queens. The system of behavioral caste differentiation into sitters, fighters, and foragers seems to permit the wasps to work for the welfare of their colonies and gain indirect fitness without entirely closing off their reproductive options.

As in many primitively eusocial species, the wasps indulge in frequent dominance-subordinate interactions. On the basis of these interactions, most wasps in a colony can be arranged in a dominance hierarchy. In *R. marginata*, the queen, unlike the queens in most other primitively eusocial species, is not at the top of the colony's dominance hierarchy. Consistent with her classification as a sitter, the queen may participate in few or no dominance interactions. But she is nevertheless completely successful in monopolizing reproduction. It seems likely therefore that she achieves this monopoly through the production of pheromones. But when she has to establish herself as a new queen, she resorts to striking levels of physical aggression quite uncharacteristic of *R. marginata* queens in mature nests but highly reminiscent of the behavior of queens of other primitively eusocial species. As might be expected in established colonies, *R. marginata* queens do not appear to regulate the activities of their workers, which appear to continue to bring food and feed larvae in the absence of the queen as efficiently as they do in her presence. There is some evidence that the dominance-subordinate interactions among the workers may serve to regulate their own activity levels by signaling hunger levels of the adults and the larvae in the colony.

Given the primitively eusocial status *of R. marginata,* there is a surprisingly well-developed age polyethism. Feeding larvae, building the nest, bringing pulp, and bringing food are four tasks performed sequentially by successively older wasps. More than the absolute age of the wasps, their relative position in the age distribution of the colony appears to strongly influence the task profile of an individual. The pattern of interindividual interactions, especially interactions not involving dominance or food exchange, constitutes a plausible mechanism for the wasps to assess their relative age in their colony. Computer simulations reveal that the so-called activator-inhibitor model proposed for the regulation of age polyethism in honey bees provides a suitable proximate mechanism for the organization of work in *R. marginata* colonies.

In an adaptation of Hamilton's rule to compare the inclusive fitness of solitary nest foundresses and workers, I computed inclusive fitness as the product of an intrinsic productivity factor, the coefficient of genetic relatedness, and a demographic correction factor. Then to make investigation easy, I measured inequalities between workers and solitary foundresses in adult-brood genetic relatedness (genetic predisposition), in intrinsic productivity levels (ecological and physiological predisposition), and in the demographic correction factors (demographic predisposition).

Haplodiploidy is expected to create genetic asymmetries such that full sisters are more closely related to each other than a mother would be to her offspring. In *R. marginata*, however, multiple mating by queens (polyandry) and frequent queen replacements (serial polygyny) break down these relatedness asymmetries to such an extent that, if anything, workers are less related to the brood they rear than a solitary foundress might be to her offspring. Thus there is no evidence for genetic predisposition for the evolution of eusociality in *R. marginata*. *R. marginata* does have a well-developed system of nestmate discrimination, but it is based on recognition labels and templates acquired by the wasps after their eclosion and from sources outside their body. This makes it unlikely that intracolony kin recognition helps the workers to be preferentially altruistic toward their close relatives. These results make it imperative to look for factors other than genetic relatedness in the evolution of eusociality.

There is considerable evidence that factors other than genetic relatedness do play a role in that evolution. During the pre-emergence phase of the colony cycle, wasps drift from one colony to other colonies. Even mature colonies readily accept young intruders from alien colonies. These phenomena must result in considerable further reduction of intracolony genetic relatedness. Nevertheless, alien conspecifics become well integrated into their foster colonies and are not discriminated against; they even go on to become foragers and replacement queens in their foster colonies. Field studies have shown that only some 4 to 8% of the wasps nest solitarily, while the remainder nest in groups. Why should such a large proportion of wasps join multiple-foundress nests when doing so means most of them will be sterile? There is evidence that all wasps are not as efficient at raising offspring as solitary foundresses. Those that are not as efficient as solitary foundresses fare better as workers in multiple-foundress colonies. Field and laboratory experiments show that a subordinate cofoundress gains fitness by a factor of 2.9 by choosing not to found her own nest and to work instead in another foundress' nest. It is this advantage that permits workers to rear brood quite distantly related to them. This is the ecological predisposition to the evolution of eusociality.

There is also evidence for considerable physiological predisposition. Under laboratory conditions, only about 50% of the wasps are capable of initiating single-foundress nests and laying eggs. Both larval and adult nutrition play a role in such pre-imaginal caste bias. Individuals that are well fed as larvae enter a developmental pathway that makes them grow into adults that are hungry, eat more, develop their ovaries, and become egg layers. Conversely, poorly fed larvae enter a developmental pathway that makes them grow into adults that are not hungry, eat less, and become non–egg layers. It is reasonable to expect that, under natural conditions, egg layers have a better chance of capitalizing on opportunities to become queens than the non–egg layers, which are more likely to become workers.

There is also evidence for strong demographic predisposition to eusociality in *R. marginata*. Delayed reproductive maturation makes it advantageous for wasps to serve as workers in another individual's nest rather than to wait until they become reproductively mature and able to raise their own offspring. Variation between individuals in the time taken to attain reproductive maturity provides a mechanism by which some individuals (the early reproducers) can be selected to become solitary nest foundresses while others (the late reproducers) are selected to become workers. There is both theoretical and empirical evidence that wasps can follow mixed reproductive strategies—behaving as workers until they become reproductively mature and raising their own offspring later. If a solitary foundress dies before she brings the brood under her care to independence, she necessarily loses all her investment in them. But a worker has more assured fitness returns because even if she dies, another worker in the colony will continue to look after the brood that were under her care. Quantitative inclusive fitness models show that the advantage of such assured fitness returns can be greater than the maximum advantage provided by haplodiploidy. In combination with delayed reproductive maturation, variation in age at reproductive maturity, and the possibility of mixed reproductive strategies, assured fitness returns leads to a significant demographic predisposition for the evolution of eusociality.

Although many factors that might tilt the inclusive fitness balance in favor of workers are yet to be considered, a preliminary unified model combining such genetic, ecological, physiological, and demographic factors as can now be quantified predicts that about 5% of the wasps should opt for the solitary nesting strategy, and that the remaining 95% should opt for the worker strategy. This result is remarkably close to the empirically observed proportions of solitary nest foundresses and workers in nature.

In addition to elucidating the relative roles of genetic, ecological, physiological, and demographic factors in promoting the evolution of eusociality, I have discussed the possible evolutionary route to eusociality, mechanisms for the evolution of caste polymorphism, and the possibility of reversal of social evolution. Perhaps the single most important point I have made is that ecological, physiological, and demographic factors can be more important in promoting the evolution of eusociality than the genetic relatedness asymmetries potentially created by haplodiploidy. Put in another way, the benefit and cost terms in Hamilton's rule deserve more attention than the relatedness term.

REFERENCES
INDEX

REFERENCES

Agosti, D., Grimaldi, D., and Carpenter, J. M. 1997. Oldest known ant fossils discovered. *Nature* 391: 447.

Alcock, J. 1974. Social interactions in the solitary wasp *Cerceris simplex* (Hymenoptera: Sphecidae). *Behaviour* 54: 142–151.

Alexander, B. A., and Michener, C. D. 1995. Phylogenetic studies of the families of short-tongued bees (Hymenoptera: Apoidea). *The Univ. Kansas Sci. Bull.* 55: 377–424.

Alexander, R. D. 1974. The evolution of social behavior. *Ann. Rev. Ecol. Syst.* 5: 325–383.

Alexander, R. D., Noonan, K. M., and Crespi, B. J. 1991. The evolution of eusociality. In *The Biology of the Naked Mole-Rat,* ed. Sherman, P. W., Jarvis, J. U. M., and Alexander, R. D, pp. 3–44.Princeton: Princeton University Press.

Altmann, J. 1974. Observational study of behavior: sampling methods. *Behaviour* 49: 227–267.

Ananthakrishnan, T. N. 1973. *Thrips: Biology and Control.* Delhi, Madras: The Macmillan Co. of India.

Ananthakrishnan, T. N., and Raman, A. 1989. *Thrips and Gall Dynamics.* New Delhi: Oxford & IBH Publishing Co.

Anderberg, M. R. 1973. *Cluster Analysis for Applications.* New York: Academic Press.

Andersson, M. 1984. The evolution of eusociality. *Ann. Rev. Ecol. Syst.* 15: 165–189.

Aoki, S. 1977a. *Colophina clematis* (Homoptera, Pemphigidae), an aphid species with "soldiers." *Kontyu* (Tokyo) 45: 276–282.

———1977b. A new species of *Colophina* (Homoptera, Aphidoidea) with soldiers. *Kontyu* (Tokyo) 45: 333–337.

Arathi, H. S., and Gadagkar, R. 1998. Cooperative nest building and brood care by nestmates and non nestmates in *Ropalidia marginata:* implications for the evolution of eusociality. *Oecologia* 117: 295–299.

Arathi, H. S., Shakarad, M., and Gadagkar, R. 1997a. Factors affecting the acceptance of alien conspecifics on nests of the primitively eusocial wasp, *Ropalidia marginata* (Hymenoptera: Vespidae). *J. Insect Behav.* 10: 343–353.

————1997b. Social organisation in experimentally assembled colonies of *Ropalidia marginata:* comparison of introduced and natal wasps. *Insectes Soc.* 44: 139–146.

Arnold, G., Quenet, B., Cornuet, J.-M., Masson, C., De Schepper, B., Estoup, A., and Gasqui, P. 1996. Kin recognition in honeybees. *Nature* 379: 498.

Atsatt, P. R. 1981a. Ant-dependent food plant selection by the mistletoe butterfly *Ogyris amaryllis* (Lycaenidae). *Oecologia* 48: 60–63.

————1981b. Lycaenid butterflies and ants: selection for enemy-free space. *Am. Nat.* 118: 638–654.

Avilés, L. 1997. Causes and consequences of cooperation and permanent sociality in spiders. In *The Evolution of Social Behavior in Insects and Arachnids,* ed. Choe, J. C., and Crespi, B. J, pp. 476–498. Cambridge: Cambridge University Press.

Bagnères, A.-G., Lorenzi, M. C., Dusticier, G., Turillazzi, S., and Clément, J.-L. 1996. Chemical usurpation of a nest by paper wasp parasites. *Science* 272: 889–892.

Baroni Urbani, C., Bolton, B., and Ward, P. S. 1992. The internal phylogeny of ants (Hymenoptera: Formicidae). *Syst. Entomol.* 17: 301–329.

Batra, S. W. T. 1966. Nests and social behavior of halictine bees of India (Hymenoptera: Halictidae). *The Indian J. Entomol.* 28: 375–393.

Belavadi, V. V., and Govindan, R. 1981. Nesting habits and behaviour of *Ropalidia (Icariola) marginata* (Hymenoptera: Vespidae) in South India. *Colemania* 1: 95–101.

Benton, T. G., and Foster, W. A. 1992. Altruistic housekeeping in a social aphid. *Proc. R. Soc. Lond. B* 247: 199–202.

Berenbaum, M. R. 1990. Coevolution between herbivorous insects and plants: tempo and orchestration. In *Insect Life Cycles: Genetics, Evolution, and Co-ordination,* ed. Gilbert, F., pp. 87–99. London: Springer-Verlag.

Bingham, C. T. 1897. *The Fauna of British India, including Ceylon and Burma—Hymenoptera.* Vol. 1: *Wasps and Bees.* London: Taylor and Francis.

Bloch, G., and Hefetz, A. 1999. Regulation of reproduction by dominant workers in bumblebee *(Bombus terrestris)* queenright colonies. *Behav. Ecol. Sociobiol.* 45: 125–135.

Bonabeau, E., Theraulaz, G., Deneubourg, J.-L., Aron, S., and Camazine, S. 1997. Self-organization in social insects. *Trends Ecol. & Evol.* 12: 188–193.

Bourke, A. F. G. 1988. Worker reproduction in the higher eusocial Hymenoptera. *Quart. Rev. Biol.* 63: 291–311.

Bourke, A. F. G., and Franks, N. R. 1995. *Social Evolution in Ants.* Princeton: Princeton University Press.

Braude, S., and Lacey, E. 1992. The underground society: The secret, communal life of the naked mole-rat. *The Sciences:* 23–28.

Breed, M. D., Butler, L., and Stiller, T. M. 1985. Kin discrimination by worker honey bees in genetically mixed groups. *Proc. Natl. Acad. Sci. USA* 82: 3058–3061.

Breed, M. D., and Gamboa, G. J. 1977. Behavioral control of workers by queens in primitively eusocial bees. *Science* 195: 694–696.

Breed, M. D., Velthuis, H. H. W., and Robinson, G. E. 1984. Do worker honey bees discriminate among unrelated and related larval phenotypes? *Ann. Entomol. Soc. Am.* 77: 737–739.

Breed, M. D., Williams, K. R, and Fewell, J. H. 1988. Comb wax mediates the acquisition of nest-mate recognition in honey bees. *Proc. Natl. Acad. Sci. USA* 85: 8766–8769.

Brockmann, H. J., and Dawkins, R. 1979. Joint nesting in a digger wasp as an evolutionarily stable preadaptation to social life. *Behaviour* 71: 203–245.

Brockmann, H. J., Grafen, A., and Dawkins, R. 1979. Evolutionarily stable nesting strategy in a digger wasp. *J. Theor. Biol.* 77: 473–496.

Brothers, D. J., and Michener, C. D. 1974. Interactions in colonies of primitively social bees. III. Ethometry of division of labor in *Lasioglossum zephyrum* (Hymenoptera: Halictidae). *J. Comp. Physiol.* 90: 129–168.

Bull, N. J., and Adams, M. 1999. Kin associations during nest founding in an allodapine bee *Exoneura robusta:* do females distinguish between relatives and familiar nestmates? *Ethology* 105: 1–13.

Bull, N. J., Mibus, A. C., Norimatsu, Y., Jarmyn, B. L., and Schwarz, M. P. 1998. Giving your daughters the edge: bequeathing reproductive dominance in a primitively social bee. *Proc. Roy. Soc. Lond. B* 265: 1411–1415.

Bull, N. J., and Schwarz, M. P. 1996. The habitat saturation hypothesis and sociality in an allodapine bee: cooperative nesting in not "making the best of a bad situation." *Behav. Ecol. Sociobiol.* 39: 267–274.

———1997. Rearing of non-descendant offspring in an allodapine bee, *Exoneura bicolor* Smith (Hymenoptera: Apidae: Zylocopinae): a preferred strategy or queen coercion? *Aust. J. Entomol.* 36: 391–394.

Burda, H., and Kawalika, M. 1993. Evolution of eusociality in the Bathyergidae: the case of the giant mole rats (*Cryptomys mechowi*). *Naturwissenschaften* 80: 235–237.

Burian, R. M. 1996. Some epistemological reflections on *Polistes* as a model organism. In *Natural History and Evolution of Paper-Wasps,* ed. Turillazzi, S., and West-Eberhard, M. J., pp. 318–337. Oxford: Oxford University Press.

Burtis, K. C. 1993. The regulation of sex determination and sexually dimorphic differentiation in *Drosophila. Current Opinion in Cell Biology* 5: 1006–1014.

Cagniant, H. 1973. Apparition d' ouvrières à Partir d'oeufs Pondus par des ouvrières chez la Fourmi *Cataglyphis cursor* Fonscolombe (Hymènoptères, Formicidae). *C. R. Acad. Sc. Paris D.* 277: 2197–2198.

———1979. La parthenogenese thelytoque et arrhenotoque chez la Fourmi *Cataglypis cursor* Fonsc. (Hym. Form.) cycle biologique en elevage des colonies avec reine et des colonies sans reine. *Insectes Soc.* 26: 51–60.

———1982. La parthénogenèse thélytoque et arrhénotoque chez la Fourmi *Cataglypis cursor* Fonscolombe (Hymenoptera, Formicidae) étude des oeufs pondus par les reines et les ouvrières: morphologie, devenir, influence sur le déterminisme de la caste reine. *Insectes Soc.* 29: 175–188.

Calderone, N. W. 1998. Proximate mechanisms of age polyethism in the honey bee, *Apis mellifera* L. *Apidologie* 29: 127–158.

Camazine, S. 1991. Self-organizing pattern formation on the combs of honey bee colonies. *Behav. Ecol. Sociobiol.* 28: 61–76.

Cameron, S. A. 1989. Temporal patterns of division of labor among workers in the primitively eusocial bumble bee, *Bombus griseocollis* (Hymenoptera: Apidae). *Ethology* 80: 137–151.

———1993. Multiple origins of advanced eusociality in bees inferred from mitochondrial DNA sequences. *Proc. Natl. Acad. Sci. USA* 90: 8687–8691.

Carl, J. 1934. *Ropalidia montana* n.sp. et son nid. Un type nouveau d'architecture vespienne. *Rev. Suisse Zool.* 41: 675–691.

Carlin, N. F. and Frumhoff, P. C. 1990. Nepotism in honey bees. *Nature* 346: 706–707.

Carpenter, J. M. 1982. The phylogenetic relationships and natural classification of the Vespoidea (Hymenoptera). *Syst. Entomol.* 7: 11–38.

———1987. Phylogenetic relationships and classification of the Vespinae (Hymenoptera: Vespidae). *Syst. Entomol.* 12: 413–431.

———1988. The phylogenetic system of the Stenogastrinae (Hymenoptera: Vespidae). *J. New York Entomol. Soc.* 96: 140–175.

———1989. Testing scenarios: wasp social behavior. *Cladistics* 5: 131–144.

———1991. Phylogenetic relationships and the origin of social behavior in the Vespidae. In *The Social Biology of Wasps*, ed. Ross, K. G., and Matthews, R. W., pp. 7–32. Ithaca: Cornell University Press.

Carpenter, J. M., Strassmann, J. E., Turillazzi, S., Hughes, C. R., Solís, C. R., and Cervo, R. 1993. Phylogenetic relationships among paper wasp social parasites and their hosts (Hymenoptera: Vespidae; Polistinae). *Cladistics* 9: 129–146.

Caspari, E. 1952. Pleiotropic gene action. *Evolution* 6: 1–18.

Chandran, S., and Gadagkar, R. 1990. Social organization in laboratory colonies of *Ropalidia marginata*. In *Social Insects and the Environment: Proceedings of the 11th International Congress of IUSSI, Bangalore*, ed. Veeresh, G. K., Mallik, B., and Viraktamath, C. A., p. 78. New Delhi: Oxford & IBH Publishing Co.

Chandrashekara, K., Bhagavan, S., Chandran, S., Nair, P., and Gadagkar, R. 1990. Perennial indeterminate colony cycle in a primitively eusocial wasp. In *Social Insects and the Environment: Proceedings of the 11th International Congress of IUSSI, Bangalore*, ed. Veeresh, G. K., Mallik, B., and Viraktamath, C. A., p. 81. New Delhi: Oxford & IBH Publishing Co.

Chandrashekara, K., and Gadagkar, R. 1990a. Behavioural castes and their correlates in the primitively eusocial wasp, *Ropalidia marginata* (Lep.) (Hymenoptera: Vespidae). In *Social Insects—An Indian Perspective:* ed. Veeresh, G. K., Kumar, A. R. V., and Shivashankar, T., pp. 153–160. Bangalore: IUSSI Indian Chapter, Department of Entomology, UAS, GKVK.

———1990b. Evolution of eusociality: lessons from social organization in *Ropalidia marginata* (Lep.) (Hymenoptera: Vespidae). In *Social Insects and the Environment: Proceedings of the 11th International Congress of IUSSI, Bangalore*, ed. Veeresh, G. K., Mallik, B., and Viraktamath, C. A., pp. 73–74. New Delhi: Oxford & IBH Publishing Co.

————1991a. Behavioural castes, dominance, and division of labour in a primitively eusocial wasp. *Ethology* 87: 269–283.

————1991b. Unmated queens in the primitively eusocial wasp *Ropalidia marginata* (Lep.) (Hymenoptera: Vespidae). *Insectes Soc.* 38: 213–216.

————1992. Queen succession in the primitively eusocial tropical wasp *Ropalidia marginata* (Lep.) (Hymenoptera: Vespidae). *J. Insect Behav.* 5: 193–209.

Choe, J. C., and Crespi, B. J., ed. 1997. *The Evolution of Social Behavior in Insects and Arachnids.* Cambridge: Cambridge University Press.

Clarke, F. M., and Faulkes, C. G. 1998. Hormonal and behavioural correlates of male dominance and reproductive status in captive colonies of the naked mole-rat, *Heterocephalus glaber. Proc. Roy. Soc. Lond. B* 265: 1391–1399.

Clarke, G. M., and McKenzie, J. A. 1987. Developmental stability of insecticide resistant phenotypes in blowfly; a result of canalizing natural selection. *Nature* 325: 345–346.

Clarke, S. R., Dani, F. R., Jones, G. R., Morgan, E. D., and Turillazzi, S. 1999. Chemical analysis of the swarming trail pheromone of the social wasp *Polybia sericea* (Hymenoptera: Vespidae). *J. Insect Physiol.* 45: 877–883.

Clutton-Brock, T. H., Albon, S. D., Gibson, R. M., and Guinness, F. E. 1979. The logical stag: adaptive aspects of fighting in Red Deer *(Cervus elaphus L.). Anim. Behav.* 27: 211–225.

Cohan, F. M., and Hoffmann, A. A. 1986. Genetic divergence under uniform selection. II. Different responses to selection for knockdown resistance to ethanol among *Drosophila melanogaster* populations and their replicate lines. *Genetics* 114: 145–163.

Cohn, J. P. 1992. Naked mole-rats—African rodents' insect-like behavior piques biologists' curiosity. *BioScience* 42: 86–89.

Corbora, B., Lachaud, J.-P., and Fresneau, D. 1989. Individual variability, social structure, and division of labour in the ponerine ant *Ectatomma ruidum* Roger (Hymenoptera, Formicidae). *Ethology* 82: 89–100.

Costa, J. T.. and Fitzgerald, T. D. 1996. Developments in social terminology: semantic battles in a conceptual war. *Trends Ecol. & Evol.* 11: 285–289.

Cowan, D. P. 1991. The solitary and presocial Vespidae. In *The Social Biology of Wasps,* ed. Ross, K. G., and Matthews, R. W., pp. 33–73. Ithaca: Cornell University Press.

Cox, D. R., and Snell, E. J. 1989. *Analysis of Binary Data.* London: Chapman and Hall.

Craig, R. 1983. Subfertility and the evolution of eusociality by kin selection. *J. Theor. Biol.* 100: 379–397.

Craig, R., and Crozier, R. H. 1978. Caste-specific locus expression in ants. *Isozyme Bull.* 11: 64–65.

Crespi, B. J. 1992. Eusociality in Australian gall thrips. *Nature* 359: 724–726.

Crespi, B. J., and Mound, L. A. 1997. Ecology and evolution of social behavior among Australian gall thrips and their allies. In *The Evolution of Social Behavior in Insects and Arachnids,* ed. Choe, J. C., and Crespi, B. J., pp. 166–180. Cambridge: Cambridge University Press.

Crespi, B. J., and Yanega, D. 1995. The definition of eusociality. *Behav. Ecol.* 6: 109–115.

Crozier, R. H. 1977. Evolutionary genetics of the *Hymenoptera*. *Ann. Rev. Entomol.* 22: 263–288.

Crozier, R. H., and Pamilo, P. 1996. *Evolution of Social Insect Colonies: Sex Allocation and Kin Selection.* Oxford: Oxford University Press.

Danforth, B. N., and Eickwort, G. C. 1997. The evolution of social behavior in the augochlorine sweat bees (Hymenoptera: Halictidae) based on a phylogenetic analysis of the genera. In *The Evolution of Social Behavior in Insects and Arachnids*, ed. Choe, J. C., and Crespi, B. J., pp. 270–292. Cambridge: Cambridge University Press.

Dani, F. R., Fratini, S., and Turillazzi, S. 1996. Behavioural evidence for the involvement of Dufour's gland secretion in nestmate recognition in the social wasp *Polistes dominulus* (Hymenoptera: Vespidae). *Behav. Ecol. Sociobiol.* 38: 311–319.

Dani, F. R., Morgan, E. D., and Turillazzi, S. 1995. Chemical analysis of sternal gland secretion of paper wasp *Polistes dominulus* (Christ) and its social parasite *Polistes sulcifer* (Zimmermann) (Hymenoptera: Vespidae). *J. Chem. Ecol.* 21: 1709–1718.

Das, B. P., and Gupta, V. K. 1983. A catalogue of the families Stenogastridae and Vespidae from the Indian subregion (Hymenoptera : Vespoidea). *Oriental Insects,* 17: 395–464.

———1989. *The Social Wasps of India and the Adjacent Countries (Hymenoptera: Vespidae).* Gainseville, Fla.: The Association for the Study of Oriental Insects.

Dawkins, R. 1976. *The Selfish Gene.* Oxford: Oxford University Press.

De Ghett, V. J. 1978. Hierarchical cluster analysis. In *Quantitative Ethology*, ed. Colgan, P. W., pp. 115–144. New York: John Wiley and Sons.

Detrain, C., Deneubourg, J.-L., and Pasteels, J. M. E. 1999. *Information Processing in Social Insects.* Basel: Birkhäuser Verlag.

Dew, H. E. 1983. Division of labor and queen influence in laboratory colonies of *Polistes metricus* (Hymenoptera; Vespidae). *Z. Tierpsychol.* 61: 127–140.

Downing, H. A. 1991a. The function and evolution of exocrine glands. In *The Social Biology of Wasps*, ed. Ross, K. G., and Matthews, R. W., pp. 540–569. Ithaca: Cornell University Press.

———1991b. A role of the Dufour's gland in the dominance interactions of the paper wasp, *Polistes fuscatus* (Hymenoptera: Vespidae). *J. Insect Behav.* 4: 557–565.

Downing, H. A., and Jeanne, R. L. 1985. Communication of status in the social wasp *Polistes fuscatus* (Hymenoptera: Vespidae). *Z. Tierpsychol.* 67: 78–96.

———1987. A comparison of nest construction behavior in two species of *Polistes* paper wasps (Insecta, Hymenoptera: Vespidae). *J. Ethol.* 5: 53–66.

———1988. Nest construction by the paper wasp, *Polistes:* a test of stigmergy theory. *Anim. Behav.* 36: 1729–1739.

———1990. The regulation of complex building behaviour in the paper wasp, *Polistes fuscatus* (Insecta, Hymenoptera, Vespidae). *Anim. Behav.* 39: 105–124.

Duffy, J. E. 1996. Eusociality in a coral-reef shrimp. *Nature* 381: 512–514.

———1998. On the frequency of eusociality in snapping shrimps (Decapoda:

Alpheidae), with description of a second eusocial species. *Bull. Marine Science* 62: 387–400.

Duffy, J. E., and Macdonald, K. S. 1999. Colony structure of the social snapping shrimp *Synalpheus filidigitus* in Belize. *J. Crustacean Biol.* 19: 283–292.

Duffy, J. E., Morrison, C. L., and Rios, R. 2000. Multiple origins of eusociality among sponge-dwelling shrimps *(Synalpheus). Evolution* 54: 503–516.

Dykhuizen, D., and Davies, M. 1980. An experimental model: bacterial specialists and generalists competing in chemostats. *Ecology* 61: 1213–1227.

Ehrlich, P. R., and Raven, P. H. 1964. Butterflies and plants: a study in coevolution. *Evolution* 18: 586–608.

Eickwort, G. C. 1981. Presocial insects. In *Social Insects*, ed. Hermann, H. R., pp. 199–280. New York: Academic Press

Eickwort, G. C., Eickwort, J. M., Gordon, J., and Eickwort, M. A. 1996. Solitary behavior in a high-altitude population of the social sweat bee *Halictus rubicundus* (Hymenoptera: Halictidae). *Behav. Ecol. Sociobiol.* 38: 227–233.

Evans, H. E. 1977. Extrinsic versus intrinsic factors in the evolution of insect sociality. *BioScience* 27: 613–617.

Evans, H. E., and O'Neill, K. M. 1978. Alternative mating strategies in the digger wasp *Philanthus zebratus* Cresson. *Proc. Natl. Acad. Sci.USA* 75: 1901–1903.

Evans, J. D., and Wheeler, D. E. 1999. Differential gene expression between developing queens and workers in the honey bee, *Apis mellifera. Proc. Natl. Acad. Sci. USA* 96: 5575–5580.

Fabricius, J. C. F. 1793. *Entomologia Systematica*. Vol. 2: *Hafniae*.

Fahrbach, S. E. 1997. Regulation of age polyethism in bees and wasps by juvenile hormone. *Adv. Study Behav.* 26: 285–316.

Fahrbach, S. E., and Robinson, G. E. 1996. Juvenile hormone, behavioral maturation, and brain structure in the honey bee. *Dev. Neurosci.* 18: 102–114.

Falconer, D. S. 1981. *Introduction to Quantitative Genetics*. London: Longman.

Field, J. 1989. Alternative nesting tactics in a solitary wasp. *Behaviour* 110: 219–243.

Field, J. and Foster, W. 1999. Helping behaviour in facultatively eusocial hover wasps: an experimental test of the subfertility hypothesis. *Anim. Behav.* 57: 633–636.

Field, J., Foster, W., Shreeves, G., and Sumner, S. 1998. Ecological constraints on independent nesting in facultatively eusocial hover wasps. *Proc. Roy. Soc. Lond. B* 265: 973–977.

Field, J., Shreeves, G., and Sumner, S. 1999. Group size, queuing, and helping decisions in facultatively eusocial hover wasps. *Behav. Ecol. Sociobiol.* 45: 378–385.

Field, J., Shreeves, G., Sumner, S., and Casiraghi, M. 2000. Insurance-based advantage to helpers in a tropical hover wasp. *Nature* 404: 869–871.

Fisher, R. A. 1958. *The Genetical Theory of Natural Selection*. New York: Dover.

Franks, N. R. 1989. Army ants: a collective intelligence. *Am. Sci.* 77: 139–145.

Franks, N. R., and Fletcher, C. R. 1983. Spatial patterns in army ant foraging and migration: *Eciton burchelli* on Barrao Colorado Island, Panama. *Behav. Ecol. Sociobiol.* 12: 261–270.

Franks, N. R., Gomez, N., Goss, S., and Deneubourg, J. L. 1991. The blind leading the blind in army ant raid patterns: testing a model of self-organisation (Hymenoptera: Formicidae). *J. Insect Behav.* 4: 583–607.

Franks, N. R., and Sendova-Franks, A. B. 1992. Brood sorting by ants: distributing the workload over the work surface. *Behav. Ecol. Sociobiol.* 30: 109–123.

Franks, N. R., and Tofts, C. 1994. Foraging for work: how tasks allocate workers. *Anim. Behav.* 48: 470–472.

Franks, N. R., Tofts, C., and Sendova-Franks, A. B. 1997. Studies of the division of labour: neither physics nor stamp collecting. *Anim. Behav.* 53: 219–224.

Free, J. B. 1987. *Pheromones of Social Bees.* London: Chapman and Hall.

Fresneau, D., and Dupuy, P. 1988. A study of polyethism in a ponerine ant: *Neoponera apicalis* (Hymenoptera, Formicidae). *Anim. Behav.* 36: 1389–1399.

Fresneau, D., Perez, J. G., and Jaisson, P. 1982. Evolution of polyethism in ants: observational results and theories. In *Social Insects in the Tropics*, ed. Jaisson, P., pp. 129–155. Paris: Presses de l'Universite Paris-Nord.

Frey, D. F., and Pimentel, R. A. 1978. Principal component analysis and factor analysis. In *Quantitative Ethology*, ed. Colgan, P. W., pp. 219–245. New York: John Wiley & Sons.

Fukuda, H., Kubo, K., Takeshi, K., Takahashi, A., Takahashi, M., Tanaka, B., Wakabayashi, M., and Shirozu, T. 1978. *Insects' Life in Japan.* Vol. 3: *Butterflies.* Tokyo: Hoikushu.

Furey, R. E. 1992. Division of labour can be morphological and/or temporal: a reply to Tsuji. *Anim. Behav.* 44: 571

Gadagkar, R. 1980. Dominance hierarchy and division of labour in the social wasp, *Ropalidia marginata* (Lep.) (Hymenoptera: Vespidae). *Curr. Sci.* 49: 772–775.

———1985a. Evolution of insect sociality: a review of some attempts to test modern theories. *Proc. Indian Acad. Sci. (Anim. Sci.)* 94: 309–324.

———1985b. Kin recognition in social insects and other animals: A review of recent findings and a consideration of their relevance for the theory of kin selection. *Proc. Indian Acad. Sci. (Anim. Sci.)* 94: 587–621.

———1987. Social structure and the determinants of queen status in the primitively eusocial wasp *Ropalidia cyathiformis.* In *Chemistry and Biology of Social Insects, Proceedings of the 10th International Congress of IUSSI, Munich*, ed. Eder, J., and Rembold, H., pp. 377–378. Munich: Verlag J. Peperny.

———1990a. Evolution of eusociality: the advantage of assured fitness returns. *Phil. Trans. Roy. Soc. Lond. B* 329: 17–25.

———1990b. Evolution of insect societies: some insights from studying tropical wasps. In *Social Insects—An Indian Perspective*: ed. Veeresh, G. K., Kumar, A. R. V., and Shivshankar, T., pp. 129–152. Bangalore: IUSSI Indian Chapter, Department of Entomology, UAS, GKVK.

———1990c. The haplodiploidy threshold and social evolution. *Curr. Sci.* 59: 374–376.

———1990d. Origin and evolution of eusociality: a perspective from studying primitively eusocial wasps. *J. Genet.* 69: 113–125.

————1990e. Social biology of *Ropalidia:* investigations into the origins of eusociality. In *Social Insects and the Environment: Proceedings of the 11th International Congress of IUSSI, Bangalore,* ed. Veeresh, G. K., Mallik, B., and Viraktamath, C. A., pp. 9–11. New Delhi: Oxford & IBH Publishing Co.

————1990f. A test of the role of haplodiploidy in the evolution of hymenopteran eusociality. In *Social Insects and the Environment: Proceedings of the 11th International Congress of IUSSI, Bangalore,* ed. Veeresh, G. K., Mallik, B., and Viraktamath, C. A., pp. 539–540. New Delhi: Oxford & IBH Publishing Co.

————1991a. *Belonogaster, Mischocyttarus, Parapolybia,* and independent founding *Ropalidia.* In *The Social Biology of Wasps* , ed. Ross, K. G., and Matthews, R. W., pp. 149–190. Ithaca: Cornell University Press.

————1991b. Demographic predisposition to the evolution of eusociality: a hierarchy of models. *Proc. Natl. Acad. Sci. USA* 88: 10993–10997.

————1991c. On testing the role of genetic asymmetries created by haplodiploidy in the evolution of eusociality in the Hymenoptera. *J. Genet.* 70: 1–31.

————1993. And now . . . eusocial thrips! *Curr. Sci.* 64: 215–216.

————1994a. The evolution of eusociality. In *Les insectes sociaux: Proceedings of the 12th Congress of the International Union for the Study of Social Insects, Sorbonne, Paris,* ed. Lenoir, A., Arnold, G., and Lepage, M., pp. 10–12. Paris: Université Paris Nord.

————1994b. Why the definition of eusociality is not helpful to understand its evolution and what should we do about it. *Oikos* 70: 485–488.

————1995. Observational study of animal behaviour: from instinct to intelligence. *Curr. Sci.* 68: 185–196.

————1996a. The evolution of eusociality, including a review of the social status of *Ropalidia marginata.* In *Natural History and Evolution of Paper-Wasps,* ed. Turillazzi, S., and West-Eberhard, M. J., pp. 248–271. Oxford: Oxford University Press.

————1996b. What's the essence of royalty—one keto group? *Curr. Sci.* 71: 975–980.

————1997a. The evolution of caste polymorphism in social insects: genetic release followed by diversifying evolution. *J. Genet.* 76: 167–179.

————1997b. The evolution of communication and the communication of evolution: the case of the honey bee queen pheromone. In *Orientation and Communication in Arthropods,* ed. Lehrer, M., pp. 375–395. Basel: Birkhäuser Verlag.

————1997c. Social evolution—has nature ever rewound the tape? *Curr. Sci.* 72: 950–956.

————1997d. *Survival Strategies: Cooperation and Conflict in Animal Societies.* Cambridge, Mass.: Harvard University Press; Hyderabad, India: Universities Press.

————1998a. Killer genes, green beards. . . . *Down to Earth* 7: 15–16.

————1998b. Red ants with green beards. *J. Biosci.* 23: 535–536.

Gadagkar, R., Bhagavan, S., Chandrashekara, K., and Vinutha, C. 1991b. The role of larval nutrition in pre-imaginal biasing of caste in the primitively eusocial wasp *Ropalidia marginata* (Hymenoptera: Vespidae). *Ecol. Entomol.* 16: 435–440.

Gadagkar, R., Bhagavan, S., Malpe, R., and Vinutha, C. 1990. On reconfirming the evidence for pre-imaginal caste bias in a primitively eusocial wasp. *Proc. Indian Acad. Sci., (Anim. Sci.)* 99: 141–150.

————1991c. Seasonal variation in the onset of egg laying in a primitively eusocial wasp: implications for the evolution of sociality. *Entomon* 16: 167–174.

Gadagkar, R., and Bonner, J. T. 1994. Social insects and social amoebae. *J. Biosci.* 19: 219–245.

Gadagkar, R., Chandrashekara, K., Chandran, S., and Bhagavan, S. 1990. Serial polygyny in *Ropalidia marginata*: implications for the evolution of eusociality. In *Social Insects and the Environment: Proceedings of the 11th International Congress of IUSSI, Bangalore,* ed. Veeresh, G. K., Mallik, B., and Viraktamath, C. A., pp. 227–228. New Delhi: Oxford & IBH Publishing Co.

————1991a. Worker-brood genetic relatedness in a primitively eusocial wasp: a pedigree analysis. *Naturwissenschaften* 78: 523–526.

————1993a. Queen success is correlated with worker-brood genetic relatedness in a primitively eusocial wasp *(Ropalidia marginata). Experientia* 49: 714–717.

————1993b. Serial polygyny in the primitively eusocial wasp *Ropalidia marginata*: implications for the evolution of sociality. In *Queen Number and Sociality in Insects,* ed. Keller, L., pp. 188–214. Oxford: Oxford University Press.

Gadagkar, R., Gadgil, M., Joshi, N. V., and Mahabal, A. S. 1982. Observations on the natural history and population ecology of the social wasp *Ropalidia marginata* (Lep.) from peninsular India (Hymenoptera: Vespidae). *Proc. Indian Acad. Sci. (Anim. Sci)* 91: 539–552.

Gadagkar, R., Gadgil, M., and Mahabal, A. S. 1982. Observations of population ecology and sociobiology of the paper wasp *Ropalidia marginata marginata* (Lep.) (Family Vespidae). *Proc. Symp. Ecol. Anim. Popul. Zool. Surv. India* 4: 49–61.

Gadagkar, R., and Joshi, N. V. 1982a. Behaviour of the Indian social wasp *Ropalidia cyathiformis* on a nest of separate combs (Hymenoptera: Vespidae). *J. Zool. Lond.* 198: 27–37.

————1982b. A comparative study of social structure in colonies of *Ropalidia*. In *The Biology of Social Insects: Proceedings of the 9th Congress of the International Union for the Study of Social Insects, Boulder, Colorado,* ed. Breed, M. D., Michener, C. D., and Evans, E., pp. 187–191. Boulder: Westview Press,

————1983. Quantitative ethology of social wasps: time-activity budgets and caste differentiation in *Ropalidia marginata* (Lep.) (Hymenoptera: Vespidae). *Anim. Behav.* 31: 26–31.

————1984. Social organization in the Indian wasp *Ropalidia cyathiformis* (Fab.) (Hymenoptera: Vespidae). *Z. Tierpsychol.* 64: 15–32.

————1985. Colony fission in a social wasp. *Curr. Sci.* 54: 57–62.

Gadagkar, R., Vinutha, C., Shanubhogue, A., and Gore, A. P. 1988. Pre-imaginal biasing of caste in a primitively eusocial insect. *Proc. Roy. Soc. Lond. B* 233: 175–189.

Gadgil, M., and Mahabal, A. S. 1974. Caste differentiation in the paper wasp *Ropalidia marginata* (Lep.). *Curr. Sci.* 43: 482

Gadgil, S., and Joshi, N. V. 1983. Climatic clusters of the Indian region. *J. Climatology* 3: 47–63.

Gamboa, G. J. 1978. Intraspecific defense: advantage of social cooperation among paper wasp foundresses. *Science* 199: 1463–1465.

Gamboa, G. J., and Dropkin, J. A. 1979. Comparisons of behaviors in early vs. late foundress associations of the paper wasp, *Polistes metricus* (Hymenoptera: Vespidae). *Can. Entomol.* 111: 919–926.

Gamboa, G. J., Klahn, J. E., Parman, A. O., and Ryan, R. E. 1987. Discrimination between nestmate and non-nestmate kin by social wasps (*Polistes fuscatus*, Hymenoptera: Vespidae). *Behav. Ecol. Sociobiol.* 21: 125–128.

Gamboa, G. J., Reeve, H. K., Ferguson, D., and Wacker, T. L. 1986. Nestmate recognition in social wasps: the origin and acquisition of recognition odours. *Anim. Behav.* 34: 685–695.

Gamboa, G. J., Reeve, H. K., and Pfennig, D. W. 1986. The evolution and ontogeny of nestmate recognition in social wasps. *Ann. Rev. Entomol.* 31: 431–454.

Gamboa, G. J., Wacker, T. L., Duffy, K. G., Dobson, S. W., and Fishwild, T. G. 1992. Defence against intraspecific usurpation by paper wasp cofoundresses (*Polistes fuscatus*, Hymenoptera: Vespidae). *Can. J. Zool.* 70: 2369–2372.

Gamboa, G. J., Wacker, T. L., Scope, J. A., Cornell, T. J., and Shellman-Reeve, J. 1990. The mechanism of queen regulation of foraging by workers in paper wasps (*Polistes fuscatus*, Hymenoptera, Vespidae). *Ethology* 85: 335–343.

Getz, W. M., and Smith, K. B. 1983. Genetic kin recognition: honey bees discriminate between full and half sisters. *Nature* 302: 147–148.

Gibo, D. L. 1978. The selective advantage of foundress associations in *Polistes fuscatus* (Hymenoptera: Vespidae): a field study of the effects of predation on productivity. *Can. Entomol.* 110: 519–540.

Gilbert, F. 1990. Size, phylogeny, and life-history in the evolution of feeding specialization in insect predators. In *Insect Life Cycles: Genetics, Evolution, and Co-ordination,* ed. Gilbert, F., pp. 101–124. London: Springer-Verlag.

Gordon, D. M. 1996. The organization of work in social insect colonies. *Nature* 380: 121–124.

———1999. *Ants at Work: How an Insect Society Is Organized.* New York: Free Press.

Gorman, M., Franke, A., and Baker, B. S. 1995. Molecular characterization of the *male-specific lethal-3* gene and investigations of the regulation of dosage compensation in *Drosophila. Development* 121: 463–475.

Gorman, M., Kuroda, M. I., and Baker, B. S. 1993. Regulation of the sex-specific binding of the maleless dosage compensation protein to the male X chromosome in *Drosophila. Cell* 72: 39–49.

Grafen, A. 1991. Modelling in behavioural ecology. In *Behavioural Ecology: An Evolutionary Approach,* 3rd ed., ed. Krebs, J. R., and Davies, N. B., pp. 5–31. Oxford: Blackwell Scientific Publications.

———1998. Green beard as death warrant. *Nature* 394: 521–523.

Haldane, J. B. S. 1932. *The Causes of Evolution*. Ithaca: Cornell University Press.

Hall, B. G. 1983. Evolution of new metabolic functions in laboratory organisms. In *Evolution of Genes and Proteins,* ed. Nei, M., and Koehn, R. K., pp. 234–257. Sunderland, Mass.: Sinauer Associates.

Hamilton, W. D. 1964a. The genetical evolution of social behaviour, I. *J. Theor. Biol.* 7: 1–16.

————1964b. The genetical evolution of social behaviour, II. *J. Theor. Biol.* 7: 17–52.

Hansell, M. 1987. Nest building as a facilitating and limiting factor in the evolution of eusociality in the Hymenoptera. In *Oxford Surveys in Evolutionary Biology,* ed. Harvey, P. H., and Patridge, L., pp. 155–181. Oxford: Oxford University Press.

Hardison, R. C. 1991. Evolution of globin gene families. In *Evolution at the Molecular Level,* ed. Selander, R. K., Clark, A. G., and Whittam, T. S., pp. 272–289. Sunderland, Mass.: Sinauer Associates.

Haridas, C. V., and Rajarshi, M. B. 1997. On computing worker-brood genetic relatedness in colonies of eusocial insects: a matrix model. *J. Genet.* 76: 43-54.

————2000. A stochastic model for evolution of sociality in insects. *Theor. Popul. Biol.,* 59: 107–117.

Henikoff, S., and Meneely, P. M. 1993. Unwinding dosage compensation. *Cell* 72: 1–2.

Henning, S. F. 1983. Biological groups within the Lycaenidae (Lepidoptera). *J. Entomol. Soc. S. Afr.* 46: 65–85.

Hepburn, H. R. 1992. Pheromonal and ovarial development covary in Cape worker honeybees, *Apis mellifera capensis. Naturwissenschaften* 79: 523–524.

Hepburn, H. R., and Crewe, R. M. 1990. Defining the Cape honeybee: reproductive traits of queenless workers. *South African J. Sci.* 86: 524–527.

————1991. Portrait of the Cape honeybee, *Apis mellifera capensis. Apidologie* 22: 567–580.

Hepburn, H. R., Magnuson, P., Herbert, L. ,and Whiffler, L. A. 1991. The development of laying workers in field colonies of the Cape honey bee. *J. Apic. Res.* 30: 107–112.

Hölldobler, B. 1984. Evolution of insect communication. In *Insect Communication (12th Symposium of the Royal Entomological Society of London),* ed. Lewis, T., pp. 349–377. New York: Academic Press.

Hölldobler, B., and Wilson, E. O. 1977. The number of queens: an important trait in ant evolution. *Naturwissenschaften* 64: 8–15.

————1990. *The Ants* Cambridge, Mass.: Harvard University Press.

Hoshikawa, T. 1979. Observations on the polygynous nests of *Polistes chinensis antennalis* PÉREZ (Hymenoptera, Vespidae) in Japan. *Kontyû* (Tokyo) 47: 239–243.

Huang, Z.-Y., Plettner, E., and Robinson, G. E. 1998. Effects of social environment and worker mandibular glands on endocrine-mediated behavioral development in honey bees. *J. Comp. Physiol. A* 183: 143–152.

Huang, Z.-Y., and Robinson, G. E. 1992. Honeybee colony integration: worker-worker interactions mediate hormonally regulated plasticity in division of labor. *Proc. Natl. Acad. Sci. USA* 89: 11726–11729.

————1996. Regulation of honey bee division of labor by colony age demography. *Behav. Ecol. Sociobiol.* 39: 147–158.

Hughes, C. R., Beck, M. O., and Strassmann, J. E. 1987. Queen succession in the social wasp, *Polistes annularis. Ethology* 76: 124–132.

Hughes, C. R., Queller, D. C., Strassmann, J. E., and Davis, S. K. 1993. Relatedness and altruism in *Polistes* wasps. *Behav. Ecol.* 4: 128–137.

Hughes, C. R., and Strassmann, J. E. 1988. Age is more important than size in determining dominance among workers in the primitively eusocial wasp, *Polistes instabilis. Behaviour* 107: 1–14.

Hung, A. C. F., Dowler, M., and Vinson, S. B. 1977. Alpha-glycerophosphate dehydrogenase isozyme of the fire ant *Solenopsis invicta. Isozyme Bull.* 10: 29.

Hung, A. C. F., and Vinson, S. B. 1977. Interspecific hybridization and caste specificity of protein in fire ant. *Science* 196: 1458–1460.

Hunt, J. H. 1984. Adult nourishment during larval provisioning in a primitively eusocial wasp, *Polistes metricus* Say. *Insectes Soc.* 31: 452–460.

————1988. Lobe erection behavior and its possible social role in larvae of *Mischocyttarus* paper wasps. *J. Insect Behav.* 1: 379–386.

————1990. Nourishment and caste in social wasps. In *Social Insects and the Environment: Proceedings of the 11th International Congress of IUSSI,* ed. Veeresh, G. K., Mallik, B., and Viraktamath, C. A., pp. 651–652. New Delhi: Oxford and IBH Publishing Co.

————1991. Nourishment and the evolution of the social Vespidae. In *The Social Biology of Wasps,* ed. Ross, K. G., and Matthews, R. W., pp. 426–450. Ithaca: Cornell University Press.

————1999. Trait mapping and salience in the evolution of eusocial vespid wasps. *Evolution* 53: 225–237.

Hunt, J. H., Jeanne, R. L., Baker, I., and Grogan, D. E. 1987. Nutrient dynamics of a warm-founding social wasp species, *Polybia occidentalis* (Hymenoptera: Vespidae). *Ethology* 75: 291–305.

Hunt, J. H., Schmidt, D. K., Mulkey, S. S., and Williams, M. A. 1996. Caste dimorphism in the wasp *Epipona guerini* (Hymenoptera: Vespidae; Polistinae, Epiponini): further evidence for larval determination. *J. Kans. Entomol.Soc.* 69: 362–369.

Hurst, G. D. D., and McVean, G. A. T. 1998. Selfish genes in a social insect. *Trends Ecol. & Evol.* 13: 434–435.

Itô, Y. 1985. A comparison of frequency of intra-colony aggressive behaviours among five species of polistine wasps (Hymenoptera: Vespidae). *Z. Tierpsychol.* 68: 152–167.

————1986. On the pleometrotic route of social evolution in the Vespidae. *Monitore zool. Ital.* (n.s.) 20: 241–262.

————1993a. *Behaviour and Social Evolution of Wasps: The Communal Aggregation Hypothesis.* Oxford: Oxford University Press.

————1993b. The evolution of polygyny in primitively eusocial polistine wasps with special reference to the genus *Ropalidia*. In *Queen Number and Sociality in Insects,* ed. Keller, L., pp. 171–187. Oxford: Oxford University Press.

Itô, Y., and Higashi, S. 1987. Spring behaviour of *Ropalidia plebeiana* (Hymenoptera: Vespidae) within a huge aggregation of nests. *Appl. Entomol. Zool.* 22: 519–527.

Itô, Y., and Iwahashi, O. 1987. An analysis of foundress group size in *Ropalidia fasciata* (Hymenoptera: Vespidae), with zero-truncated distributions. *Res. Popul. Ecol.* 29: 189–194.

Itô, Y., Katada, S., Tsuchida, K., and Rowe, R. J. 1996. Social structure of the primitively eusocial paper wasp *Ropalidia gregaria spilocephala* (Cameron) (Hymenoptera: Vespidae) in Queensland. *Aust. J. Entomol.* 35: 231–233.

Itô, Y., Yamane, S., and Spradbery, J. P. 1988. Population consequences of huge nesting aggregations of *Ropalidia plebeiana* (Hymenoptera: Vespidae). *Res. Popul. Ecol.* 30: 279–295.

Itow, T., Kobayashi, K., Kubota, M., Ogata, K., Imai, H. T., and Crozier, R. H. 1984. The reproductive cycle of the queenless ant *Pristomyrmex pungens. Insectes Soc.* 31: 87–102.

Jackson, L. L., and Bartlet, R. J. 1986. Cuticular hydrocarbons of *Drosophila virilis:* comparison by age and sex. *Insect Biochem.* 16: 433–439.

Jarvis, J. U. M. 1981. Eusociality in a mammal: cooperative breeding in naked mole-rat colonies. *Science* 212: 571–573.

Jarvis, J. U. M., and Bennett, N. C. 1993. Eusociality has evolved independently in two genera of bathyergid mole-rats—but occurs in other subterranean mammals. *Behav. Ecol. Sociobiol.* 33: 253–260.

Jarvis, J. U. M., O'Riain, M. J., Bennett, N. C., and Sherman, P. W. 1994. Mammalian eusociality: a family affair. *Trends Ecol. & Evol.* 9: 47–51.

Jaycox, E. R. 1976. Behavioral changes in worker honey bees (*Apis mellifera* L.) after injection with synthetic juvenile hormone (Hymenoptera: Apidae). *J. Kans. Entomol. Soc.* 49: 165–170.

Jeanne, R. L. 1970. Chemical defense of brood by a social wasp. *Science* 168: 1465–1466.

———1972. Social biology of the neotropical wasp *Mischocyttarus drewseni. Bull. Mus. Comp. Zool.* 144: 63–150.

———1975. The adaptiveness of social wasp nest architecture. *Quart. Rev. Biol.* 50: 267–287.

———1980. Evolution of social behavior in the Vespidae. *Ann. Rev. Entomol.* 25: 371–396.

———1981. Chemical communication during swarm emigration in the social wasp *Polybia sericea* (Olivier). *Anim. Behav.* 29: 102–113.

———1986a. The evolution of the organization of work in social insects. *Monitore Zool. Ital.* 20: 119–133.

———1986b. The organization of work in *Polybia occidentalis:* costs and benefits of specialization in a social wasp. *Behav. Ecol. Sociobiol.* 19: 333–341.

———1991a. Polyethism. In *The Social Biology of Wasps,* ed. Ross, K. G., and Matthews, R. W., pp. 389–425. Ithaca: Cornell University Press.

————1991b. The swarm-founding Polistinae. In *The Social Biology of Wasps*, ed. Ross, K. G., and Matthews, R. W., pp. 191–231. Ithaca: Cornell University Press.

————1996. Regulation of nest construction behaviour in *Polybia occidentalis. Anim. Behav.* 52: 473–488.

————1999. Group size, productivity, and information flow in social wasps. In *Information Processing in Social Insects*, ed. Detrain, C., Denenbourg, J. L., and Pasteels, J., pp. 3–30. Basel: Birkhäuser Verlag.

Jeanne, R. L., Downing, H. A., and Post, D. C. 1983. Morphology and function of sternal glands in polistine wasps (Hymenoptera: Vespidae). *Zoomorphology* 103: 149–164.

————1988. Age polyethism and individual variation in *Polybia occidentalis*, an advanced eusocial wasp. In *Interindividual Behavioural Variability in Social Insects*, ed. Jeanne, R. L., pp. 323–357. Boulder, Colo.: Westview Press.

Jeanne, R. L., and Hunt, J. H. 1992. Observations on the social wasp *Ropalidia montana* from peninsular India. *J. Biosci.* 17: 1–14.

Jeanne, R. L., and Nordheim, E. V. 1996. Productivity in a social wasp: per capita output increases with swarm size. *Behav. Ecol.* 7: 43–48.

Judd, T. M., and Sherman, P. W. 1996. Naked mole-rats recruit colony mates to food sources. *Anim. Behav.* 52: 957–969.

Karsai, I., and Wenzel, J. W. 1998. Productivity, individual-level and colony-level flexibility, and organization of work as consequences of colony size. *Proc. Natl. Acad. Sci. USA* 95: 8665–8669.

Kasuya, E. 1981. Polygyny in the Japanese paper wasp *Polistes jadwigae* Dalla Torre (Hymenoptera, Vespidae). *Kontyu* (Tokyo) 49: 306–313.

————1983. Behavioral ecology of Japanese paper wasps, *Polistes* spp. (Hymenoptera: Vespidae). *Z. Tierpsychol.* 63: 303–317.

Keeping, M. G. 1990. Rubbing behaviour and morphology of van der Vecht's gland in *Belonogaster petiolata* (Hymenoptera: Vespidae). *J. Insect Behav.* 3: 85–104.

————1992. Social organization and division of labour in colonies of the polistine wasp, *Belonogaster petiolata. Behav. Ecol. Sociobiol.* 31: 211–224.

Keeping, M. G., Crewe, R. M., and Kojima, J. 1995. Ant-repellent allomones of four genera of independent-founding paper wasps (Hymenoptera: Vespidae). In *Proceedings of the 10th Entomological Congress of the Entomological Society of Southern Africa*, p. 78. Grahamstown, South Africa.

Keller, L., and Nonacs, P. 1993. The role of queen pheromones in social insects: queen control or queen signal? *Anim. Behav.* 45: 787–794.

Keller, L., and Ross, K. G. 1998. Selfish genes: a green beard in the red fire ant. *Nature* 394: 573–575.

Kelley, R. L., Solovyeva, I., Lyman, L. M., Richman, R., Solovyev, V., and Kuroda, M. I. 1995. Expression of msl-2 causes assembly of dosage compensation regulators on the X chromosomes and female lethality in *Drosophila. Cell* 81: 867–877.

Kent, D. S., and Simpson, J. A. 1992. Eusociality in the beetle *Austroplatypus incompertus* (Coleoptera: Curculionidae). *Naturwissenschaften* 79: 86–87.

Kirkendall, L. R., Kent, D. S., and Raffa, K. F. 1997. Interactions among males, females, and offspring in bark and ambrosia beetles: the significance of living in tunnels for the evolution of social behavior. In *The Evolution of Social Behavior in Insects and Arachnids*, ed. Choe, J. C., and Crespi, B. J, pp. 181–215. Cambridge: Cambridge University Press.

Klahn, J. 1988. Intraspecific comb usurpation in the social wasp *Polistes fuscatus. Behav. Ecol. Sociobiol.* 23: 1–8.

Knerer, G., and Plateaux-Quénu, C. 1966a. Sur l'importance de l'ouverture des cellules à couvain dans l'évolution des Halictinae (insectes Hyménoptères) sociaux. *Compte Rendu de l'Académie des Sciences* (Paris) 263: 1622–1625.

———1966b. Sur la polygynie chez les Halictinae (insectes Hyménoptéres). *Compte Rendu de l'Académie des Sciences* (Paris) 263: 2014–2017.

———1966c. Sur le polymorphisme des femelles chez quelques Halictinae (insectes Hyménoptéres) paléarctiques. *Compte Rendu de l'Académie des Sciences* (Paris) 263: 1759–1761.

———1967a. Sur la production continue ou périodique de couvain chez les Halictinae (insectes Hyménoptéres). *Compte Rendu de l'Académie des Sciences* (Paris) 264: 651–653.

———1967b. Sur la production de mâles chez les Halictinae (insectes Hyménoptéres) sociaux. *Compte Rendu de l'Académie des Sciences* (Paris) 264: 1096–1099.

Kojima, J. 1982a. Nest architecture of three *Ropalidia* species (Hymenoptera: Vespidae) on Leyte Island, the Philippines. *Biotropica* 14: 272–280.

———1982b. Notes on rubbing behavior in *Ropalidia gregaria* (Hymenoptera, Vespidae). *New Entomol.* 31: 17–19.

———1982c. Taxonomic revision of the subgenus *Icarielia* of the genus *Ropalidia* (Vespidae) in the Philippines. *Kontyu* (Tokyo) 40: 108–124.

———1983a. Defense of the pre-emergence colony against ants by means of a chemical barrier in *Ropalidia fasciata* (Hymenoptera, Vespidae). *Jap. J. Ecol.* 33: 213–223.

———1983b. Peritrophic sac extraction in *Ropalidia fasciata* (Hymenoptera, Vespidae). *Kontyu* (Tokyo) 51: 502–508.

———1984. *Ropalidia* wasps in the Philippines (Hymenoptera, Vespidae). I, Subgenus *Icariola. Kontyu* (Tokyo) 52: 522–532.

———1989. Growth and survivorship of preemergence colonies of *Ropalidia fasciata* in relation to foundress group size in the subtropics (Hymenoptera: Vespidae). *Insectes Soc.* 36: 197–218.

———1992. Temporal relationships of rubbing behavior with foraging and petiole enlargement in *Parapolybia indica* (Hymenoptera: Vespidae). *Insectes Soc.* 39: 275–284.

———1996a. Colony cycle of an Australian swarm-founding paper wasp, *Ropalidia romandi* (Hymenoptera: Vespidae). *Insectes Soc.* 43: 411–420.

———1996b. Meconium egestion by larvae of *Ropalidia* Guérin (Hymenoptera: Vespidae) without adult aid, with a note on the evolution of meconium extraction behaviour in the tribe Ropalidiini. *Aust. J. Entomol.* 35: 73–75.

————1997. Abandonment of the subgeneric concept in the Old World polistine genus *Ropalidia* Guérin-Méneville, 1831 (Insecta: Hymenoptera: Vespidae). *Nat. Hist. Bull. Ibaraki Univ.* 1: 93–106.

Kojima, J., and Jeanne, R. L. 1986. Nests of *Ropalidia (Icarielia) nigrescens* and *R. (I.) extrema* from the Philippines, with reference to the evolutionary radiation in nest architecture within the subgenus *Icarielia* (Hymenoptera: Vespidae). *Biotropica* 18: 324–336.

Kojima, J., and Tano, T. 1985. *Ropalidia* wasps in the Philippines (Hymenoptera, Vespidae). II. A new species from Palawan, with brief notes on the distribution of four species. *Kontyu* (Tokyo) 53: 520–526.

Kukuk, P. F. 1994. Replacing the terms "primitive"and "advanced": new modifiers for the term "eusocial." *Anim. Behav.* 47: 1475–1478.

Kukuk, P. F., Eickwort, G. C., Raveret-Richter, M., Alexander, B., Gibson, R., Morse, R. A., and Ratnieks, F. 1989. Importance of the sting in the evolution of sociality in the Hymenoptera. *Ann. Entomol. Soc. Am.* 82: 1–5.

Lenoir, A., and Cagniant, H. 1986. Role of worker thelytoky in colonies of the ant *Cataglyphis cursor* (Hymenoptera: Formicidae). *Entomologia Generalis* 11: 153–157.

Lenski, R. E. 1988a. Experimental studies of pleiotropy and epistasis in *Escherichia coli*. I. Variation in competitive fitness among mutants resistant to virus T4. *Evolution* 42: 425–432.

————1988b. Experimental studies of pleiotropy and epistasis in *Escherichia coli*. II. Compensation for maladaptive effects associated with resistance to virus T4. *Evolution* 42: 433–440.

Lenski, R. E., and Levin, B. R. 1985. Constraints on the coevolution of bacteria and virulent phage: a model, some experiments, and predictions for natural communities. *Am. Nat.* 125: 585–602.

Lepeletier, A. L. M. 1836. *Histoire naturelle des insectes, Hyménoptères,* vol.1. Paris: Roret's Suites à Bufforn.

Li, W.-H., and Graur, D. 1991. *Fundamentals of Molecular Evolution.* Sunderland, Mass.: Sinaeur Associates.

Lin, N., and Michener, C. D. 1972. Evolution of sociality in insects. *Quart. Rev. Biol.* 47: 131–159.

Litte, M. 1977. Behavioral ecology of the social wasp, *Mischocyttarus mexicans. Behav. Ecol. Sociobiol.* 2: 229–246.

————1979. *Mischocyttarus flavitarsis* in Arizona: social and nesting biology of a polistine wasp. *Z. Tierpsychol.* 50: 282–312.

————1981. Social biology of the polistine wasp *Mischocyttarus labiatus:* survival in a Colombian rain forest. *Smithsonian Contrib. Zool.* 327: 1–27.

Macalintal, E. A., and Starr, C. K. 1996. Comparative morphology of the stinger in the social wasp genus *Ropalidia* (Hymenoptera: Vespidae). *Memoirs Entomol. Soc. Washington* 17: 108–115.

Makino, S., Yamane, S., Itô, Y., and Spradbery, J. P. 1994. Process of comb division of re used nests in the Australian paper wasp *Ropalidia plebeiana* (Hymenoptera, Vespidae). *Insectes Soc.* 41: 411–422.

Marino Piccioli, M. T., and Pardi, L. 1970. Studi sulla biologia di *Belonogaster* (Hymenoptera, Vespidae). 1. Sull'etogramma di *Belonogaster griseus* (Fab.). *Monitore Zool. Ital.* (n.s.) 9: 197–225.

———1978. Studies on the biology of *Belonogaster* (Hymenoptera Vespidae). 3. The nest of *Belonogaster griseus* (Fab). *Monitore Zool. Ital.* (n.s.) Supplement 10: 179–228.

Matsuura, M., and Yamane, S. 1990. *Biology of the Vespine Wasps*. New York: Springer-Verlag.

Matthews, R. W. 1968a. *Microstigmus comes:* sociality in a sphecid wasp. *Science* 160: 787–788.

———1968b. Nesting biology of the social wasp *Microstigmus comes* (Hymenoptera: Sphecidae, Pemphredoninae). *Psyche* 75: 23–45.

Maynard Smith, J., Burian, R., Kauffman, S., Alberch, P., Campbell, J., Goodwin, B., Lande, R., Raup, D., and Wolpert, L. 1985. Developmental constraints and evolution. *Quart. Rev. Biol.* 60: 265–287.

McKenzie, J. A., Whitten, M. J., and Adena, M. A. 1982. The effect of genetic background on the fitness of Diazinon resistance genotypes of the Australian sheep blowfly, *Lucilia cuprina. Heredity* 49: 1–9.

Michener, C. D. 1958. The evolution of social behaviour in bees. *Proceedings of the 10th International Congress of Entomology* (Montreal) 2: 441–448.

———1964a. The bionomics of *Exoneurella,* a solitary relative of *Exoneura. Pacific Insects* 6: 411–426.

———1964b. Reproductive efficiency in relation to colony size in hymenopterous societies. *Insectes Soc.* 11: 317–342.

———1965. The life cycle and social organization of bees of the genus *Exoneura* and their parasite, *Inquilina. The Univ. Kans. Sci. Bull.* 46: 317–358.

———1969. Comparative social behavior of bees. *Ann. Rev. Entomol.* 14: 299–342.

———1974. *The Social Behavior of the Bees: A Comparative Study*. Cambridge, Mass.: The Belknap Press of Harvard University Press.

———1990. Reproduction and castes in social halictine bees. In *Social Insects: An Evolutionary Approach to Castes and Reproduction,* ed. Engels, W., pp. 77–121. Berlin: Springer-Verlag.

———2000. *The Bees of the World*. Baltimore: The John Hopkins University Press.

Moffett, M. W. 1987. Sociobiology of the ants of the genus *Pheidologeton*. Phd. Dissertation, *Harvard University*

Morel, L., Vander Meer, R. K., and Lavine, B. K. 1988. Ontogeny of nestmate recognition cues in the red carpenter ant *(Camponotus floridanus):* behavioral and chemical evidence for the role of age and social experience. *Behav. Ecol. Sociobiol.* 22: 175–183.

Mueller, U. G., Rehner, S. A., and Schultz, T. R. 1998. The evolution of agriculture in ants. *Science* 281: 2034–2038.

Muller, H. J. 1935. The origination of chromatin deficiencies as minute deletions subject to insertion elsewhere. *Genetica* 17: 237–252.

Muralidharan, K., Shaila, M. S., and Gadagkar, R. 1986. Evidence for multiple mating in the primitively eusocial wasp *Ropalidia marginata* (Lep.) (Hymenoptera: Vespidae). *J. Genet.* 65: 153–158.

Myles, T. G. 1988. Resource inheritance in social evolution from termites to man. In *The Ecology of Social Behaviour,* ed., Slobodchikoff, C. N., pp. 379–423. New York: Academic Press.

Nair, P., Bose, P., and Gadagkar, R. 1990. The determinants of dominance in a primitively eusocial wasp. In *Social Insects and the Environment: Proceedings of the 11th International Congress of IUSSI, Bangalore,* ed. Veeresh, G. K., Mallik, B., and Viraktamath, C. A., p. 79. New Delhi: Oxford & IBH Publishing Co.

Nalepa, C. A. 1991. Ancestral transfer of symbionts between cockroaches and termites: an unlikely scenario. *Proc. Roy. Soc. Lond. B* 246: 185–189.

Naug, D., and Gadagkar, R. 1998a. Division of labor among a cohort of young individuals in a primitively eusocial wasp. *Insectes Soc.* 45: 247–254.

————1998b. The role of age in temporal polyethism in a primitively eusocial wasp. *Behav. Ecol. Sociobiol.* 42: 37–47.

————1999. Flexible division of labor mediated by social interactions in an insect colony: a simulation model. *J. Theor. Biol.* 197: 123–133.

Noirot, C. 1989. Social structure in termite societies. *Ethol. Ecol. Evol.* 1: 1–17.

Nonacs, P. 1991. Alloparental care and eusocial evolution: the limits of Queller's head-start advantage. *Oikos* 61: 122–125.

Nonacs, P., and Reeve, H. K. 1993. Opportunistic adoption of orphaned nests in paper wasps as an alternative reproductive strategy. *Behavioural Processes* 30: 47–60.

————1995. The ecology of cooperation in wasps: causes and consequences of alternative reproductive decisions. *Ecology* 76: 953–967.

Noonan, K. C. 1986. Recognition of queen larvae by worker honey bees (*Apis mellifera*). *Ethology* 73: 295–306.

O'Donnell, S. 1996. RAPD markers suggest genotypic effects on forager specialization in a eusocial wasp. *Behav. Ecol. Sociobiol.* 38: 83–88.

————1998a. Dominance and polyethism in the eusocial wasp *Mischocyttarus mastigophorus* (Hymenoptera: Vespidae). *Behav. Ecol. Sociobiol.* 43: 327–331.

————1998b. Effects of experimental forager removals on division of labour in the primitively eusocial wasp *Polistes instabilis* (Hymenoptera: Vespidae). *Behaviour* 135: 173–193.

————1998c. Genetic effects on task performance, but not on age polyethism, in a swarm-founding eusocial wasp. *Anim. Behav.* 55: 417–426.

————1998d. Reproductive caste determination in eusocial wasps (Hymenoptera: Vespidae). *Ann. Rev. Entomol.* 43: 323–346.

O'Donnell, S., and Jeanne, R. L. 1990. Forager specialization and the control of nest repair in *Polybia occidentalis* Oliver (Hymenoptera: Vespidae). *Behav. Ecol. Sociobiol.* 27: 359–364.

————1993. Methoprene accelerates age polyethism in workers of a social wasp *(Polybia occidentalis)*. *Physiol. Entomol.* 18: 189–194.

————1995a. Implications of senescence patterns for the evolution of age polyethism in eusocial insects. *Behav. Ecol.* 6: 269–273.

————1995b. Worker lipid stores decrease with outside-nest task performance in wasps: implications for the evolution of age polyethism. *Experientia* 51: 749–752.

Ohno, S. 1970. *Evolution by Gene Duplication,* Berlin: Springer-Verlag.

Oldroyd, B. P., Rinderer, T. E., and Buco, S. M. 1990. Nepotism in honey bees: Page *et al.* reply. *Nature* 346: 707–708.

Oldroyd, B. P., Rinderer, T. E., Schwenke, J. R., and Buco, S. M. 1994. Subfamily recognition and task specialisation in honey bees *(Apis mellifera* L.) (Hymenoptera: Apidae). *Behav. Ecol. Sociobiol.* 34: 169–173.

O'Riain, M. J., and Jarvis, J. U. M. 1997. Colony member recognition and xenophobia in the naked mole-rat. *Anim. Behav.* 53: 487–498.

O'Riain, M. J., Jarvis, J. U. M., and Faulkes, C. G. 1996. A dispersive morph in the naked mole-rat. *Nature* 380: 619–621.

Oster, G. F., and Wilson, E. O. 1978. *Caste and Ecology in the Social Insects.* Princeton: Princeton University Press.

Packer, L. 1990. Solitary and eusocial nests in a population of *Augochlorella striata* (Provancher) (Hymenoptera; Halictidae) at the northern edge of its range. *Behav. Ecol. Sociobiol.* 27: 339–344.

————1991. The evolution of social behavior and nest architecture in sweat bees of the subgenus *Evylaeus* (Hymenoptera: Halictidae): a phylogenetic approach. *Behav. Ecol. Sociobiol.* 29: 153–160.

Packer, L., and Knerer, G. 1985. Social evolution and its correlates in bees of the subgenus *Evylaeus* (Hymenoptera; Halictidae). *Behav. Ecol. Sociobiol.* 17: 143–149.

Page, R. E., Jr. 1997. The evolution of insect societies. *Endeavour* 21: 114–120.

Page, R. E., Jr., Breed, M. D., and Getz, W. M. 1990. Nepotism in honey bees: Page *et al.* reply. *Nature* 346: p. 707

Page, R. E., Jr., and Metcalf, R. A. 1982. Multiple mating, sperm utilization, and social evolution. *Am. Nat.* 119: 263–281.

Page, R. E., Jr., and Robinson, G. E. 1990. Nepotism in honey bees: Page and Robinson reply. *Nature* 346: p. 708.

————1991. The genetics of division of labour in honey bee colonies. *Adv. Insect Physiol.* 23: 118–169.

Page, R. E., Jr., Robinson, G. E., and Fondrk, M. K. 1989. Genetic specialists, kin recognition, and nepotism in honey-bee colonies. *Nature* 338: 576–579.

Pardi, L. 1948. Dominance order in *Polistes* wasps. *Physiol. Zool.* 21: 1–13.

Pardi, L., and Marino Piccioli, M. T. 1970. Studi ulla biologia di *Belonogaster* (Hymenoptera, Vespidae). 2. Differenziamento castale incipiente in *B. griseus* (Fab.). *Monitore Zool. Ital.* (n.s.) Supplement 3: 235–265.

————1981. Studies on the biology of *Belonogaster* (Hymenoptera Vespidae). 4. On caste differences in *Belonogaster griseus* (Fab.) and the position of this genus among social wasps. *Monitore Zool. Ital.* (n.s.) Supplement 14: 131–146.

Parker, B. L., Skinner, M., and Lewis, T. E. 1994. *Thrips: Biology and Management.* New York: Plenum Press.

Peeters, C. 1987. The diversity of reproductive systems in ponerine ants. In *Chemistry and Biology of Social Insects,* ed. Eder, J., and Rembold, H., pp. 253–254. Munich: Verlag J. Peperny.

———1997. Morphologically "primitive" ants: comparative review of social characters, and the importance of queen-worker dimorphism. In *The Evolution of Social Behavior in Insects and Arachnids,* ed. Choe, J., and Crespi, B., pp. 372–391. Cambridge: Cambridge University Press.

Peeters, C., and Crewe, R. 1984. Insemination controls the reproductive division of labour in a ponerine ant. *Naturwissenschaften* 71: 50–51.

———1985. Worker reproduction in the ponerine ant *Ophthalmopone berthoudi:* an alternative form of eusocial organization. *Behav. Ecol. Sociobiol.* 18: 29–37.

Pfennig, D. W., and Klahn, J. E. 1985. Dominance as a predictor of cofoundress disappearance order in social wasps (*Polistes fuscatus*). *Z. Tierpsychol.* 67: 198–203.

Pierce, N. E., and Elgar, M. A. 1985. The influence of ants on host plant selection by *Jalmenus evagoras,* a myrmecophilous lycaenid butterfly. *Behav. Ecol. Sociobiol.* 16: 209–222.

Plettner, E., Otis, G. W., Wimalaratne, P. D. C., Winston, M. L., Slessor, K. N., Pankiw, T., and Punchihewa, P. W. K. 1997. Species- and caste-determined mandibular gland signals in honey bee (*Apis*). *J. Chem. Ecol.* 23: 363–377.

Plettner, E., Slessor, K. N., Winston, M. L., and Oliver, J. E. 1996. Caste-selective pheromone biosynthesis in honeybees. *Science* 271: 1851–1853.

Post, D. C., Mohamed, M. A., Coppel, H. C., and Jeanne, R. L. 1984. Identification of ant repellent allomone produced by social wasp *Polistes fuscatus* (Hymenoptera: Vespidae). *J. Chem. Ecol.* 10: 1799–1807.

Pratte, M. 1989. Foundress association in the paper wasp *Polistes dominulus* Christ (Hymen. Vesp.). Effects of dominance hierarchy on the division of labour. *Behaviour* 111: 208–219.

Premnath, S., Chandrashekara, K., Chandran, S., and Gadagkar, R. 1990. Constructing dominance hierarchies in a primitively eusocial wasp. In *Social Insects and the Environment: Proceedings of the 11th International Congress of IUSSI, Bangalore,* ed. Veeresh, G. K., Mallik, B., and Viraktamath, C. A., p. 80. New Delhi: Oxford & IBH Publishing Co.

Premnath, S., Sinha, A., and Gadagkar, R. 1994. Dominance behaviour and the resolution of intra-colonial conflicts in the primitively eusocial wasp, *Ropalidia marginata.* In *Les insectes sociaux: Proceedings of the 12th Congress of the International Union for the Study of Social Insects, Sorbonne, Paris,* ed. Lenoir, A., Arnold, G., and Lepage, M., p. 242. Paris: Université Paris Nord.

———1995. Regulation of worker activity in a primitively eusocial wasp, *Ropalidia marginata. Behav. Ecol.* 6: 117–123.

———1996a. Dominance relationships in the establishment of reproductive division of labour in a primitively eusocial wasp (*Ropalidia marginata*). *Behav. Ecol. Sociobiol.* 39: 125–132.

————1996b. How is colony activity regulated in *Ropalidia marginata?* In *Readings in Behaviour,* ed. Ramamurthi, R., and Geethabali, pp. 160–167. New Delhi: New Age International.

Queller, D. C. 1989. The evolution of eusociality: reproductive head starts of workers. *Proc. Natl. Acad. Sci. USA* 86: 3224–3226.

————1994. Extended parental care and the origin of eusociality. *Proc. Roy. Soc. Lond. B* 256: 105–111.

————1996. The origin and maintenance of eusociality: the advantage of extended parental care. In *Natural History and Evolution of Paper-Wasps,* ed. Turillazzi, S., and West-Eberhard, M. J., pp. 218–234. Oxford: Oxford University Press.

Queller, D. C., Hughes, C. R., and Strassmann, J. E. 1990. Wasps fail to make distinctions. *Nature* 344: 388.

Ratnieks, F. L. W. 1988. Reproductive harmony via mutual policing by workers in eusocial Hymenoptera. *Am. Nat.* 132: 217–236.

Ratnieks, F. L. W., and Visscher, P. K. 1989. Worker policing in the honeybee. *Nature* 342: 796–797.

Rau, P. 1933. *The Jungle Bees and Wasps of Barro Colorado Island.* Kirkwood, St. Louis Co., Mo.: Phil Rau.

Rau, P., and Rau, N. 1918. *Wasp Studies Afield.* Princeton: Princeton University Press.

Reeve, H. K. 1991. Polistes. In *The Social Biology of Wasps,* ed. Ross, K. G., and Matthews, R. W., pp. 99–148. Ithaca: Cornell University Press.

————1992. Queen activation of lazy workers in colonies of the eusocial naked mole-rat. *Nature* 358: 147–149.

————1993. Haplodiploidy, eusociality, and absence of male parental and alloparental care in Hymenoptera: a unifying genetic hypothesis distinct from kin selection theory. *Phil. Trans. Roy. Soc. Lond. B* 342: 335–352.

Reeve, H. K., and Gamboa, G. J. 1983. Colony activity integration in primitively eusocial wasps: the role of the queen (*Polistes fuscatus,* Hymenoptera: Vespidae). *Behav. Ecol. Sociobiol.* 13: 63–74.

————1987. Queen regulation of worker foraging in paper wasps: a social feedback control system (*Polistes fuscatus,* Hymenoptera: Vespidae). *Behaviour* 102: 147–167.

Reeve, H. K., and Nonacs, P. 1992. Social contracts in wasp societies. *Nature* 359: 823–825.

————1997. Within-group aggression and the value of group members: theory and a field test with social wasps. *Behav. Ecol.* 8: 75–82.

Reeve, H. K., and Sherman, P. W. 1991. Intracolonial aggression and nepotism by the breeding female naked mole-rat. In *The Biology of the Naked Mole-Rat,* ed. Sherman, P. W., Jarvis, J. U. M., and Alexander, R. D., pp. 337–357. Princeton: Princeton University Press.

Reeve, H. K., Westneat, D. F., Noon, W. A., Sherman, P. W., and Aquadro, C. F. 1990. DNA "fingerprinting" reveals high levels of inbreeding in colonies of the eusocial naked mole-rat. *Proc. Natl. Acad. Sci. USA* 87: 2496–2500.

Richards, O. W. 1978. The Australian social wasps (Hymenoptera: Vespidae). *Aust. J. Zoology* 61: 1–132.

Richards, O. W., and Richards, M. J. 1951. Observations on the social wasps of South America (Hymenoptera: Vespidae). *Trans. Roy. Ent. Soc. Lond.* 102: 1–168.

Robinson, G. E. 1987. Regulation of honey bee age polyethism by juvenile hormone. *Behav. Ecol. Sociobiol.* 20: 329–338.

———1992. Regulation of division of labor in insect societies. *Ann. Rev. Entomol.* 37: 637–665.

Robinson, G. E., Page, R. E., Jr., and Huang, Z.-Y. 1994. Temporal polyethism in social insects is a developmental process. *Anim. Behav.* 48: 467–469.

Robinson, G. E., Page, R. E., Jr., Strambi, C., and Strambi, A. 1989. Hormonal and genetic control of behavioral integration in honey bee colonies. *Science* 246: 109–112.

———1992. Colony integration in honey bees: mechanisms of behavioral reversion. *Ethology* 90: 336–348.

Robinson, G. E., and Vargo, E. L. 1996. Juvenile hormone in adult eusocial Hymenoptera: gonadotropin and behavioral pacemaker. *Arch. Ins. Biochem. Physiol.* 35: 559–583.

Robson, S. K. A., Bean, K., Hansen, J., Norling, K., Rowe, R. J., and White, D. 2000. Social and spatial organisation in colonies of primitively eusocial wasp, *Ropalidia revolutionalis* (de Saussure) (Hymenoptera: Vespidae). *Aust. J. Entomol.* 39: 20–24.

Robson, S. K., and Beshers, S. N. 1997. Division of labour and "foraging for work": simulating reality versus the reality of simulations. *Anim. Behav.* 53: 214–218.

Roff, D. A. 1990. Understanding the evolution of insect life-cycles: the role of genetic analysis. In *Insect Life Cycles: Genetics, Evolution, and Co-ordination*, ed. Gilbert, F., pp. 5–27. London: Springer-Verlag.

Roig-Alsina, A., and Michener, C. D. 1993. Studies of the phylogeny and classification of long-tongued bees (Hymenoptera: Apoidea). *The Univ. Kans. Sci. Bull.* 55: 123–162.

Roisin, Y. 1993. Selective pressures on pleometrosis and secondary polygyny: a comparison of termites and ants. In *Queen Number and Sociality in Insects*, ed. Keller, L., pp. 402–421. Oxford: Oxford University Press.

———1994. Intragroup conflicts and the evolution of sterile castes in termites. *Am. Nat.* 143: 751–765.

———1999. Philopatric reproduction: a prime mover in the evolution of termite sociality? *Insectes Soc.* 46: 297–305.

Rose, M. R. 1982. Antagonistic pleiotropy, dominance, and genetic variation. *Heredity* 48: 63–78.

Rose, M. R., Graves, J. L., and Hutchinson, E. W. 1990. The use of selection to probe patterns of pleiotropy in fitness characters. In *Insect Life Cycles: Genetics, Evolution, and Co-ordination*, ed. Gilbert, F., pp. 29–42. London: Springer-Verlag.

Röseler, P.-F. 1991. Reproductive competition during colony establishment. In *The So-*

cial Biology of Wasps, ed. Ross, K. G., and Matthews, R. W., pp. 309–335. Ithaca: Cornell University Press.

Röseler, P.-F., Röseler, I., and Strambi, A. 1980. The activity of corpora allata in dominant and subordinated females of the wasp *Polistes gallicus. Insectes Soc.* 27: 97–107.

———1985. Role of ovaries and ecdysteroids in dominance hierarchy establishment among foundresses of the primitively social wasp, *Polistes gallicus. Behav. Ecol. Sociobiol.* 18: 9–13.

Röseler, P.-F., Röseler, I., Strambi, A., and Augier, R. 1984. Influence of insect hormones on the establishment of dominance hierarchies among foundresses of the paper wasp, *Polistes gallicus. Behav. Ecol. Sociobiol.* 15: 133–142.

Röseler, P.-F., and van Honk, G. J. 1990. Castes and reproduction in bumblebees. In *Social Insects: An Evolutionary Approach to Castes and Reproduction,* ed. Engels, W., pp. 147–166. Berlin: Springer-Verlag.

Ross, K. G., and Carpenter, J. M. 1991. Population genetic structure, relatedness, and breeding systems. In *The Social Biology of Wasps,* ed. Ross, K. G., and Matthews, R. W., pp. 451–479. Ithaca: Cornell University Press.

Ross, K. G., and Matthews, R. W. 1989a. New evidence for eusociality in the sphecid wasp *Microstigmus comes. Anim. Behav.* 38: 613–619.

———1989b. Population genetic structure and social evolution in the sphecid wasp *Microstigmus comes. Am. Nat.* 134: 574–598.

Rossi, A. M., and Hunt, J. H. 1988. Honey supplementation and its developmental consequences: evidence for food limitation in a paper wasp, *Polistes metricus. Ecol. Entomol.* 13: 437–442.

Roubaud, E. 1916. Recherches biologiques sur les guêpes solitaires et sociales d'Afrique. La genèse de la vie sociale et l'évolution de l'instinct maternel chez les vespides. *Annales des Sciences Naturelles* 10: 1–160.

Roubik, D. W. 1989. *Ecology and Natural History of Tropical Bees.* Cambridge: Cambridge University Press.

Rutz, W., Gerig, L., Wille, H., and Lüscher, M. 1976. The function of juvenile hormone in adult worker honeybees, *Apis mellifera. J. Insect Physiol.* 22: 1485–1491.

Saito, Y. 1994. Is sterility by deleterious recessives an origin of inequalities in the evolution of eusociality? *J. Theor. Biol.* 166: 113–115.

Sakagami, S. F., and Maeta, Y. 1982. Further experiments on the artificial induction of multifemale association in the principally solitary bee genus *Ceratina.* In *The Biology of Social Insects,* ed. Breed, M. D., Michener, C. D., and Evans, H. E., pp. 171–174. Boulder, Colo.: Westview Press.

———1987. Sociality, induced and/or natural, in the basically solitary small carpenter bees (*Ceratina*). In *Animal Societies: Theories and Facts,* ed. Itô, Y., Brown, J. L., and Kikkawa, J., pp. 1–16. Tokyo: Japan Scientific Societies Press.

Saussure, H. de 1853–1858. *Monographie des guêpes sociales ou de la tribu des vespiens, ouvrage faisant suite à la monographie des guêpes solitaires,* Paris: V. Masson.

———1862. Sur divers vespides asiatiques et africaines du Musée de Leyden. *Stettin. Entomol. Ztg.* 23: 129–141.

Schmitz, J., and Moritz, R. F. A. 1998. Molecular phylogeny of Vespidae (Hymenoptera) and the evolution of sociality in wasps. *Molecular Phylogenetics and Evol.* 9: 183–191.

Schwarz, M. P. 1988. Intra-specific mutualism and kin-association of cofoundresses in allodapine bees (Hymenoptera: Anthophoridae). *Monitore Zool. Ital.* 22: 245–254.

Scott, S. M., and Dingle, H. 1990. Developmental programmes and adaptive syndromes in insect life-cycles. In *Insect Life Cycles: Genetics, Evolution, and Co-ordination,* ed. Gilbert, F., pp. 69–85. London: Springer-Verlag.

Seeley, T. D. 1982. Adaptive significance of the age polyethism schedule in honeybee colonies. *Behav. Ecol. Sociobiol.* 11: 287–293.

————1985. *Honeybee Ecology: A Study of Adaptation in Social Life.* Princeton: Princeton University Press.

————1995. *The Wisdom of the Hive: The Social Physiology of Honey Bee Colonies.* Cambridge, Mass.: Harvard University Press.

Seeley, T. D., and Kolmes, S. A. 1991. Age polyethism for hive duties in honey bees: illusion or reality? *Ethology* 87: 284–297.

Seger, J. 1991. Cooperation and conflict in social insects. In *Behavioural Ecology: An Evolutionary Approach,* 3rd ed. ed. Krebs, J. R., and Davies, N. B., pp. 338–373. Oxford: Blackwell Scientific Publications.

Sendova-Franks, A. B., and Franks, N. R. 1995a. Demonstrating new social interactions in ant colonies through randomization tests: separating seeing from believing. *Anim. Behav.* 50: 1683–1696.

————1995b. Spatial relationships within nests of the ant *Leptothorax unifasciatus* (Latr.) and their implications for the division of labour. *Anim. Behav.* 50: 121–136.

Service, P. M., and Lenski, R. E. 1982. Aphid genotypes, plant phenotypes, and genetic diversity: a demographic analysis of experimental data. *Evolution* 36: 1276–1282.

Shakarad, M., and Gadagkar, R. 1995. Colony founding in the primitively eusocial wasp, *Ropalidia marginata* (Lep.) (Hymenoptera: Vespidae). *Ecol. Entomol.* 20: 273–282.

————1996. Why are there multiple-foundress colonies in *Ropalidia marginata?* In *Readings in Behaviour,* ed. Ramamurthi, R., and Geethabali, pp. 145–152. New Delhi: New Age International.

————1997. Do social wasps choose nesting strategies based on their brood rearing abilities? *Naturwissenschaften* 84: 79–82.

Shanubhogue, A., and Gore, A. P. 1987. Using logistic regression in ecology. *Curr. Sci.* 56: 933–936.

Shaw, C. R., and Prasad, R. 1970. Starch gel electrophoresis of enzymes: a compilation of recipes. *Biochem. Genet.* 4: 297–320.

Shellman, J. S., and Gamboa, G. J. 1982. Nestmate discrimination in social wasps: the role of exposure to nest and nestmates (*Polistes fuscatus,* Hymenoptera: Vespidae). *Behav. Ecol. Sociobiol.* 11: 51–53.

Shellman-Reeve, J. S. 1997. The spectrum of eusociality in termites. In *The Evolution of Social Behavior in Insects and Arachnids,* ed. Choe, J. C., and Crespi, B. J., pp. 52–93. Cambridge: Cambridge University Press.

Sherman, P. W. 1979. Insect chromosome numbers and eusociality. *Am. Nat.* 113: 925–935.

Sherman, P. W., Jarvis, J. U. M., and Alexander, R. D. 1991. *The Biology of the Naked Mole-Rat.* Princeton: Princeton University Press.

Sherman, P. W., Jarvis, J. U. M., and Braude, S. H. 1992. Naked mole rats. *Sci. Amer.* 267: 72–78.

Sherman, P. W., Lacey, E. A., Reeve, H. K., and Keller, L. 1995. The eusociality continuum. *Behav. Ecol.* 6: 102–108.

Singer, T. L. 1998. Roles of hydrocarbons in the recognition systems of insects. *Am. Zool.* 38: 394–405.

Singer, T. L., and Espelie, K. E. 1992. Social wasps use nest paper hydrocarbons for nestmate recognition. *Anim. Behav.* 44: 63–68.

————1996. Nest surface hydrocarbons facilitate nestmate recognition for the social wasp, *Polistes metricus* Say (Hymenoptera: Vespidae). *J. Insect Behav.* 9: 857–870.

————1997. Exposure to nest paper hydrocarbons is important for nest recognition by a social wasp, *Polistes metricus* Say (Hymenoptera, Vespidae). *Insectes Soc.* 44: 245–254.

Sinha, A., Premnath, S., Chandrashekara, K., and Gadagkar, R. 1993. *Ropalidia rufoplagiata*: a polistine wasp society probably lacking permanent reproductive division of labour. *Insectes Soc.* 40: 69–86.

Sokal, R. R., and Rohlf, F. J. 1981. *Biometry.* San Francisco: W. H. Freeman and Co.

Solís, C. R., Hughes, C. R., Klingler, C. J., Strassmann, J. E., and Queller, D. C. 1998. Lack of kin discrimination during wasp colony fission. *Behav. Ecol.* 9: 172–176.

Solís, C. R., and Strassmann, J. E. 1990. Presence of brood affects caste differentiation in the social wasp, *Polistes exclamans* Viereck (Hymenoptera: Vespidae). *Functional Ecol.* 4: 531–541.

Sommeijer, M. J. 1984. Distribution of labor among workers of *Melipona favosa* F: age-polyethism and worker oviposition. *Insectes Soc.* 31: 171–184.

Spradbery, J. P. 1975. The biology of *Stenogaster concina* van der Vecht, with comments on the phylogeny of the Stenogastrinae (Hymenoptera: Vespidae). *J. Aust. Ent. Soc.* 14: 309–318.

Spradbery, J. P., and Kojima, J. 1989. Nest descriptions and colony populations of eleven species of *Ropalidia* (Hymenoptera, Vespidae) in New Guinea. *Jpn. J. Entomol.* 57: 632–653.

Starr, C. K. 1979. Origin and evolution of insect sociality: a review of modern theory. In *Social Insects,* ed. Hermann, H. R., pp. 35–79. Orlando: Academic Press.

————1984. Sperm competition, kinship, and sociality in the aculeate Hymenoptera. In *Sperm Competition and the Evolution of Animal Mating Systems,* ed. Smith, R. L., pp. 427–464. Orlando: Academic Press.

————1985. Enabling mechanisms in the origin of sociality in the Hymenoptera: the sting's the thing. *Ann. Entomol. Soc. Am.* 78: 836–840.

————1989. In reply: is the sting the thing? *Ann. Entomol. Soc. Am.* 82: 6–8.

Stern, D. L., and Foster, W. A. 1997. The evolution of sociality in aphids: a clone's-eye view. In *The Evolution of Social Behavior in Insects and Arachnids*, ed. Choe, J. C., and Crespi, B. J., pp. 150–165. Cambridge: Cambridge University Press.

Strassmann, J. E. 1981. Wasp reproduction and kin selection: reproductive competition and dominance hierarchies among *Polistes annularis* foundresses. *Florida Entomol.* 64: 74–88.

———1996. Selective altruism towards closer over more distant relatives in colonies of the primitively eusocial wasp, *Polistes*. In *Natural History and Evolution of Paper-Wasps*, ed. Turillazzi, S., and West-Eberhard, M. J., pp. 190–201. Oxford: Oxford University Press.

Strassmann, J. E., Hughes, C. R., Queller, D. C., Turillazzi, S., Cervo, R., Davis, S. K., and Goodnight, K. F. 1989. Genetic relatedness in primitively eusocial wasps. *Nature* 342: 268–269.

Strassmann, J. E., Klingler, C. J., Arévalo, E., Zacchi, F., Husain, A., Williams, J., Seppä, P., and Queller, D. C. 1997. Absence of within-colony kin discrimination in behavioural interactions of swarm-founding wasps. *Proc. Roy. Soc. Lond. B* 264: 1565–1570.

Strassmann, J. E., and Meyer, D. C. 1983. Gerontocracy in the social wasp, *Polistes exclamans*. *Anim. Behav.* 31: 431–438.

Strassmann, J. E., Meyer, D. C., and Matlock, R. L. 1984. Behavioral castes in the social wasp, *Polistes exclamans* (Hymenoptera: Vespidae). *Sociobiology* 8: 211–224.

Strassmann, J. E., and Queller, D. C. 1989. Ecological determinants of social evolution. In *The Genetics of Social Evolution*, ed. Breed, M. D., and Page, R. E., Jr., pp. 81–101. Boulder: Westview Press.

Strassmann, J. E., Queller, D. C., and Hughes, C. R. 1988. Predation and the evolution of sociality in the paper wasp *Polistes bellicosus*. *Ecology* 69: 1497–1505.

Strassmann, J. E., Queller, D. C., Solís, C. R., and Hughes, C. R. 1991. Relatedness and queen number in the Neotropical wasp, *Parachartergus colobopterus*. *Anim. Behav.* 42: 461–470.

Suzuki, H., and Murai, M. 1980. Ecological studies of *Ropalidia fasciata* in Okinawa island. I. Distribution of single- and multiple-foundress colonies. *Res. Popul. Ecol.* 22: 184–195.

Theraulaz, G., Gervet, J., Thon, B., Pratte, M., and Semenoff-Tian-Chanski, S. 1992. The dynamics of colony organization in the primitively eusocial wasp *Polistes dominulus* Christ. *Ethology* 91: 177–202.

Thompson, J. N. 1989. Concepts of coevolution. *Trends Ecol. & Evol.* 4: 179–183.

———1994. *The Coevolutionary Process*. Chicago: University of Chicago Press.

Thorne, B. L. 1990. A case for ancestral transfer of symbionts between cockroaches and termites. *Proc. Roy. Soc. Lond. B* 241: 37–41.

Thorne, B. L., and Carpenter, J. M. 1992. Phylogeny of the Dictyoptera. *Syst. Entomol.* 17: 253–268.

Tinbergen, N. 1932. Über die Orientierung des Bienenwolfes (*Philanthus triangulum* Fabr.). *Zs. vergl. Physiol.* 16: 305–334.

————1935. Über die Orientierung des Bienenwolfes (*Philanthus triangulum* Fabr.). II. Die Bienenjagd. *Zs. vergl. Physiol.* 21: 699–716.

Tinbergen, N., and Kruyt, W. 1938. Über die Orientierung des Bienenwolfes (*Philanthus triangulum* Fabr.). III. Die Bevorzugung bestimmter Wegmarken. *Zs. vergl. Physiol.* 25: 292–334.

Tofts, C., and Franks, N. R. 1992. Doing the right thing: ants, honeybees, and naked mole-rats. *Trends Ecol. & Evol.* 7: 346–349.

Traniello, J. F. A., and Rosengaus, R. B. 1997. Ecology, evolution, and division of labour in social insects. *Anim. Behav.* 53: 209–213.

Trivers, R. 1985. *Social Evolution.* Menlo Park, Calif.: The Benjamin/Cummings Publishing Co.

Trivers, R. L., and Hare, H. 1976. Haplodiploidy and the evolution of the social insects. *Science* 191: 249–263.

Tsuji, K. 1988. Obligate parthenogenesis and reproductive division of labor in the Japanese queenless ant *Pristomyrmex pungens:* comparison of intranidal and extranidal workers. *Behav. Ecol. Sociobiol.* 23: 247–255.

————1990. Reproductive division of labour related to age in the Japanese queenless ant, *Pristomyrmex pungens. Anim. Behav.* 39: 843–849.

————1992. Sterility for life: applying the concept of eusociality. *Anim. Behav.* 44: 572–573.

Tsuji, K., and Itô, Y. 1986. Territoriality in a queenless ant, *Pristomyrmex pungens* (Hymenoptera: Myrmicinae). *Appl. Entomol. Zool.* 21: 377–381.

Turillazzi, S. 1989. The origin and evolution of social life in the Stenogastrinae (Hymenoptera, Vespidae). *J. Insect Behav.* 2: 649–661.

————1991. The Stenogastrinae. In *The Social Biology of Wasps,* ed. Ross, K. G., and Matthews, R. W., pp. 74–98. Ithaca: Cornell University Press.

Uyenoyama, M. K. 1986. Pleiotropy and the evolution of genetic systems conferring resistance to pesticides. In *Pesticide Resistance: Strategies and Tactics for Management,* ed. National Research Council Committee on Strategies for the Management of Pesticide Resistant Pest Populations, pp. 207–221. Washington D.C.: National Academy Press.

Vecht, J., van der 1941. The Indo-Australian species of the genus *Ropalidia* (= *Icaria*) (Hym., Vespidae) (first part). *Treubia Deel* 18: 103–190.

————1962. The Indo-Australian Species of the genus *Ropalidia* (Icaria) (Hymenoptera, Vespidae) (second part). *Zool. Verhandelingen* 57: 1–71.

————1977. Studies of oriental Stenogastrinae (Hymenoptera Vespoidea). *Tijd. Entomol.* 120: 55–75.

Venkataraman, A. B., and Gadagkar, R. 1992. Kin recognition in a semi-natural context: behaviour towards foreign conspecifics in the social wasp *Ropalidia marginata* (Lep.) (Hymenoptera: Vespidae). *Insectes Soc.* 39: 285–299.

————1993. Differential aggression towards alien conspecifics in a primitively eusocial wasp. *Curr. Sci.* 64: 601–603.

————1995. Age-specific acceptance of unrelated conspecifics on nests of the primitively eusocial wasp, *Ropalidia marginata*. *Proc. Indian Natl. Sci. Acad. B* 61: 299–314.

Venkataraman, A. B., Swarnalatha, V. B., Nair, P., and Gadagkar, R. 1988. The mechanism of nestmate discrimination in the tropical social wasp *Ropalidia marginata* and its implications for the evolution of sociality. *Behav. Ecol. Sociobiol.* 23: 271–279.

Venkataraman, A. B., Swarnalatha, V. B., Nair, P., Vinutha, C., and Gadagkar, R. 1990. Nestmate discrimination in the social wasp *Ropalidia marginata*. In *Social Insects and the Environment: Proceedings of the 11th International Congress of IUSSI, Bangalore*, ed. Veeresh, G. K., Kumar, A. R. V., and Shivashankar, T., pp. 161–171. New Delhi: Oxford and IBH Publishing Co.

Verma, S., and Ruttner, F. 1983. Cytological analysis of the thelytokous parthenogenesis in the Cape honeybee (*Apis mellifera capensis* Escholtz). *Apidologie* 14: 41–57.

Via, S. 1984. The quantitative genetics of polyphagy in an insect herbivore. II. Genetic correlations in larval performance within and among host plants. *Evolution* 38: 896–905.

Visscher, P. K. 1986. Kinship discrimination in queen rearing by honey bees *(Apis mellifera)*. *Behav. Ecol. Sociobiol.* 18: 453–460.

Vollrath, F. 1986. Eusociality and extraordinary sex ratios in the spider *Anelosimus eximius* (Araneae: Theridiidae). *Behav. Ecol. Sociobiol.* 18: 283–287.

Waldman, B. 1988. The ecology of kin recognition. *Ann. Rev. Ecol. Syst.* 19: 543–571.

Wcislo, W. T. 1996. Commentary: solitary behavior in social bees. *Behav. Ecol. Sociobiol.* 38: 235–236.

———1997. Are behavioral classifications blinders to studying natural variation? In *The Evolution of Social Behavior in Insects and Arachnids*, ed. Choe, J. C., and Crespi, B. J., pp. 8–13. Cambridge: Cambridge University Press.

Wcislo, W. T., and Danforth, B. N. 1997. Secondarily solitary: the evolutionary loss of social behavior. *Trends Ecol. & Evol.* 12: 468–474.

Wcislo, W. T., West-Eberhard, M. J., and Eberhard, W. G. 1988. Natural history and behavior of a primitively social wasp, *Auplopus semialatus*, and its parasite, *Irenangelus eberhardi* (Hymenoptera; Pompilidae). *J. Insect Behav.* 1: 247–260.

Wenzel, J. W. 1987. *Ropalidia formosa*, a nearly solitary paper wasp from Madagascar (Hymenoptera: Vespidae). *J. Kans. Entomol. Soc.* 60: 549–556.

———1992. Extreme queen-worker dimorphism in *Ropalidia ignobilis*, a small-colony wasp (Hymenoptera: Vespidae). *Insectes Soc.* 39: 31–43.

———1998. A generic key to the nest of hornets, yellowjackets, and paper wasps worldwide (Vespidae: Vespinae, Polistinae). *Am. Mus. Novitates* 3224: 1–39.

Wenzel, J. W., and Carpenter, J. M. 1994. Comparing methods: adaptive traits and tests of adaptation. In *Phylogenetics and Ecology*, ed. Eggleton, P., and Vane-Wright, R. I., pp. 79–101. London: Academic Press.

Wenzel, J. W., and Pickering, J. 1991. Cooperative foraging, productivity, and the central limit theorem. *Proc. Natl. Acad. Sci. USA* 88: 36–38.

West, M. J. 1967. Foundress associations in polistine wasps: dominance hierarchies and the evolution of social behavior. *Science* 157: 1584–1585.

West-Eberhard, M. J. 1969. The social biology of polistine wasps. *Misc. Publ. Mus. Zool. Univ. Mich.* 140: 1–101.

———1975. The evolution of social behavior by kin selection. *Quart. Rev. Biol.* 50: 1–33.

———1977. The establishment of reproductive dominance in social wasp colonies. In *Proceedings of the 8th International Congress of the International Union for the Study of Social Insects*, pp. 223–227. Wageningen, Holland.

———1978a. Polygyny and the evolution of social behavior in wasps. *J. Kans. Entomol. Soc.* 51: 832–856.

———1978b. Temporary queens in *Metapolybia* wasps: nonreproductive helpers without altruism? *Science* 200: 441–443.

———1979. Sexual selection, social competition, and evolution. *Proc. Am. Phil. Soc.* 123: 222–234.

———1982a. Communication in social wasps: predicted and observed patterns, with a note on the significance of behavioral and ontogenetic flexibility for theories of worker "altruism". In *La communication chez les sociétiés d'insectes*, ed. De Haro, A., and Espalader, X., pp. 13–36. Bellaterra: Universidad Autonoma de Barcelona.

———1982b. Diversity of dominance displays in *Polistes* and its possible evolutionary significance. In *The Biology of Social Insects*, ed. Breed, M. D., Michener, C. D., and Evans, H. E., pp. 222–223. Boulder: Westview Press.

———1986a. Alternative adaptations, speciation, and phylogeny (a review). *Proc. Natl .Acad. Sci. USA* 83: 1388–1392.

———1986b. Dominance relations in *Polistes canadensis* (L.), a tropical social wasp. *Monitore Zool. Ital.* (n.s.) 20: 263–281.

———1987. Flexible strategy and social evolution. In *Animal Societies: Theories and Facts*, ed. Itô, Y., Brown, J. L., and Kikkawa, J., pp. 35–51. Tokyo: Japan Scientific Societies Press.

———1989. Phenotypic plasticity and the origins of diversity. *Ann. Rev. Ecol. Syst.* 20: 249–278.

———1990. The genetic and social structure of polygynous social wasp colonies (Vespidae: Polistinae). In *Social Insects and the Environment: Proceedings of the 11th International Congress of IUSSI*, ed. Veeresh, G. K., Mallik, B., and Viraktamath, C. A., pp. 254–255. New Delhi: Oxford and IBH Publishing Co.

———1992. Behavior and evolution. In *Molds, Molecules, And Metazoa: Growing Points in Evolutionary Biology*, ed. Grant, P. R., and Horn, H. S., pp. 57–75. Princeton: Princeton University Press.

———1996. Wasp societies as microcosms for the study of development and evolution. In *Natural History and Evolution of Paper-Wasps*, ed. Turillazzi, S., West-Eberhard, M.J., pp. 290–317. Oxford: Oxford University Press.

Wheeler, D. E. 1986. Developmental and physiological determinants of caste in social Hymenoptera: evolutionary implications. *Am. Nat.* 128: 13–34.

Wheeler, W. M. 1910. *Ants: Their Structure, development, and Behavior.* New York: Columbia University Press.

———1923. *Social Life among the Insects.* New York: Harcourt, Brace.

———1928. *The Social Insects: Their Origin and Evolution.* London: Kegan, Paul, Trench, Trubner and Co.

Wilson, E. O. 1953. The origin and evolution of polymorphism in ants. *Quart. Rev. Biol.* 28: 136–156.

———1968. The ergonomics of caste in the social insects. *Am. Nat.* 102: 41–66.

———1971. *The Insect Societies.* Cambridge, Mass.: The Belknap Press of Harvard University Press.

———1975. *Sociobiology: The New Synthesis.* Cambridge, Mass.: The Belknap Press of Harvard University Press.

———1990. *Success and Dominance in Ecosystems: The Case of the Social Insects.* Oldendorf/Luhe, Germany: Ecology Institute.

Winston, M. L. 1987. *The Biology of the Honey Bee.* Cambridge, Mass.: Harvard University Press.

Winston, M. L., and Michener, C. D. 1977. Dual origin of highly social behavior among bees. *Proc. Natl. Acad. Sci. USA* 74: 1135–1137.

Winston, M. L., and Slessor, K. N. 1992. The essence of royalty: honey bee queen pheromone. *Am. Sci.* 80: 374–385.

Wright, S. 1968. *Evolution and Genetics of Populations.* Vol.1: *Genetics and Biometric Foundations.* Chicago: University of Chicago Press.

Yamane, S. 1984. Nest architecture of two oriental paper wasps, *Parapolybia varia* and *P. nodosa,* with notes on its adaptive significance (Vespidae, Polistinae). *Zool. Jb. Syst.* 111: 119–141.

———1986. The colony cycle of the sumatran paper wasp *Ropalidia (Icariola) variegata jacobsoni* (Buysson), with reference to the possible occurrence of serial polygyny (Hymenoptera Vespidae). *Monitore Zool. Ital.* (n.s.) 20: 135–161.

Yamane, S., and Itô, Y. 1994. Nest architecture of the Australian paper wasp *Ropalidia romandi Cabeti,* with a note on its developmental process (Hymenoptera, Vespidae). *Psyche* 101: 145–158.

Yamane, S., Kojima, J., and Yamane, Sk. 1983. Queen/worker size dimorphism in an oriental polistine wasp, *Ropalidia montana* Carl (Hymenoptera: Vespidae). *Insectes Soc.* 30: 416–422.

Yamane, S., and Yamane, Sk. 1979. Polistine wasps from Nepal (Hymenoptera, Vespidae). *Insecta Matsumurana* 15: 1–37.

Yamane, Sk. 1973. Discovery of a pleometrotic association in *Polistes chinensis antennalis* PÉREZ (Hymenoptera: Vespidae). *Life Study, Fukui* (Japan) 17: 3–4.

Yanega, D. 1988. Social plasticity and early-diapausing females in a primitively social bee. *Proc. Natl. Acad. Sci. USA* 85: 4374–4377.

———1989. Caste determination and differential diapause within the first brood of *Halictus rubicundus* in New York (Hymenoptera: Halictidae). *Behav. Ecol. Sociobiol.* 24: 97–107.

INDEX